AF381390

Operator Extensions, Interpolation of Functions and Related Topics

14th International Conference on Operator Theory,
Timişoara (Romania), June 1–5, 1992

Edited by
A. Gheondea
D. Timotin
F.-H. Vasilescu

Springer Basel AG

Editors

A. Gheondea
D. Timotin
F.-H. Vasilescu
Institute of Mathematics
of the Romanian Academy
C.P. 1–764
70700 Bucharest
Romania

A CIP catalogue record for this book is available from the Library of Congress, Washington D.C., USA

Deutsche Bibliothek Cataloging-in-Publication Data
Operator extensions, interpolation of functions and related
topics / 14th International Conference on Operator Therory,
Timişoara (Romania), June 1–5, 1992. Ed. by A. Gheondea ... –
Basel ; Boston ; Berlin : Birkhäuser, 1993
   (Operator theory ; Vol. 61)
      ISBN 978-3-0348-9687-0   ISBN 978-3-0348-8575-1 (eBook)
      DOI 10.1007/978-3-0348-8575-1
NE: Gheondea, Aurelian [Hrsg.]; International Conference on Operator
   Theory <14, 1992, Timişoara>; GT

© 1993 Springer Basel AG
Originally published by Birkhäuser Verlag,Basel in 1993
Camera-ready copy prepared by the editors
Printed on acid-free paper produced from chlorine-free pulp
Cover design: Heinz Hiltbrunner, Basel

ISBN 978-3-0348-9687-0

9 8 7 6 5 4 3 2 1

CONTENTS

Foreword

Since 1976 the Institute of Mathematics of the Romanian Academy (formerly the Department of Mathematics of INCREST) and the Faculty of Mathematics (formerly the Faculty of Sciences) of the University of Timişoara have organized several Conferences on Operator Theory. These Conferences were held yearly in Timişoara (or in Timişoara and Herculane) and beginning with 1985 they were held in Bucharest (1985, 1986), in Timişoara (1988) and in Predeal (1990).

At the beginning, these Conferences answered the need of a part of the Romanian Mathematical Community of exploring other forms of survival, after the dissolution of the Institute of Mathematics in 1975. Soon, these meetings evolved to International Conferences with a broad participation and where important results in Operator Theory and Operator Algebras and their interplay with Complex Function Theory, Differential Equations, Mathematical Physics, System Theory, etc. were presented.

The *14th Conference on Operator Theory* was held between June 1st and June 5th 1992, at the University of Timişoara. It was partially supported by the Institute of Mathematics of the Romanian Academy and by the Faculty of Mathematics of the University of Timişoara. Another important contribution towards covering the costs of this meeting came from *The Soros Foundation for an Open Society*. Without this generous help the organizing of this event would be impossible.

Since 1980, the Proceedings of OT Conferences were published by Birkhäuser Verlag in the series Operator Theory: Advances and Applications. The abstracts of the talks were collected in the Conference Report, published by the University of Timişoara.

This volume consists of a careful selection of contributed papers of the participants to OT 14 Conference. The problems of extensions of operators and their connections with interpolation of analytic functions and with the spectral theory of differential operators were frequent topics in the lectures of this Conference and consequently they are reflected in most of the papers. The other topics concern operator inequalities, spectral theory in general spaces and operator theory in Krein spaces. Original results of new research in fast developing areas are included.

We are indebted to Professor Israel Gohberg for including these Proceedings in the OT Series and for valuable advice in the editing process. Birkhäuser Verlag was very cooperative in publishing this volume.

The Editors

Programme of the Conference

Timişoara, June 1–5, 1992

Monday June 1st

Morning Session

Chairman: D. Gaşpar

8:30– 9:30   *Registration.*

9:30– 9:55   *Opening.*

9:55–10:35   M. Martin: *Some Comparability Results in Inductive C^*-algebras.*

11:00–11:40   D. Alpay: *Reproducing Kernel Pontryagin Spaces, Operator Colligations and Operator Models.*

11:55–12:25   P. A. Cojuhari: *On the Spectrum of a Singular Nonselfadjoint Differential Operator.*

Afternoon Session

Chairman: T. Furuta

16:00–16:30   L. Klotz: *Different Definitions for Operator-Valued Kernels.*

17:10–17:30   E. J. Ionaşcu: *Some New Proofs in Connection with Jordan Operators.*

17:35–17:55   L. Burlando: *Approximation by Semi-Fredholm and Semi-α-Fredholm Operators with Fixed Index in a Hilbert Space.*

18:00–18:20   C. Gr. Ambrozie: *A Stability Result Concerning the Fredholm Index in Banach Spaces.*

Tuesday June 2nd

Morning Session

Chairman: V. M. Adamyan

9:00– 9:40   T. Furuta: *Order Preserving Operator Inequality and its Applications.*

9:55–10:35   B. Sz-Nagy: *On the Monograph "Harmonic Analysis of Operators on Hilbert Space".*

11:00–11:40   H. Neidhardt: *On the Spectra of Selfadjoint Extensions.*

11:55–12:25   G. Nævdal: *Completions of Partial Positive Semidefinite Toeplitz Matrices.*

Afternoon Session

Chairman: H. Langer

16:00–16:30   S. A. M. Marcantognini: *The Commutant Lifting Theorem for Contractions on Krein Spaces.*

16:35–16:55   M. Zajac: *Hyperreflexivity and Reflexivity of Operators in Hilbert Spaces.*

17:35–18:05   V. Matache: *Spectral Properties of Operators with Dense Orbits.*

18:10–18:30   P. Găvruţă: *On Subnormal Pairs.*

Wednesday June 3rd

Morning Session

Chairman: B. Sz.-Nagy

9:00– 9:40   R. Mennicken: *On Boundary Eigenvalue Problems from Magnetohydrodynamics.*

9:55–10:35   H. Langer: *Sturm-Liouville Problems which are Rational in the Spectral Parameter.*

11:00–11:40   M. A. Dritschel: *A Method for Constructing Invariant Subspaces for Normal Operators on Krein Spaces.*

11:55–12:25   R. Levy: *On the Functional Models and Principal Functions of Commuting Operators.*

Afternoon Session

Chairman: A. A. Nudelman

16:00–16:30   M. D. Morán: *Unitary Extension of Commuting Isometric Operators.*

16:35–16:55   P. Jonas: *On Selfadjoint Extensions of Nonnegative Operators with Defect 1 in Krein Spaces.*

17:10–17:30   N. Buyukliev: *Existence of Cross-Sections and an Index Formula for some Extensions of Groupoid C^*-Algebra.*

17:35–17:55   N. Cotfas: *A class of O_7^n-Invariant Linear Operators with Applications to Crystal-Physics.*

18:00–18:20   S. Hassi: *Antitonicity of the Inverse and J-Selfadjoint Operators.*

18:25–18:45   A. Gheondea: *Rank Possibilities in One-Step Completions of Partial Matrices.*

Thursday June 4th

Morning Session

Chairman: **P. Masani**

9:00– 9:40   V. M. Adamyan: *Schur Problems with Incomplete Data.*

9:55–10:35   L. Waelbroeck: *The Quotient Bornological Spaces.*

11:00–11:40   B. Chevreau: *Dual Algebras, Dilation Theory, Invariant Subspaces.*

11:55–12:25   Şt. Frunză: *The Regularity Problem for Generalized Scalar Operators.*

Friday June 5th

Morning Session

Chairman: **L. Waelbroeck**

9:00– 9:40   I. Suciu: *The Kobayashi Distance between Two Contractions.*

9:55–10:35   A. A. Nudelman: *Some Generalizations of Classical Interpolation Problems.*

List of Participants*

Vadim M. ADAMYAN	*University of Odessa, Ukraine*
Daniel ALPAY	*Ben-Gurion University of the Negev, Israel*
Călin-Grigore AMBROZIE	*Institutul de Matematică, Bucureşti*
Laura BURLANDO	*Università di Genova, Italia*
Constantin BUŞE	*Universitatea din Timişoara*
Nikolai P. BUYUKLIEV	*Gabrovo, Bulgaria*
Traian CEAUŞU	*Universitatea Tehnică Timişoara*
Bernard CHEVREAU	*Université de Bordeaux I, France*
Nicolae COFAN	*Universitatea Tehnică Timişoara*
Petru A. COJUHARI	*Universitatea de Stat din Chişinău, Moldova*
Ion COLOJOARĂ	*Universitatea din Bucureşti*
Nicolae COTFAS	*Universitatea din Bucureşti*
Mircea CRAIOVEANU	*Universitatea din Timişoara*
Octavian CRĂSNARU	*Liceul Jimbolia*
Michael A. DRITSCHEL	*University of Groningen, The Netherlands*
Dorin DUMITRAŞCU	*Universitatea din Craiova*
Gheorghe ECKSTEIN	*Universitatea din Timişoara*
Ştefan FRUNZĂ	*Universitatea A. I. Cuza din Iaşi*
Takayuki FURUTA	*University of Tokyo, Japan*
Dumitru GAŞPAR	*Universitatea din Timişoara*
Paşc GĂVRUŢĂ	*Universitatea Tehnică Timişoara*
Aurelian GHEONDEA	*Institutul de Matematică, Bucureşti*
Seppo HASSI	*University of Helsinki, Finland*
Dumitru HĂRĂGUŞ	*Universitatea din Timişoara*
Viorel HIRIŞ	*Universitatea din Timişoara*
Sorin HOARĂ	*Universitatea Aurel Vlaicu din Arad*
Eugen-Julien IONAŞCU	*Institutul de Matematică, Bucureşti*

---

*The Romanian participants are listed only with affiliation.

Peter JONAS   *Mathematische Institut, Berlin, Germany*
Francisc KLEPP   *Universitatea din Timişoara*
Lutz KLOTZ   *Universtität Leipzig, Germany*
Marek KOSIEK   *Jagellonian University, Poland*
Jean-Philippe LABROUSSE   *Université de Nice, France*
Heinz LANGER   *Technische Universität Wienn, Austria*
Rony Nissim LEVY   *Sofia University, Bulgaria*
Stefania MARCANTOGNINI   *University of Groningen, The Netherlands*
Mircea MARTIN   *Institutul de Matematică, Bucureşti*
Pesi MASANI   *University of Pittsburgh, USA*
Valentin MATACHE   *Universitatea din Timişoara*
Mihai MEGAN   *Universitatea din Timişoara*
Reinhard MENNICKEN   *Universität Regensburg, Germany*
Dobrica MIHAILOV   *Universitatea Tehnică Timişoara*
María Dolores MORÁN   *Universidad Central de Venezuela*
Sorin NĂDĂBAN   *Universitatea din Timişoara*
Geir NÆVDAL   *Statens Sikkerhetshøgskole Skåregt, Norway*
Hagen NEIDHARDT   *Technische Universität Berlin, Germany*
Adolf A. NUDELMAN   *Civil Engineering Institute, Odessa, Ukraine*
Doru PĂUNESCU   *Universitatea Tehnică Timişoara*
George POPESCU   *Universitatea din Craiova*
Dan POPOVICI   *Universitatea din Timişoara*
Gabriel PRĂJITURĂ   *Universitatea din Craiova*
Ilie STAN   *Universitatea Tehnică Timişoara*
Ion SUCIU   *Institutul de Matematică, Bucureşti*
Nicolae SUCIU   *Universitatea din Timişoara*
Bela SZÖKEFALVI-NAGY   *Széged University, Hungary*
Alexandru TERESCENCO   *Universitatea din Timişoara*
Todor TODOROV   *University of Sofia, Bulgaria*
Paul TOPUZU   *Universitatea din Timişoara*
Ilie VALUŞESCU   *Institutul de Matematică, Bucureşti*
Lucien WAELBROECK   *Université Libre de Bruxelles, Belgium*
Michal ZAJAC   *EFSTU Bratislava, Slovakia*

Operator Theory:
Advances and Applications, Vol. 61
© 1993 Birkhäuser Verlag Basel

On Some Operator Colligations and Associated Reproducing Kernel Hilbert Spaces

Daniel Alpay, Vladimir Bolotnikov, Aad Dijksma and Henk de Snoo

Abstract. We introduce the notion of unitary, isometric and coisometric realizations for a class of operator valued functions, and prove the existence of such realizations for a class of functions which includes as a particular case the family of Schur functions.

1. Introduction

In this work we study realization theorems for operator valued functions which are analytic and contractive in some open subset of the complex plane. Let us first recall some definitions from [AD1] and [AD2]. Let Ω be an open and connected subset of $\mathbb{C}$; the class D_Ω consists of the functions ρ which can be written as

$$\rho_\omega(z) = a(z)a(\omega)^* - b(z)b(\omega)^*, \tag{1.1}$$

where a and b are analytic in Ω and such that the sets

$$\Omega_+ = \left\{\omega \in \Omega; |a(\omega)| > |b(\omega)|\right\}$$

and

$$\Omega_- = \left\{\omega \in \Omega; |a(\omega)| < |b(\omega)|\right\}$$

are nonempty. Then (see [AD2]) the set $\Omega_0 = \left\{\omega \in \Omega; |a(\omega)| = |b(\omega)|\right\}$ contains an element ω with $a(\omega) \neq 0$.

The representation (1.1) is essentially unique: if $\rho_\omega(z) = c(z)c(\omega)^* - d(z)d(\omega)^*$ is another representation of ρ, then

$$\begin{pmatrix} c(z) & d(z) \end{pmatrix} = \begin{pmatrix} a(z) & b(z) \end{pmatrix} M$$

for some 2×2 matrix M which is $\begin{pmatrix} 1 & 0 \\ 0 & -1 \end{pmatrix}$-unitary.

The function $\rho_\omega(z) = 1 - z\omega^*$ is clearly of the form (1.1) with $a(z) = 1$ and $b(z) = z$. The sets Ω_+, Ω_- and Ω_0 are equal to the open unit disk $\mathbb{D}$, the complement in $\mathbb{C}$ of the closed unit disk and the unit circle $\mathbb{T}$, respectively. Two other instances of interest are $\rho_\omega(z) = -i(z - \omega^*)$ and $\rho_\omega(z) = -i(z - w^*)(1 - zw^*)$. In the first case, $a(z) = \frac{1+iz}{\sqrt{2}}$ and $b(z) = \frac{1-iz}{\sqrt{2}}$; Ω_+ is equal to the open upper half plane $\mathbb{C}_+$, Ω_- is the lower open half plane $\mathbb{C}_-$ and $\Omega_0 = \mathbb{R}$. In the second case, the functions a and b are equal to $a(z) = \frac{z+i(z^2+1)}{\sqrt{2}}$ and $b(z) = \frac{z-i(z^2+1)}{\sqrt{2}}$. Moreover, $\Omega_+ = (\mathbb{D} \cap \mathbb{C}_+) \cup (\mathbb{C}_- \setminus \mathbb{D})$, $\Omega_- = (\mathbb{D} \cap \mathbb{C}_-) \cup (\mathbb{C}_+ \setminus \mathbb{D})$ and $\Omega_0 = \mathbb{R} \cup \mathbb{D}$. Note that in this last example neither Ω_+ nor Ω_- are connected. For more on these and other examples we refer to [AD1].

Let us take a function ρ in D_Ω, a point α in Ω_+ and two Hilbert spaces $\mathcal{F}$ and $\mathcal{G}$. An $\mathcal{L}(\mathcal{F}, \mathcal{G})$ valued function Θ which is analytic in a neighbourhood U_α of α is said to belong to the class $\mathcal{S}_\alpha(\rho, \mathcal{F}, \mathcal{G})$ if the kernel

$$\sigma_\Theta(z, \omega) = \frac{I - \Theta(z)\Theta(\omega)^*}{\rho_\omega(z)}$$

is nonnegative (positive definite or of positive type) in U_α, i.e., if for every positive integer r, every choice of points $\omega_1, \ldots, \omega_r$ in U_α and vectors $g_1, \ldots, g_r$ in $\mathcal{G}$, the $r \times r$ hermitian matrix with (i, j)–th entry

$$\left[\sigma_\Theta(\omega_i, \omega_j) g_j, g_i \right]_\mathcal{G}$$

is nonnegative.

By a theorem of Aronszjan and Moore [Ar] for such a function Θ there exists a (unique) reproducing kernel Hilbert space of $\mathcal{G}$-valued functions with reproducing kernel σ_Θ. This space will be denoted by $\mathcal{H}(\Theta)$. It is uniquely determined by the following: for every choice of $\omega \in U_\alpha$ and $g \in \mathcal{G}$ the function $z \to \sigma_\Theta(z, \omega) g$ belongs to $\mathcal{H}(\Theta)$ and

$$\left[x, \sigma_\Theta(\cdot, \omega) g \right]_{\mathcal{H}(\Theta)} = \left[x(\omega), g \right]_\mathcal{G}$$

for every element $x \in \mathcal{H}(\Theta)$.

When $\Omega = \mathbb{C}$, $U_\alpha = \mathbb{D} = \Omega_+$ and $\rho_\omega(z) = 1 - zw^*$, the functions Θ are the so-called Schur functions and the space $\mathcal{H}(\Theta)$ was first introduced by L. de Branges and J. Rovnyak [dBR1], [dBR2]; it provides a coisometric state space realization of Θ, i.e., the functions $\Theta(z)$ can be written as

$$\Theta(z) = H_0 + zG_0(I - zT_0)^{-1}F_0,$$

where the operators H_0, F_0, T_0 and G_0 are defined by

$$\begin{aligned}
(T_0 x)(z) &= \tfrac{x(z)-x(0)}{z}, & T_0 &: \mathcal{H}(\Theta) \to \mathcal{H}(\Theta), \\
F_0 f(z) &= \tfrac{\Theta(z)-\Theta(0)}{z} f, & F_0 &: \mathcal{F} \to \mathcal{H}(\Theta), \\
G_0 x &= x(0), & G_0 &: \mathcal{H}(\Theta) \to \mathcal{G},
\end{aligned}$$

and
$$H_0 f = \Theta(0) f, \qquad H_0 : \mathcal{F} \to \mathcal{G},$$

and where the operator matrix $\begin{pmatrix} T_0 & F_0 \\ G_0 & H_0 \end{pmatrix}$ is coisometric from $\mathcal{H}(\Theta) \oplus \mathcal{F}$ into $\mathcal{H}(\Theta) \oplus \mathcal{G}$.

If Θ is a Schur function so is the function $\tilde{\Theta}(z) = \Theta(z^*)^*$; the corresponding space $\mathcal{H}(\tilde{\Theta})$ provides an isometric realization for Θ. Moreover, the kernel

$$D_\Theta(z,\omega) = \begin{pmatrix} \sigma_\Theta(z,\omega) & \frac{\Theta(z)-\Theta(\omega^*)}{z-\omega^*} \\ \frac{\tilde{\Theta}(z)-\tilde{\Theta}(\omega^*)}{z-\omega^*} & \sigma_{\tilde{\Theta}}(z,\omega) \end{pmatrix}$$

is nonnegative in $\mathbb{D}$; the associated reproducing kernel Hilbert space $\mathcal{D}(\Theta)$ provides a unitary realization of the function Θ (see [dBS]).

The questions we address here are the following: Given an element Θ in a class $S_\alpha(\rho, \mathcal{F}, \mathcal{G})$, does the associated space $\mathcal{H}(\Theta)$ still allow us to define a coisometric realization of Θ (in a sense to be made precise)? Also, are there analogues of the function $\tilde{\Theta}$, the space $\mathcal{D}(\Theta)$ and of the associated unitary realization?

In this paper we look for realizations of Θ of the form

$$\Theta(z) = H + \big(b(z)a(\alpha) - a(z)b(\alpha)\big)G\big(a(z)A - b(z)B\big)^{-1}F, \qquad (1.2)$$

where A and B are bounded operators in some Hilbert space $\mathcal{H}$ such that $a(\alpha)A - b(\alpha)B$ is invertible and F (G) belongs to $\mathcal{L}(\mathcal{F}, \mathcal{H})$ $(\mathcal{L}(\mathcal{H}, \mathcal{G})$, respectively). Clearly, $H = \Theta(\alpha)$. Such nonstandard realizations were introduced and studied in [AD1] and [AD3], where all the indicated spaces are finite dimensional. When $\rho_\omega(z) = 1 - z\omega^*$, the realization (1.2) reduces to the nonstandard realizations of the form considered in [GK] and [GKR]. In these papers also the operator A is not necessarily boundedly invertible. As a general reference on realizations we mention the book [Fu] by P.Fuhrmann.

In the second section of this paper we define coisometric, isometric and unitary realizations for expressions of the form (1.2). We also define the function $\tilde{\Theta}$ and introduce some minimality conditions on (1.2). In Section 3 we prove that $\mathcal{H}(\Theta)$ is indeed the state space for a coisometric realization, while in Section 4 we study a unitary realization in terms of the (new) space $\mathcal{D}(\Theta)$.

One can study the same questions when $\mathcal{F}$ and $\mathcal{G}$ are Krein spaces and the kernel (1.1) has a finite number of negative squares. This has been done in [ADPS], [vdP] and [ADS] for $\rho_\omega(z) = 1 - z\omega^*$ and essentially for the case where $\mathcal{F}$ and $\mathcal{G}$ are Pontryagin spaces with the same index.

We will present similar results for general ρ in future publications.

The notation will be quite standard and introduced where needed; we conclude this introduction with one more definition. If ρ belongs to D_Ω, then the kernel $\frac{1}{\rho}$ is nonnegative in Ω_+ and therefore there is an associated reproducing kernel Hilbert space, which we denote by H_ρ. For $\rho_\omega(z) = 1 - z\omega^*$, the space H_ρ is just the classical Hardy space on the disk. For ρ of the form (1.1), H_ρ was studied in [AD1] and its main properties are gathered in the next theorem.

Theorem 1.1 *The reproducing kernel Hilbert space with reproducing kernel $\frac{1}{\rho}$ consists of functions of the form*

$$f(z) = \frac{1}{a(z)} \sum_{n=0}^{\infty} f_n \left(\frac{b(z)}{a(z)} \right)^n$$

with $f_n \in \mathbb{C}$ and $\|f\|_{H_\rho}^2 = \sum |f_n|^2 < \infty$.

When the f_n are taken to be in the Hilbert space $\mathcal{F}$ rather than $\mathbb{C}$ the absolute value of f_n is replaced by its norm and we will denote the corresponding space by $H_\rho(\mathcal{F})$.

2. Colligations and characteristic functions

Let $\rho_\omega(z)$ be a function of the form (1.1) and let α be in Ω_+. In particular $|a(\alpha)| > 0$ and without loss of generality we assume $a(\alpha)$ and $b(\alpha)$ are real. (Indeed, if $a(\alpha) = \rho_1 e^{i\theta_1}$ and, if $b(\alpha) \neq 0$, $b(\alpha) = \rho_2 e^{i\theta_2}$ are the polar decompositions of $a(\alpha)$ and $b(\alpha)$, then the functions $\tilde{a}(z) = a(z)e^{-i\theta_1}$ and $\tilde{b}(z) = b(z)e^{-i\theta_2}$ represent the same $\rho_\omega(z)$ and their values in α are real.)

A *colligation* is a collection of the form

$$\Delta = (\alpha, \rho, \mathcal{H}, \mathcal{F}, \mathcal{G}, A, B, H, G, F) \tag{2.1}$$

consisting of three Krein spaces $\mathcal{H}$ (*the state space*), $\mathcal{F}$ (*the incoming space*), $\mathcal{G}$ (*the outgoing space*) and of operators $H \in \mathcal{L}(\mathcal{F}, \mathcal{G})$, $G \in \mathcal{L}(\mathcal{H}, \mathcal{G})$, $F \in \mathcal{L}(\mathcal{F}, \mathcal{H})$ and $A, B \in \mathcal{L}(\mathcal{H})$ with

$$a(\alpha)A - b(\alpha)B = I_{\mathcal{H}}. \tag{2.2}$$

The colligation is called *isometric* if

$$A^*A = B^*B, \tag{2.3}$$

where $\mathcal{A} \in \mathcal{L}(\mathcal{H} \oplus \mathcal{F})$ and $\mathcal{B} \in \mathcal{L}(\mathcal{H} \oplus \mathcal{F}, \mathcal{H} \oplus \mathcal{G})$ are given by

$$\mathcal{A} = \begin{pmatrix} A & b(\alpha)F \\ 0 & I_{\mathcal{F}} \end{pmatrix} : \begin{pmatrix} \mathcal{H} \\ \mathcal{F} \end{pmatrix} \to \begin{pmatrix} \mathcal{H} \\ \mathcal{F} \end{pmatrix}, \qquad \mathcal{B} = \begin{pmatrix} B & a(\alpha)F \\ G & H \end{pmatrix} : \begin{pmatrix} \mathcal{H} \\ \mathcal{F} \end{pmatrix} \to \begin{pmatrix} \mathcal{H} \\ \mathcal{G} \end{pmatrix}. \tag{2.4}$$

The colligation is called *coisometric* if

$$\tilde{\mathcal{A}}\tilde{\mathcal{A}}^* = \tilde{\mathcal{B}}\tilde{\mathcal{B}}^*, \tag{2.5}$$

where

$$\tilde{\mathcal{A}} = \begin{pmatrix} A & 0 \\ b(\alpha)G & I_{\mathcal{G}} \end{pmatrix} : \begin{pmatrix} \mathcal{H} \\ \mathcal{G} \end{pmatrix} \to \begin{pmatrix} \mathcal{H} \\ \mathcal{G} \end{pmatrix}, \qquad \tilde{\mathcal{B}} = \begin{pmatrix} B & F \\ a(\alpha)G & H \end{pmatrix} : \begin{pmatrix} \mathcal{H} \\ \mathcal{F} \end{pmatrix} \to \begin{pmatrix} \mathcal{H} \\ \mathcal{G} \end{pmatrix},$$
$$\tag{2.6}$$

and it is called *unitary* if (2.3) and (2.5) hold simultaneously.

The colligation (2.1) is called *closely innerconnected* if

$$\mathcal{H} = \bigvee_{z \in U_\alpha} \operatorname{ran}\left(\left(a(z)^* A - b(z)^* B\right)^{-1} F\right) = \bigvee \{B^n F f;\ n = 0, \ldots,\ f \in \mathcal{F}\},$$

closely outerconnected if

$$\mathcal{H} = \bigvee_{z \in U_\alpha} \operatorname{ran}\left(\left(a(z) A - b(z) B\right)^{*-1} G^*\right) = \bigvee \{B^{*n} G^* g;\ n = 0, \ldots,\ g \in \mathcal{G}\},$$

and it is called *closely connected* if

$$\mathcal{H} = \bigvee_{z, \omega \in U_\alpha} \left\{ \operatorname{ran}\left(\left(a(z)^* A - b(z)^* B\right)^{-1} F\right),\ \operatorname{ran}\left(\left(a(\omega) A - b(\omega) B\right)^{*-1} G^*\right) \right\}.$$

In these definitions U_α stands for a small neighbourhood around α and $\bigvee_{z \in U}$ denotes the closed linear span of the sets with index $z \in U$. These definitions are independent of the choice of U_α. We call an isometric (coisometric, unitary) colligation *minimal* if it is closely innerconnected (closely outerconnected, closely connected, respectively).

Associated with a colligation Δ is the so-called *characteristic function*

$$\Theta(z) = \Theta_\Delta(z) = H + \delta_\alpha(z) G \left(a(z) A - b(z) B\right)^{-1} F, \tag{2.7}$$

where

$$\delta_\omega(z) = b(z) a(\omega) - a(z) b(\omega). \tag{2.8}$$

In view of (2.2), Θ is analytic in some neighbourhood U_α of α.

If for an $\mathcal{L}(\mathcal{F}, \mathcal{G})$ valued function $\Theta(z)$ analytic in U_α there exists a colligation Δ of the form (2.1) such that $\Theta(z) = \Theta_\Delta(z)$ for all z in U_α, then this colligation is called a *realization* of Θ.

Together with Θ defined by (2.7) we consider the $\mathcal{L}(\mathcal{G}, \mathcal{F})$ valued function $\tilde{\Theta}(z)$ given by

$$\tilde{\Theta}(z) = \tilde{\Theta}_\Delta(z) = H^* + \delta_\alpha(z) F^* \left(a(z) A^* - b(z) B^*\right)^{-1} G^*, \tag{2.9}$$

which is the characteristic function of the so-called *adjoint colligation*

$$\Delta^* = (\alpha, \rho, \mathcal{H}, \mathcal{G}, \mathcal{F}, A^*, B^*, H^*, F^*, G^*). \tag{2.10}$$

The next lemma follows directly from the definitions.

Lemma 2.1 *The colligation* (2.1) *is isometric (coisometric) if and only if the adjoint colligation* (2.10) *is coisometric (isometric, respectively). Moreover, Δ is minimal if and only if Δ^* is minimal.*

We associate with Θ the kernels

$$\sigma_\Theta(z, \omega) = \frac{I_\mathcal{G} - \Theta(z)\Theta(\omega)^*}{\rho_\omega(z)}, \qquad \sigma_{\tilde{\Theta}}(z, \omega) = \frac{I_\mathcal{F} - \tilde{\Theta}(z)\tilde{\Theta}(\omega)^*}{\rho_\omega(z)} \tag{2.11}$$

with values in $\mathcal{L}(\mathcal{G})$ and $\mathcal{L}(\mathcal{F})$, respectively, and the kernel

$$D_\Theta(z,\omega) = \begin{pmatrix} \sigma_\Theta(z,\omega) & \dfrac{\Theta(z)-\widetilde{\Theta}(\omega)^*}{\varphi_\omega(z)} \\[2mm] \dfrac{\widetilde{\Theta}(z)-\Theta(\omega)^*}{\varphi_\omega(z)} & \sigma_{\widetilde{\Theta}}(z,\omega) \end{pmatrix} \tag{2.12}$$

with values in $\mathcal{L}(\mathcal{G} \oplus \mathcal{F})$, where

$$\varphi_\omega(z) = b(z)a(\omega)^* - a(z)b(\omega)^*. \tag{2.13}$$

Lemma 2.2 *Let Δ be a colligation of the form* (2.1), *and let Θ and $\widetilde{\Theta}$ be functions given by* (2.7) *and* (2.9), *respectively.*

(i) *If Δ is isometric then*

$$\sigma_{\widetilde{\Theta}}(z,\omega) = F^*\big(a(z)^*A - b(z)^*B\big)^{*-1}\big(a(\omega)^*A - b(\omega)^*B\big)^{-1}F. \tag{2.14}$$

(ii) *If Δ is coisometric then*

$$\sigma_\Theta(z,\omega) = G\big(a(z)A - b(z)B\big)^{-1}\big(a(\omega)A - b(\omega)B\big)^{*-1}G^*. \tag{2.15}$$

(iii) *If Δ is unitary then*

$$D_\Theta(z,\omega) = \begin{pmatrix} G\big(a(z)A - b(z)B\big)^{-1} \\[2mm] F^*\big(a(z)^*A - b(z)^*B\big)^{*-1} \end{pmatrix}$$
$$\times \Big(\big(a(\omega)A - b(\omega)B\big)^{*-1}G^*, \ \big(a(\omega)^*A - b(\omega)^*B\big)^{-1}F\Big). \tag{2.16}$$

Proof. Let Δ be isometric. Then, in view of (2.3) and (2.4), we have

$$A^*A - B^*B = G^*G, \tag{2.17}$$

$$F^*\big(b(\alpha)A - a(\alpha)B\big) = H^*G, \tag{2.18}$$

and

$$I_\mathcal{F} - \rho_\alpha(\alpha)F^*F = H^*H. \tag{2.19}$$

Using (2.2), (2.9) and (2.17)–(2.19) we obtain

$$I - \widetilde{\Theta}(z)\widetilde{\Theta}(\omega)^* = I - \Big(H^* + \delta_\alpha(z)F^*\big(a(z)A^* - b(z)B^*\big)^{-1}G^*\Big)$$
$$\times \Big(H + \delta_\alpha(\omega)^*G\big(a(\omega)^*A - b(\omega)^*B\big)^{-1}F\Big) \tag{2.20}$$
$$= F^*\big(a(z)A^* - b(z)B^*\big)^{-1}N(z,\omega)\big(a(\omega)^*A - b(\omega)^*B\big)^{-1}F,$$

where

$$\begin{aligned}
N(z,\omega) &= \rho_\alpha(\alpha)\big(a(z)A^* - b(z)B^*\big)\big(a(\omega)^*A - b(\omega)^*B\big) \\
&\quad - \delta_\alpha(z)\big(b(\alpha)A^* - a(\alpha)B^*\big)\big(a(\omega)^*A - b(\omega)^*B\big) \\
&\quad - \delta_\alpha(\omega)^*\big(a(z)A^* - b(z)B^*\big)\big(b(\alpha)A - a(\alpha)B\big) \\
&\quad - \delta_\alpha(z)\delta_\alpha(\omega)^*(A^*A - B^*B) \\
&= \rho_\omega(z)\big(a(\alpha)A - b(\alpha)B\big)^*\big(a(\alpha)A - b(\alpha)B\big) \\
&= \rho_\omega(z)I_{\mathcal{H}}.
\end{aligned} \tag{2.21}$$

Substituting (2.21) into (2.20) and taking into account (2.11) we obtain (2.14). Equality (2.15) can be obtained from (2.2) and (2.7) in the same way. Using (2.7), (2.9) and (2.13) we have

$$\begin{aligned}
\Theta(z) - \tilde{\Theta}(\omega)^* &= G\Big\{\delta_\alpha(z)\big(a(z)A - b(z)B\big)^{-1} - \delta_\alpha(\omega)^*\big(a(\omega)^*A - b(\omega)^*B\big)^{-1}\Big\}F \\
&= \varphi_\omega(z)G\big(a(z)A - b(z)B\big)^{-1}\big(a(\omega)^*A - b(\omega)^*B\big)^{-1}F,
\end{aligned}$$

which together with (2.14) and (2.15) implies (2.16). ∎

In the following we denote by $\mathrm{ind}_\pm \mathcal{H}$ the dimension of the space $\mathcal{H}_\pm$ in a fundamental decomposition $\mathcal{H} = \mathcal{H}_- + \mathcal{H}_+$ of the Krein space $\mathcal{H}$, and by $\mathrm{sq}_-(\sigma_\Theta)$ ($\mathrm{sq}_+(\sigma_\Theta)$) the number of negative (positive, respectively) squares of the kernel σ_Θ.

Corollary 2.3 *Let Δ be an isometric (coisometric or unitary) colligation and let $\Theta = \Theta_\Delta$ be its characteristic function. Then*

(i) $\mathrm{ind}_\pm \mathcal{H} \geq \mathrm{sq}_\pm(\sigma_{\tilde{\Theta}})$ $(\mathrm{ind}_\pm \mathcal{H} \geq \mathrm{sq}_\pm(\sigma_\Theta)$ or $\mathrm{ind}_\pm \mathcal{H} \geq \mathrm{sq}_\pm(D_\Theta))$,

respectively);

(ii) *if the colligation is minimal then in* (i) *equality prevails.*

Proof. The assertions of the corollary follow from (2.14)–(2.16), which imply respectively

$$\big[\sigma_{\tilde{\Theta}}(z,\omega)f_1, f_2\big]_{\mathcal{F}} = \Big[\big(a(\omega)^*A - b(\omega)^*B\big)^{-1}Ff_1,\ \big(a(z)^*A - b(z)^*B\big)^{-1}Ff_2\Big]_{\mathcal{H}},$$

$$\big[\sigma_\Theta(z,\omega)g_1, g_2\big]_{\mathcal{G}} = \Big(\big(a(\omega)A - b(\omega)B\big)^{*-1}G^*g_1,\ \big(a(z)A - b(z)B\big)^{*-1}G^*g_2\Big]_{\mathcal{H}},$$

$$\begin{aligned}
\Big[D_\Theta(z,\omega)\begin{pmatrix} g_1 \\ f_1 \end{pmatrix}, \begin{pmatrix} g_2 \\ f_2 \end{pmatrix}\Big]_{\mathcal{G}\oplus\mathcal{F}} &= \Big[\big(a(\omega)A - b(\omega)B\big)^{*-1}G^*g_1 \\
&\quad + \big(a(\omega)^*A - b(\omega)^*B\big)^{-1}Ff_1,\ \big(a(z)A - b(z)B\big)^{*-1}G^*g_2 \\
&\quad + \big(a(z)^*A - b(z)^*B\big)^{-1}Ff_2\Big]_{\mathcal{H}}
\end{aligned}$$

for arbitrary $f_1, f_2 \in \mathcal{F}$ and $g_1, g_2 \in \mathcal{G}$. ∎

Theorem 2.4 *Let Θ belong to $S_\alpha(\rho, \mathcal{F}, \mathcal{G})$, where ρ is of the form (1.1). Then Θ can be uniquely extended to an $\mathcal{L}(\mathcal{F}, \mathcal{G})$ valued function $\hat{\Theta}$ which is analytic in Ω_+ and such that the kernel $\sigma_{\hat{\Theta}}(z, \omega)$ is nonnegative in Ω_+.*

Proof. Let $H_\rho(\mathcal{F})$ and $H_\rho(\mathcal{G})$ be the Hardy spaces defined after Theorem 1.1. Assume that the kernel $\sigma_\Theta(z, \omega)$ is nonnegative in a neighbourhood $U_\alpha \subset \Omega_+$ of α. Then for any choice of $g_1, \ldots, g_n \in \mathcal{G}$ and $\omega_1, \ldots, \omega_n \in U_\alpha$,

$$\left[\sum_{i=1}^n \frac{g_i}{\rho_{\omega_i}}, \sum_{j=1}^n \frac{g_j}{\rho_{\omega_j}} \right]_{H_\rho(\mathcal{G})} - \left[\sum_{i=1}^n \frac{\Theta(\omega_i)^* g_i}{\rho_{\omega_i}}, \sum_{j=1}^n \frac{\Theta(\omega_j)^* g_j}{\rho_{\omega_j}} \right]_{H_\rho(\mathcal{F})}$$

$$= \sum_{i,j=1}^n \left[\sigma_\Theta(\omega_j, \omega_i) g_i, g_j \right]_{\mathcal{G}} \geq 0,$$

which implies that the formula

$$\left(T \frac{g}{\rho_\omega} \right)(z) = \frac{\Theta(\omega)^* g}{\rho_\omega(z)}, \qquad g \in \mathcal{G}, \ \omega \in U_\alpha, \ z \in \Omega_+, \tag{2.22}$$

gives rise to a contraction T which is well defined and densely defined in $H_\rho(\mathcal{G})$ and has values in $H_\rho(\mathcal{F})$. Therefore T admits a continuous extension to all of $H_\rho(\mathcal{G})$, and its adjoint T^* mapping $H_\rho(\mathcal{F})$ to $H_\rho(\mathcal{G})$ is also a contraction.

For every $f \in \mathcal{F}$ the function $z \to \frac{f}{a(z)}$ belongs to $H_\rho(\mathcal{F})$ and hence $\left(T^* \frac{f}{a} \right)(z)$ belongs to $H_\rho(\mathcal{G})$. In view of Theorem 1.1, there exist vectors $g_0, g_1, \ldots$ in $\mathcal{G}$ such that

$$\left(T^* \frac{f}{a} \right)(z) = \frac{1}{a(z)} \sum_{i=0}^\infty g_i \left(\frac{b(z)}{a(z)} \right)^i, \tag{2.23}$$

where the convergence is both in $H_\rho(\mathcal{G})$ and pointwise. Since T^* is a contraction,

$$\left\| T^* \frac{f}{a} \right\|^2_{H_\rho(\mathcal{G})} = \sum_{i=0}^\infty \|g_i\|^2_{\mathcal{G}} \leq \|f\|^2_{\mathcal{F}},$$

and therefore

$$\|g_i\|_{\mathcal{G}} \leq \|f\|_{\mathcal{F}}, \qquad i = 0, 1, \ldots . \tag{2.24}$$

Hence the map $f \to g_i$ defines a linear operator $G_i \colon \mathcal{F} \to \mathcal{G}$

$$G_i f = g_i, \qquad i = 0, 1, \ldots, \tag{2.25}$$

which, in view of (2.24), is a contraction:

$$\|G_i\| \leq 1, \qquad i = 0, 1, \ldots . \tag{2.26}$$

The last inequalities imply that for every z, $|z| < 1$,

$$\left\| \sum_{i=0}^\infty G_i z^i \right\| \leq \sum_{i=0}^\infty |z|^i = \frac{1}{1 - |z|},$$

and therefore the series

$$M(z) = \sum_{i=0}^{\infty} G_i z^i \tag{2.27}$$

defines an $\mathcal{L}(\mathcal{F}, \mathcal{G})$ valued function which is analytic in the unit disc $\mathbb{D}$. Substituting (2.25) and (2.26) into (2.23) we obtain

$$\left(T^* \frac{f}{a}\right)(\omega) = \frac{M\left(\frac{b(\omega)}{a(\omega)}\right)}{a(\omega)} f, \qquad f \in \mathcal{F}. \tag{2.28}$$

It follows from (2.22), (2.28) and the reproducing properties of the kernels of $H_\rho(\mathcal{G})$ and $H_\rho(\mathcal{F})$ that for arbitrary $f \in \mathcal{F}$, $g \in \mathcal{G}$ and $\omega \in U_\alpha$,

$$\left[\frac{M\left(\frac{b(\omega)}{a(\omega)}\right)}{a(\omega)} f, g\right]_{\mathcal{G}} = \left[T^* \frac{f}{a}, \frac{g}{\rho_\omega}\right]_{H_\rho(\mathcal{G})} = \left[\frac{f}{a}, T \frac{g}{\rho_\omega}\right]_{H_\rho(\mathcal{F})} = \left[\frac{\Theta(\omega)}{a(\omega)} f, g\right]_{\mathcal{G}},$$

which implies $\Theta(\omega) = M\left(\frac{b(\omega)}{a(\omega)}\right)$, $\omega \in U_\alpha$. Since the kernel σ_Θ is nonnegative in U_α, it follows by a change of variable that the kernel

$$\frac{I - M(z)M(\nu)^*}{1 - z\nu^*}$$

is nonnegative for z, ν in the neighbourhood $\frac{b}{a}(U_\alpha)$ of $\frac{b(\alpha)}{a(\alpha)}$. This means that $M(z)$ is a contraction for all z in this neighbourhood, and therefore M defined by (2.27) is a Schur function on all of $\mathbb{D}$. (For a detailed acount of this conclusion we refer to [Br] Theorem 3.3 and Section 3.5, where further relevant references are given.) Clearly

$$\widehat{\Theta}(\omega) = M\left(\frac{b(\omega)}{a(\omega)}\right), \qquad \omega \in \Omega_+, \tag{2.29}$$

is the unique extension of Θ to all of Ω_+ with the required property that the kernel $\sigma_{\widehat{\Theta}}(z, \omega)$ is nonnegative on Ω_+. ∎

Let Θ be in $S_\alpha(\rho, \mathcal{F}, \mathcal{G})$, and assume that it is extended to all of Ω_+ according to (2.29), in which M is the function defined via (2.27). Then we define the *associated function* $\widetilde{\Theta}$ by

$$\widetilde{\Theta}(z) = M\left(\frac{b(z)^*}{a(z)^*}\right)^*, \qquad z \in \Omega_+. \tag{2.30}$$

Note that

$$\widetilde{\Theta}(\alpha) = \Theta(\alpha)^*.$$

The following result is easy to verify; we omit the proof.

Lemma 2.5 *If $\Theta \in S_\alpha(\rho, \mathcal{F}, \mathcal{G})$ is of the form (2.7), then (2.30) coincides with (2.9) in a neighbourhood of α .*

As a corollary of Theorem 2.4 we obtain

Theorem 2.6 *Let $\Theta \in S_\alpha(\rho, \mathcal{F}, \mathcal{G})$, so that the kernel $\sigma_\Theta(z, \omega)$ is nonnegative in some neighbourhood U_α of α. Then the kernels $\sigma_{\widetilde{\Theta}}(z, \omega)$ and $D_\Theta(z, \omega)$ given by (2.11) and (2.12), respectively, are also nonnegative in U_α.*

Proof. This follows from the corresponding known fact for the classical case $\rho_\omega(z) = 1 - z\omega^*$, $\Omega_+ = \mathbb{D}$, $\alpha = 0$ and change of variables. $\blacksquare$

We conclude that for every $\Theta \in S_\alpha(\rho, \mathcal{F}, \mathcal{G})$ there exist three reproducing kernel Hilbert spaces $\mathcal{H}(\Theta)$, $\mathcal{H}(\widetilde{\Theta})$ and $\mathcal{D}(\Theta)$ with reproducing kernels σ_Θ, $\sigma_{\widetilde{\Theta}}$ and D_Θ, respectively. In view of Theorem 2.4, the functions in these spaces have unique analytic extensions to all of Ω_+. Hence $\mathcal{H}(\Theta)$, $\mathcal{H}(\widetilde{\Theta})$ and $\mathcal{D}(\Theta)$ do not depend on the choice of the point $\alpha \in \Omega_+$.

In the next two sections we show that these spaces provide state spaces for coisometric, isometric and unitary realizations of Θ.

3. Isometric and coisometric realizations

In this section we prove that every element in $S_\alpha(\rho, \mathcal{F}, \mathcal{G})$ admits isometric and coisometric realizations. We first describe the various operators defining the corresponding colligations.

Theorem 3.1 *Let $\mathcal{F}$ and $\mathcal{G}$ be two Hilbert spaces, let $\Theta \in S_\alpha(\rho, \mathcal{F}, \mathcal{G})$ be analytic in the neighbourhood U_α of α and such that the kernel $\sigma_\Theta(z, \omega)$ is nonnegative in U_α, and let $\mathcal{H}(\Theta)$ be the reproducing kernel Hilbert space with the reproducing kernel σ_Θ. Then the formulas*

$$A\sigma_\Theta(\cdot, \omega)g = \frac{b(\omega)^*\sigma_\Theta(\cdot, \omega) - b(\alpha)\sigma_\Theta(\cdot, \alpha)}{\delta_\alpha(\omega)^*}g, \tag{3.1}$$

$$B\sigma_\Theta(\cdot, \omega)g = \frac{a(\omega)^*\sigma_\Theta(\cdot, \omega) - a(\alpha)\sigma_\Theta(\cdot, \alpha)}{\delta_\alpha(\omega)^*}g, \tag{3.2}$$

$$G\sigma_\Theta(\cdot, \omega)g = \frac{\Theta(\omega)^* - \Theta(\alpha)^*}{\delta_\alpha(\omega)^*}g, \tag{3.3}$$

$$Fg = \sigma_\Theta(\cdot, \alpha)g, \tag{3.4}$$

$$Hg = \Theta(\alpha)^*g, \tag{3.5}$$

where $g \in \mathcal{G}$ and $\omega \in U_\alpha$, uniquely define bounded operators $A, B \in \mathcal{L}\big(\mathcal{H}(\Theta)\big)$, $G \in \mathcal{L}\big(\mathcal{H}(\Theta), \mathcal{F}\big)$, $F \in \mathcal{L}\big(\mathcal{G}, \mathcal{H}(\Theta)\big)$, $H \in \mathcal{L}(\mathcal{G}, \mathcal{F})$ such that

$$a(\alpha)A - b(\alpha)B = I_{\mathcal{H}(\Theta)} \tag{3.6}$$

and

$$(Ah)(z) = \frac{a(z)}{\rho_\alpha(z)}h(z) - \frac{b(\alpha)}{\rho_\alpha(z)}\Theta(z)Gh, \tag{3.7}$$

$$(Bh)(z) = \frac{b(z)}{\rho_\alpha(z)}h(z) - \frac{a(\alpha)}{\rho_\alpha(z)}\Theta(z)Gh, \qquad h \in \mathcal{H}(\Theta). \tag{3.8}$$

Their adjoints are given by

$$(A^*h)(z) = \frac{b(z)h(z) - b(\alpha)h(\alpha)}{\delta_\alpha(z)}, \tag{3.9}$$

$$(B^*h)(z) = \frac{a(z)h(z) - a(\alpha)h(\alpha)}{\delta_\alpha(z)}, \tag{3.10}$$

$$(G^*f)(z) = \frac{\Theta(z) - \Theta(\alpha)}{\delta_\alpha(z)}f, \tag{3.11}$$

$$F^*h = h(\alpha), \tag{3.12}$$

$$H^*f = \Theta(\alpha)f, \tag{3.13}$$

where $h \in \mathcal{H}(\Theta)$ *and* $f \in \mathcal{F}$.

The operators defined in (3.9) and (3.10) are introduced in [AD2], where by different methods it is shown that A^* and B^* are bounded operators on $\mathcal{H}(\Theta)$.

Proof of Theorem 3.1. First we show that A, B and G introduced in (3.1)–(3.3) are well defined operators on the set

$$\mathcal{H}_0 = \text{l.s.} \left\{ \sigma(\cdot, \omega)g;\ g \in \mathcal{G},\ \omega \in U_\alpha \right\}, \tag{3.14}$$

which clearly is dense in $\mathcal{H}(\Theta)$.

We consider finite sums of the form

$$\sum_\omega \sigma_\Theta(\cdot, \omega)g_\omega,$$

$$\sum_\omega \frac{a(\omega)^*\sigma_\Theta(\cdot, \omega) - a(\alpha)\sigma_\Theta(\cdot, \alpha)}{\delta_\alpha(\omega)^*}g_\omega,$$

$$\sum_\omega \frac{b(\omega)^*\sigma_\Theta(\cdot, \omega) - b(\alpha)\sigma_\Theta(\cdot, \alpha)}{\delta_\alpha(\omega)^*}g_\omega$$

and

$$\sum_\omega \frac{\Theta(\omega)^* - \Theta(\alpha)^*}{\delta_\alpha(\omega)^*}g_\omega,$$

where $g_\omega \in \mathcal{G}$ and the finitely many ω's are in U_α. We denote these sums by $S(g_\omega)$, $S_a(g_\omega)$, $S_b(g_\omega)$ and $S_r(g_\omega)$, respectively, to simplify the notation. Note that the first three sums belong to $\mathcal{H}(\Theta)$,

$$a(\alpha)S_b(g_\omega) - b(\alpha)S_a(g_\omega) = S(g_\omega) \tag{3.15}$$

12   D. Alpay et al.

and that in this notation $\mathcal{H}_0 = \left\{ S(g_\omega);\ g_\omega \in \mathcal{G} \right\}$.

In $(\mathcal{H}(\Theta) \oplus \mathcal{G}) \times (\mathcal{H}(\Theta) \oplus \mathcal{F})$ we define the relation

$$R = \left\{ \left\{ \begin{pmatrix} S_b(g_\omega) + b(\alpha)\sigma_\Theta(\cdot, \alpha)g \\ g \end{pmatrix}, \begin{pmatrix} S_a(g_\omega) + a(\alpha)\sigma_\Theta(\cdot, \alpha)g \\ S_r(g_\omega) + \Theta(\alpha)^*g \end{pmatrix} \right\}; g_\omega, g \in \mathcal{G} \right\}.$$

Clearly, R is linear in the sense that for $\{u_1, v_1\}, \{u_2, v_2\} \in R$ and $\lambda \in \mathbb{C}$, $\{u_1 + \lambda u_2, v_1 + \lambda v_1\}$ also belongs to R. We claim that R is isometric, that is, in the same notation,

$$[u_1, u_2]_{\mathcal{H}(\Theta) \oplus \mathcal{G}} = [v_1, v_2]_{\mathcal{H}(\Theta) \oplus \mathcal{F}}. \tag{3.16}$$

We postpone the calculations to substantiate the claim until the end of the proof.

Formula (3.16) implies in particular that

$$\|S_b(g_\omega)\|^2_{\mathcal{H}(\Theta)} = \|S_a(g_\omega)\|^2_{\mathcal{H}(\Theta)} + \|S_r(g_\omega)\|^2_{\mathcal{F}}$$
$$\geq \|S_a(g_\omega)\|^2_{\mathcal{H}(\Theta)}.$$

On the other hand, (3.15) gives that

$$\|S_b(g_\omega)\|_{\mathcal{H}(\Theta)} = \|\frac{b(\alpha)}{a(\alpha)} S_a(g_\omega) + \frac{1}{a(\alpha)} S(g_\omega)\|_{\mathcal{H}(\Theta)}$$
$$\leq \frac{b(\alpha)}{a(\alpha)} \|S_a(g_\omega)\|_{\mathcal{H}(\Theta)} + \frac{1}{a(\alpha)} \|S(g_\omega)\|_{\mathcal{H}(\Theta)}.$$

Combining these inequalities we obtain that

$$\|S_a(g_\omega)\|_{\mathcal{H}(\Theta)} \leq \frac{1}{a(\alpha) - b(\alpha)} \|S(g_\omega)\|_{\mathcal{H}(\Theta)},$$
$$\|S_b(g_\omega)\|_{\mathcal{H}(\Theta)} \leq \frac{1}{a(\alpha) - b(\alpha)} \|S(g_\omega)\|_{\mathcal{H}(\Theta)}$$

and

$$\|S_r(g_\omega)\|_{\mathcal{H}(\Theta)} \leq \frac{1}{a(\alpha) - b(\alpha)} \|S(g_\omega)\|_{\mathcal{H}(\Theta)}.$$

Since the linear space $\mathcal{H}_0$ is dense in $\mathcal{H}(\Theta)$, it follows that the closures of the linear relations

$$R_A = \{\{S(g_\omega), S_b(g_\omega)\}; g_\omega \in \mathcal{G}\},$$
$$R_B = \{\{S(g_\omega), S_a(g_\omega)\}; g_\omega \in \mathcal{G}\}$$

and

$$R_G = \{\{S(g_\omega), S_r(g_\omega)\}; g_\omega \in \mathcal{G}\}$$

are the graphs of bounded operators $A, B \in \mathcal{L}(\mathcal{H}(\Theta))$ and $G \in \mathcal{L}(\mathcal{H}(\Theta), \mathcal{G})$, respectively. Clearly, they satisfy, and are uniquely determined by, the formulas (3.1), (3.2) and (3.3). From the boundedness of A and B and from (3.15) we obtain further that (3.6) is valid.

Let us show now that B has the form (3.8). Using the identity

$$a(\omega)^* \rho_\alpha(z) = b(z)\delta_\alpha(\omega)^* + a(\alpha)\rho_\omega(z), \tag{3.17}$$

which follows from (1.1) and (2.8), we first establish (3.8) for all $h \in \mathcal{H}_0$. Let h be an arbitrary element in $\mathcal{H}_0$, that is, of the form $S(g_\omega)$. Then, in view of (3.2), (3.3) and (3.17),

$$(Bh)(z) = S_a(g_\omega)$$

$$= \frac{b(z)}{\rho_\alpha(z)} \sum_\omega \sigma_\Theta(z,\omega)g_\omega + \frac{a(\alpha)}{\rho_\alpha(z)} \sum_\omega \frac{I - \Theta(z)\Theta(\omega)^* - \left(I - \Theta(z)\Theta(\alpha)^*\right)}{\delta_\alpha(\omega)^*} g_\omega$$

$$= \frac{b(z)}{\rho_\alpha(z)} h(z) - \frac{a(\alpha)}{\rho_\alpha(z)}\Theta(z) \sum_\omega \frac{\Theta(\omega)^* - \Theta(\alpha)^*}{\delta_\alpha(\omega)^*} g_\omega$$

$$= \frac{b(z)}{\rho_\alpha(z)} h(z) - \frac{a(\alpha)}{\rho_\alpha(z)}\Theta(z)Gh.$$

Let now h be an element in $\mathcal{H}(\Theta)$. Then $h = \lim_{n\to\infty} h_n$, for some sequence h_n in $\mathcal{H}_0$, and using the reproducing kernel property of σ_Θ we find that for all $g \in \mathcal{G}$,

$$\left[(Bh)(z), g\right]_\mathcal{G} = \lim_{n\to\infty}\left[Bh_n(z), g\right]_\mathcal{G}$$

$$= \lim_{n\to\infty}\left\{ \frac{b(z)}{\rho_\alpha(z)}\left[h_n(z), g\right]_\mathcal{G} - \frac{a(\alpha)}{\rho_\alpha(z)}\left[\Theta(z)Gh_n, g\right]_\mathcal{G}\right\}$$

$$= \lim_{n\to\infty}\left\{ \frac{b(z)}{\rho_\alpha(z)}\left[h_n, \sigma_\Theta(.,z)g\right]_\mathcal{G} - \frac{a(\alpha)}{\rho_\alpha(z)}\left[\Theta(z)Gh_n, g\right]_\mathcal{G}\right\}$$

$$= \left\{ \frac{b(z)}{\rho_\alpha(z)}\left[h, \sigma_\Theta(.,z)g\right]_\mathcal{G} - \frac{a(\alpha)}{\rho_\alpha(z)}\left[\Theta(z)Gh, g\right]_\mathcal{G}\right\}$$

$$= \left[\frac{b(z)}{\rho_\alpha(z)} h(z) - \frac{a(\alpha)}{\rho_\alpha(z)}\Theta(z)Gh, g\right]_\mathcal{G}.$$

This proves (3.8). The equalities (3.6) and (3.8) imply (3.7).

We now compute the adjoints of A, B and G. In view of (3.1), we have for every $g \in \mathcal{G}$ and $h \in \mathcal{H}(\Theta)$,

$$\left[(A^*h)(z), g\right]_\mathcal{G} = \left[A^*h,\ \sigma_\Theta(\cdot,\omega)g\right]_{\mathcal{H}(\Theta)} = \left[h, A\sigma_\Theta(\cdot,z)g\right]_{\mathcal{H}(\Theta)}$$

$$= \left[h, \frac{b(z)^*\sigma_\Theta(\cdot,z) - b(\alpha)\sigma_\Theta(\cdot,\alpha)}{\delta_\alpha(z)^*}g\right]_{\mathcal{H}(\Theta)} = \left[\frac{b(z)h(z) - b(\alpha)h(\alpha)}{\delta_\alpha(z)}, g\right]_\mathcal{G},$$

and therefore

$$(A^*h)(z) = \frac{b(z)h(z) - b(\alpha)h(\alpha)}{\delta_\alpha(z)},$$

14 *D. Alpay et al.*

which coincides with (3.9). Similarly, for every $g \in \mathcal{G}$, $h \in \mathcal{H}(\Theta)$,

$$\left[(B^* h)(z), g\right]_{\mathcal{G}} = \left[\frac{a(z)h(z) - a(\alpha)h(\alpha)}{\delta_\alpha(z)}, g\right]_{\mathcal{G}},$$

and therefore B^* is the operator given by (3.10). Taking into account (3.3) we obtain that for $f \in \mathcal{F}$ and $g \in \mathcal{G}$,

$$\left[(G^* f)(z), g\right]_{\mathcal{G}} = \left[G^* f, \sigma_\Theta(\cdot, z)g\right]_{\mathcal{H}(\Theta)} = \left[f, G\sigma_\Theta(\cdot, z)g\right]_{\mathcal{F}}$$
$$= \left[f, \frac{\Theta(z)^* - \Theta(\alpha)^*}{\delta_\alpha(z)^*} g\right]_{\mathcal{F}} = \left[\frac{\Theta(z) - \Theta(\alpha)}{\delta_\alpha(z)} f, g\right]_{\mathcal{G}},$$

which proves (3.11).

Finally, let F be defined by (3.4). Then we have for $g \in \mathcal{G}$,

$$\|Fg\|_{\mathcal{H}(\Theta)}^2 = \left[\sigma_\Theta(\cdot, \alpha)g, \ \sigma_\Theta(\cdot, \alpha)g\right]_{\mathcal{H}(\Theta)} = \left[\sigma_\Theta(\alpha, \alpha)g, g\right]_{\mathcal{G}} \leq \|\sigma_\Theta(\alpha, \alpha)\| \|g\|^2,$$

which shows that $F \in \mathcal{L}\left(\mathcal{G}, \mathcal{H}(\Theta)\right)$ and $\|F\| \leq \sqrt{\|\sigma_\Theta(\alpha, \alpha)\|}$. Again using (3.4) we obtain that for $g \in \mathcal{G}$, $h \in \mathcal{H}(\Theta)$,

$$[F^* h, g]_{\mathcal{G}} = [h, Fg]_{\mathcal{H}(\Theta)} = \left[h, \sigma_\Theta(\cdot, \alpha)g\right]_{\mathcal{H}(\Theta)} = \left[h(\alpha), g\right]_{\mathcal{G}},$$

and therefore the adjoint operator $F^* \in \mathcal{L}\left(\mathcal{H}(\Theta), \mathcal{G}\right)$ is given by (3.12). Since $\Theta \in S_\alpha(\rho, \mathcal{F}, \mathcal{G})$, the operator H defined by (3.5) is bounded, and (3.13) holds for all $f \in \mathcal{F}$.

It remains to substantiate the claim (3.16). Let g_ω, g_z, g and h belong to $\mathcal{G}$, so that

$$\left\{\begin{pmatrix} S_b(g_\omega) + b(\alpha)\sigma_\Theta(\cdot, \alpha)g \\ g \end{pmatrix}, \begin{pmatrix} S_a(g_\omega) + a(\alpha)\sigma_\Theta(\cdot, \alpha)g \\ S_r(g_\omega) + \Theta(\alpha)^* g \end{pmatrix}\right\}$$

and

$$\left\{\begin{pmatrix} S_b(g_z) + b(\alpha)\sigma_\Theta(\cdot, \alpha)h \\ h \end{pmatrix}, \begin{pmatrix} S_a(g_z) + a(\alpha)\sigma_\Theta(\cdot, \alpha)h \\ S_r(g_z) + \Theta(\alpha)^* h \end{pmatrix}\right\}$$

are elements in the relation R. We have to show that

$$\left[S_b(g_\omega) + b(\alpha)\sigma_\Theta(\cdot, \alpha)g, S_b(g_z) + b(\alpha)\sigma_\Theta(\cdot, \alpha)h\right]_{\mathcal{H}(\Theta)} + [g, h]_{\mathcal{G}}$$
$$= \left[S_a(g_\omega) + b(\alpha)\sigma_\Theta(\cdot, \alpha)g, S_a(g_z) + b(\alpha)\sigma_\Theta(\cdot, \alpha)h\right]_{\mathcal{H}(\Theta)}$$
$$+ \left[S_r(g_\omega) + \Theta(\alpha)^* g, S_r(g_z) + \Theta(\alpha)^* h\right]_{\mathcal{F}}.$$

The lefthand side of this equality can be rewritten as

$$\sum_{\omega, z} \frac{1}{\delta_\alpha(\omega)^* \delta_\alpha(z)} \left\{\left[b(\omega)^* b(z)\sigma_\Theta(z, \omega)g_\omega, h_z\right]_{\mathcal{G}}\right.$$
$$\left. - \left[b(\alpha)b(z)\sigma_\Theta(z, \alpha)g_\omega, h_z\right]_{\mathcal{G}} - \left[b(\alpha)b(w)^*\sigma_\Theta(\alpha, \omega)g_\omega, h_z\right]_{\mathcal{G}} + \left[b(\alpha)^2 \sigma_\Theta(\alpha, \alpha)g_\omega, h_z\right]_{\mathcal{G}}\right\}$$

$$+ \sum_\omega \frac{1}{\delta_\alpha(\omega)^*} \left\{ b(\alpha)b(\omega)^* \big[\sigma_\Theta(\alpha,\omega)g_\omega, h_z\big]_\mathcal{G} - b(\alpha)^2\big[\sigma_\Theta(\alpha,\alpha)g_\omega, h_z\big]_\mathcal{G} \right\}$$

$$+ \sum_\omega \frac{1}{\delta_\alpha(z)} \left\{ b(\alpha)b(z) \big[\sigma_\Theta(z,\alpha)g_\omega, h_z\big]_\mathcal{G} - b(\alpha)^2\big[\sigma_\Theta(\alpha,\alpha)g_\omega, h_z\big]_\mathcal{G} \right\}$$

$$+ b(\alpha)^2 \big[\sigma_\Theta(\alpha,\alpha)g_\omega, h_z\big]_\mathcal{G} + [g_\omega, h_z]_\mathcal{G}.$$

As for the righthand side, it can be rewritten as

$$\sum_{\omega,z} \frac{1}{\delta_\alpha(\omega)^*\delta_\alpha(z)} \left\{ \big[a(\omega)^*a(z)\sigma_\Theta(z,\omega)g_\omega, h_z\big]_\mathcal{G} \right.$$

$$- \big[a(\alpha)a(z)\sigma_\Theta(z,\alpha)g_\omega, h_z\big]_\mathcal{G} - \big[a(\alpha)a(w)^*\sigma_\Theta(\alpha,\omega)g_\omega, h_z\big]_\mathcal{G} + \big[a(\alpha)^2\sigma_\Theta(\alpha,\alpha)g_\omega, h_z\big]_\mathcal{G} \bigg\}$$

$$+ \sum_\omega \frac{1}{\delta_\alpha(\omega)^*} \left\{ a(\alpha)a(\omega)^* \big[\sigma_\Theta(\alpha,\omega)g_\omega, h_z\big]_\mathcal{G} - a(\alpha)^2\big[\sigma_\Theta(\alpha,\alpha)g_\omega, h_z\big]_\mathcal{G} \right\}$$

$$+ \sum_\omega \frac{1}{\delta_\alpha(z)} \left\{ a(\alpha)a(z) \big[\sigma_\Theta(z,\alpha)g_\omega, h_z\big]_\mathcal{G} - a(\alpha)^2\big[\sigma_\Theta(\alpha,\alpha)g_\omega, h_z\big]_\mathcal{G} \right\}$$

$$+ a(\alpha)^2 \big[\sigma_\Theta(\alpha,\alpha)g_\omega, h_z\big]_\mathcal{G} + [\Theta(\alpha)\Theta(\alpha)^*g_\omega, h_z]_\mathcal{G}.$$

Therefore the difference between the righthand side and the lefthand side is equal to

$$\sum_{z,\omega} \frac{1}{\delta_\alpha(\omega)^*\delta_\alpha(z)} \left\{ \big[(I - \Theta(z)\Theta(\omega)^*)g_\omega, h_z\big]_\mathcal{G} - \big[((I - \Theta(z)\Theta(\alpha)^*)g_\omega, h_z\big]_\mathcal{G} \right.$$

$$\left. - \big[(I - \Theta(\alpha)\Theta(\omega)^*)g_\omega, h_z\big]_\mathcal{G} + (I - \Theta(\alpha)\Theta(\alpha)^*)g_\omega, h_z\big]_\mathcal{G} \right]$$

$$+ \sum_\omega \frac{1}{\delta_\alpha(\omega)^*} \left[\big[(I - \Theta(\alpha)\Theta(\omega)^*)g_\omega, h_z\big]_\mathcal{G} - (I - \Theta(\alpha)\Theta(\alpha)^*)g_\omega, h_z\big]_\mathcal{G} \right]$$

$$+ \sum_z \frac{1}{\delta_\alpha(z)} \left[\big[(I - \Theta(z)\Theta(\alpha)^*)g_\omega, h_z\big]_\mathcal{G} - (I - \Theta(\alpha)\Theta(\alpha)^*)g_\omega, h_z\big]_\mathcal{G} \right],$$

which is easily seen to be zero. Thus the relation R is isometric. ∎

Lemma 3.2 *The operator $a(\omega)A^* - b(\omega)B^*$ is invertible for every $\omega \in U_\alpha$ and, moreover,*

$$\left(\big(a(\omega)A^* - b(\omega)B^*\big)^{-1}h\right)(z) = \frac{\delta_\alpha(z)h(z) - \delta_\alpha(\omega)h(\omega)}{\delta_\omega(z)}. \tag{3.18}$$

Proof. Let ω be a fixed point in U_α and let h be in $\mathcal{H}(\Theta)$. Then, in view of (3.9), (3.10) and (2.8),

$$\left(\big(a(\omega)A^* - b(\omega)B^*\big)h\right)(z) = \frac{\delta_\omega(z)h(z) - \delta_\omega(\alpha)h(\alpha)}{\delta_\alpha(z)} \in \mathcal{H}(\Theta). \tag{3.19}$$

In Theorem 3.1 we replace α by ω. Then we have to replace a by $\tilde{a}(z) = a(z) \cdot \frac{\overline{a(\omega)}}{|a(\omega)|}$ and b by $\tilde{b}(z) = b(z) \cdot \frac{\overline{b(\omega)}}{|b(\omega)|}$, where we set $\frac{\overline{b(\omega)}}{|b(\omega)|} = 1$ if $b(\omega) = 0$. Now $\tilde{a}(\omega)$ and $\tilde{b}(\omega)$ are real numbers. We obtain that the operators A_ω and B_ω defined by

$$
\begin{aligned}
(A_\omega h)(z) &= \frac{|a(\omega)|}{a(\omega)} \cdot \frac{\tilde{b}(z)h(z) - \tilde{b}(\omega)h(\omega)}{\tilde{b}(z)\tilde{a}(\omega) - \tilde{a}(z)\tilde{b}(\omega)} \\
&= \frac{b(z)h(z) - b(\omega)h(\omega)}{\delta_\omega(z)},
\end{aligned}
\tag{3.20}
$$

$$
\begin{aligned}
(B_\omega h)(z) &= \frac{|b(\omega)|}{b(\omega)} \cdot \frac{\tilde{a}(z)h(z) - \tilde{a}(\omega)h(\omega)}{\tilde{b}(z)\tilde{a}(\omega) - \tilde{a}(z)\tilde{b}(\omega)} \\
&= \frac{a(z)h(z) - a(\omega)h(\omega)}{\delta_\omega(z)},
\end{aligned}
\tag{3.21}
$$

map $\mathcal{H}(\Theta)$ into $\mathcal{H}(\Theta)$ and are bounded. On account of (3.20) and (3.21),

$$
\left(\big(a(\alpha)A_\omega - b(\alpha)B_\omega \big) h \right)(z) = \frac{\delta_\alpha(z)h(z) - \delta_\alpha(\omega)h(\omega)}{\delta_\omega(z)} \in \mathcal{H}(\Theta).
\tag{3.22}
$$

Combining (3.19) and (3.22) we obtain that for every $h \in \mathcal{H}(\Theta)$,

$$
\begin{aligned}
&\left(\big(a(\omega)A^* - b(\omega)B^* \big) \big(a(\alpha)A_\omega - b(\alpha)B_\omega \big) h \right)(z) \\
&= \frac{\delta_\alpha(z)h(z) - \delta_\alpha(\omega)h(\omega) - \delta_\alpha(\alpha)h(\alpha) + \delta_\alpha(\omega)h(\omega)}{\delta_\alpha(z)} = h(z).
\end{aligned}
$$

Similarly,

$$
\big(a(\alpha)A_\omega - b(\alpha)B_\omega \big) \big(a(\omega)A^* - b(\omega)B^* \big) = I,
$$

and hence

$$
\big(a(\omega)A^* - b(\omega)B^* \big)^{-1} = a(\alpha)A_\omega - b(\alpha)B_\omega \in \mathcal{L}\big(\mathcal{H}(\Theta) \big).
\tag{3.23}
$$

Finally, (3.18) follows from (3.22) and (3.23). ∎

Corollary 3.3 *The formulas*

$$
(R_\omega f)(z) = \frac{\Theta(z) - \Theta(\omega)}{\delta_\omega(z)} f, \qquad f \in \mathcal{F},
\tag{3.24}
$$

$$
(E_\omega h)(z) = h(\omega), \qquad h \in \mathcal{H}(\Theta),
\tag{3.25}
$$

define bounded operators $R_\omega \in \mathcal{L}\big(\mathcal{F}, \mathcal{H}(\Theta) \big)$ and $E_\omega \in \mathcal{L}\big(\mathcal{H}(\Theta), \mathcal{G} \big)$ for all ω in U_α. In fact, they are given by

$$
R_\omega f = \big(a(\omega)A^* - b(\omega)B^* \big)^{-1} G^* f,
\tag{3.26}
$$

$$
E_\omega h = F^* \big(a(\omega)A^* - b(\omega)B^* \big)^{-1} h.
\tag{3.27}
$$

Proof. We define the bounded operators R_ω and E_ω by (3.26) and (3.27), and show that they satisfy (3.24), (3.25). Using (3.18) we obtain

$$
\begin{aligned}
(R_\omega f(z) &= \left(\left(a(\omega)A^* - b(\omega)B^*\right)^{-1}G^*f\right)(z) \\
&= \frac{\delta_\alpha(z)(G^*f)(z) - \delta_\alpha(\omega)(G^*f)(\omega)}{\delta_\omega(z)} = \frac{\Theta(z) - \Theta(\omega)}{\delta_\omega(z)}f,
\end{aligned}
$$

$$
(E_\omega h)(z) = \left(F^*\left(a(\omega)A^* - b(\omega)B^*\right)^{-1}h\right)(z) = F^*\frac{\delta_\alpha(z)h(z) - \delta_\alpha(\omega)h(\omega)}{\delta_\omega(z)} = h(\omega).
$$

∎

Theorem 3.4 *Let Θ belong to $S_\alpha(\rho, \mathcal{F}, \mathcal{G})$, where $\mathcal{F}$ and $\mathcal{G}$ are Hilbert spaces. Let $\mathcal{H}(\Theta)$ be the reproducing kernel Hilbert space with reproducing kernel $\sigma_\Theta(z, \omega)$, and let A, B, G, F and H be the operators described in Theorem 3.1. Then the colligation*

$$
\Delta = \left(\alpha, \rho, \mathcal{H}(\Theta), \mathcal{F}, \mathcal{G}, A^*, B^*, H^*, F^*, G^*\right) \tag{3.28}
$$

is coisometric and closely outerconnected, and $\Theta_\Delta = \Theta$ on U_α. If the colligation

$$
\Delta_1 = (\alpha, \rho, \mathcal{K}, \mathcal{F}, \mathcal{G}, A_1^*, B_1^*, H_1^*, F_1^*, G_1^*) \tag{3.29}
$$

has the same properties, then

$$
H_1 = H, \qquad WF_1 = F, \qquad G_1 = GW, \qquad WB_1 = BW, \qquad WA_1 = AW,
$$

where $W: \mathcal{K} \to \mathcal{H}(\Theta)$ is the unitary mapping which assigns to each $k \in \mathcal{K}$ the function

$$
(Wk)(z) = F_1^*\left(a(z)A_1^* - b(z)B_1^*\right)^{-1}k.
$$

Proof. The first part of the theorem follows from the arguments of the proof of Theorem 3.1. They imply that the closure $\bar{R}$ of the linear isometric relation R in the space $(\mathcal{H}(\Theta) \oplus \mathcal{G}) \oplus (\mathcal{H}(\Theta) \oplus \mathcal{F})$ can be written as

$$
\bar{R} = \left\{\left\{\mathcal{A}\binom{h}{g}, \mathcal{B}\binom{h}{g}\right\}; h \in \mathcal{H}(\Theta), g \in \mathcal{G}\right\},
$$

where $\mathcal{A}$ and $\mathcal{B}$ are given by (2.4). The isometric property of R implies (2.3). By Lemma 2.1, the colligation Δ defined by (3.28) is coisometric. We now show that the colligation is closely outerconnected. It follows from (3.1), (3.2) that for every choice of $g \in \mathcal{G}$ and $\omega \in U_\alpha$,

$$
\left(a(\omega)^*A - b(\omega)^*B\right)\sigma_\Theta(\cdot, \omega)g = \sigma_\Theta(\cdot, \alpha)g,
$$

18   D. Alpay et al.

which together with (3.4) implies

$$\left(a(\omega)^*A - b(\omega)^*B\right)^{-1}Fg = \sigma_\Theta(\cdot,\omega)g.$$

The inverse on the lefthand side exists for every $\omega \in U_\alpha$ because of Lemma 3.2. Since $\mathcal{H}_0$ defined by (3.14) is dense in $\mathcal{H}(\Theta)$ we obtain

$$\bigvee_{\omega\in U_\alpha} \mathrm{ran}\left(a(\omega)^*A - b(\omega)^*B\right)^{-1}F = \mathcal{H}(\Theta),$$

which proves that the colligation (3.28) is closely outerconnected. By (3.9)–(3.13), (3.24) and (3.26), we have for arbitrary $f \in \mathcal{F}$ and $z \in U_\alpha$,

$$\begin{aligned}
\Theta_\Delta(z)f &= \left(H^* + \delta_\alpha(z)F^*\left(a(z)A^* - b(z)B^*\right)^{-1}G^*\right)f \\
&= \Theta(\alpha)f + \delta_\alpha(z)F^*\left(\frac{\Theta(z) - \Theta(\omega)}{\delta_\omega(z)}f\right) \\
&= \Theta(\alpha)f + \delta_\alpha(z)\frac{\Theta(z) - \Theta(\alpha)}{\delta_\alpha(z)}f = \Theta(z)f,
\end{aligned}$$

which yields $\Theta_\Delta = \Theta$ on U_α.

We now prove the essential uniqueness of Δ. Let Δ_1 of the form (3.29) be another closely outerconnected isometric colligation with $\Theta = \Theta_{\Delta_1}$. Then, on account of Lemma 2.2,

$$\begin{aligned}
&F^*\left(a(z)A^* - b(z)B^*\right)^{-1}\left(a(\omega)^*A - b(\omega)^*B\right)^{-1}F \\
&= \sigma_\Theta(z,\omega) = F_1^*\left(a(z)A_1^* - b(z)B_1^*\right)^{-1}\left(a(\omega)^*A_1 - b(\omega)^*B_1\right)^{-1}F_1.
\end{aligned}$$

Let S be the linear span of all elements of $\mathcal{K} \times \mathcal{H}(\Theta)$ which have the form

$$\left\{\left(a(\omega)^*A_1 - b(\omega)^*B_1\right)^{-1}F_1g, \ \left(a(\omega)^*A - b(\omega)^*B\right)^{-1}Fg\right\},$$

where g and ω vary over $\mathcal{G}$ and U_α, respectively. Then $\mathrm{dom}\, S$ is dense in $\mathcal{K}$, because of the closely outerconnectedness of Δ_1, and the above equalities show that for arbitrary $g_1, g_2 \in \mathcal{G}$ and $z,\omega \in U_\alpha$,

$$\begin{aligned}
&\left[\left(a(\omega)^*A_1 - b(\omega)^*B_1\right)^{-1}F_1g_1, \ \left(a(z)^*A_1 - b(z)^*B_1\right)^{-1}F_1g_2\right]_\mathcal{K} \\
&= \left[\left(a(\omega)^*A - b(\omega)^*B\right)^{-1}Fg_1, \ \left(a(z)^*A - b(z)^*B\right)^{-1}Fg_2\right]_{\mathcal{H}(\Theta)}.
\end{aligned}$$

So S is a densely defined isometric relation, and hence it defines a bounded isometric mapping $W\colon\mathcal{K} \to \mathcal{H}(\Theta)$. In view of the closely outerconnectedness of Δ, $\mathrm{ran}\, W$ is dense in $\mathcal{H}(\Theta)$, and hence W is unitary.

By definition, $W\big(a(\omega)^*A_1 - b(\omega)^*B_1\big)^{-1}F_1 = \big(a(\omega)^*A - b(\omega)^*B\big)^{-1}F$. If in this equality we put $\omega = \alpha$ we obtain, on account of (3.6),

$$WF_1 = F.$$

Using this equality and the identity

$$\delta_\alpha(\omega)^*B\big(a(\omega)^*A - b(\omega)^*B\big)^{-1} = -a(\alpha)I_{\mathcal{H}} + a(\omega)^*\big(a(\omega)^*A - b(\omega)^*B\big)^{-1}$$

we obtain

$$\begin{aligned}
\delta_\alpha(\omega)^*WB_1&\big(a(\omega)^*A_1 - b(\omega)^*B_1\big)^{-1}F_1 \\
&= -a(\alpha)WF_1 + a(\omega)^*W\big(a(\omega)^*A_1 - b(\omega)^*B_1\big)^{-1}F_1 \\
&= -a(\alpha)F + a(\omega)^*\big(a(\omega)^*A - b(\omega)^*B\big)^{-1}F \\
&= \delta_\alpha(\omega)^*B\big(a(\omega)^*A - b(\omega)^*B\big)^{-1}F \\
&= \delta_\alpha(\omega)^*BW\big(a(\omega)^*A_1 - b(\omega)^*B_1\big)^{-1}F_1,
\end{aligned}$$

which means that the equality

$$WB_1 k = BW k \tag{3.30}$$

holds for all $k \in \mathcal{K}$ of the form

$$k = \big(a(\omega)^*A_1 - b(\omega)^*B_1\big)^{-1}F_1 g, \qquad g \in \mathcal{G}.$$

Since such vectors form a total set in $\mathcal{K}$ and W is bounded, (3.30) is valid for all $k \in \mathcal{K}$, and

$$WB_1 = BW. \tag{3.31}$$

Using (3.31) and (3.6) we have $WA_1 = AW$.

Finally, it follows from

$$\begin{aligned}
H + \delta_\alpha(z)^*G&\big(a(z)^*A - b(z)^*B\big)^{-1}F = \Theta(z)^* \\
&= H_1 + \delta_\alpha(z)^*G_1\big(a(z)^*A_1 - b(z)^*B_1\big)^{-1}F_1, \qquad z \in U_\alpha,
\end{aligned}$$

that $H = H_1$ and that $G_1\big(a(z)^*A_1 - b(z)^*B_1\big)^{-1}F_1 = G\big(a(z)^*A - b(z)^*B\big)^{-1}F$, which, again by the continuity of W, yields $GW = G_1$. ∎

Corollary 3.5 *Let Θ belong to $S_\alpha(\rho, \mathcal{F}, \mathcal{G})$, where $\mathcal{F}$ and $\mathcal{G}$ are Hilbert spaces. Then there exists a closely interconnected isometric colligation Δ of the form (2.1) such that $\Theta(z) = \Theta_\Delta(z)$ for all $z \in U_\alpha$.*

Proof. Since $\Theta \in S_\alpha(\rho, \mathcal{F}, \mathcal{G})$, Theorem 3.3 implies that Θ admits a representation

$$\Theta(z) = H^* + \delta_\alpha(z)F^*\big(a(z)A^* - b(z)B^*\big)^{-1}G^*.$$

By Lemma 2.5, $\tilde{\Theta}$ can be written as

$$\tilde{\Theta}(z) = H + \delta_\alpha(z)G\big(a(z)A - b(z)B\big)^{-1}F,$$

and by Theorem 2.6, $\tilde{\Theta} \in S_\alpha(\rho, \mathcal{F}, \mathcal{G})$. We recall that $\mathcal{H}(\tilde{\Theta})$ stands for the reproducing kernel Hilbert space with reproducing kernel $\sigma_{\tilde{\Theta}}(z, \omega)$ defined in (2.11).

Replacing in Theorems 3.1 and 3.4 and in their proofs Θ by $\tilde{\Theta}$ we obtain that
(i) the formulas

$$\tilde{A}\sigma_{\tilde{\Theta}}(\cdot, \omega)f = \frac{b(\omega)^*\sigma_{\tilde{\Theta}}(\cdot, \omega) - b(\alpha)\sigma_{\tilde{\Theta}}(\cdot, \alpha)}{\delta_\alpha(\omega)^*}f,$$

$$\tilde{B}\sigma_{\tilde{\Theta}}(\cdot, \omega)f = \frac{a(\omega)^*\sigma_{\tilde{\Theta}}(\cdot, \omega) - a(\alpha)\sigma_{\tilde{\Theta}}(\cdot, \alpha)}{\delta_\alpha(\omega)^*}f,$$

$$\tilde{G}\sigma_{\tilde{\Theta}}(\cdot, \omega)f = \frac{\tilde{\Theta}(\omega)^* - \tilde{\Theta}(\alpha)^*}{\delta_\alpha(\omega)^*}f,$$

$$\tilde{F}f = \sigma_{\tilde{\Theta}}(\cdot, \alpha)f, \quad \widetilde{H}f = \tilde{\Theta}(\alpha)^*f, \qquad f \in \mathcal{F},$$

uniquely define bounded linear operators $\tilde{A}, \tilde{B} \in \mathcal{L}\big(\mathcal{H}(\tilde{\Theta})\big)$, $\tilde{G} \in \mathcal{L}\big(\mathcal{H}(\tilde{\Theta}), \mathcal{F}\big)$, $\tilde{F} \in \mathcal{L}\big(\mathcal{G}, \mathcal{H}(\tilde{\Theta})\big)$ and $\widetilde{H} \in \mathcal{L}(\mathcal{F}, \mathcal{G})$;

(ii) the colligation $\Delta = \big(\alpha, \rho, \mathcal{H}(\tilde{\Theta}), \mathcal{F}, \mathcal{G}, \tilde{A}, \tilde{B}, \widetilde{H}, \tilde{G}, \tilde{F}\big)$ is isometric and closely innerconnected;

(iii) $\Theta(z) = \Theta_\Delta(z)$ for all $z \in U_\alpha$. ∎

4. Unitary realization

In this section we prove the following theorem.

Theorem 4.1 *Let Θ belong to $S_\alpha(\rho, \mathcal{F}, \mathcal{G})$, where $\mathcal{F}$ and $\mathcal{G}$ are Hilbert spaces. Then there exists a closely connected unitary colligation Δ of the form (2.1):*

$$\Delta = (\alpha, \rho, \mathcal{D}(\Theta), \mathcal{F}, \mathcal{G}, A, B, H, G, F),$$

such that $\Theta(z) = \Theta_\Delta(z)$ for all $z \in U_\alpha$. Here the operators A, B and G are related by the formulas

$$(Ah)(z) = \begin{pmatrix} \frac{b(z)}{\delta_\alpha(z)}I & 0 \\ 0 & \frac{a(z)}{\rho_\alpha(z)}I \end{pmatrix} h(z) - \begin{pmatrix} \frac{b(\alpha)}{\delta_\alpha(z)}I \\ \frac{b(\alpha)}{\rho_\alpha(z)}\Theta(z) \end{pmatrix} Gh, \tag{4.1}$$

$$(Bh)(z) = \begin{pmatrix} \frac{a(z)}{\delta_\alpha(z)}I & 0 \\ 0 & \frac{b(z)}{\rho_\alpha(z)}I \end{pmatrix} h(z) - \begin{pmatrix} \frac{b(\alpha)}{\delta_\alpha(z)}I \\ \frac{a(\alpha)}{\rho_\alpha(z)}\Theta(z) \end{pmatrix} Gh, \qquad h \in D(\Theta). \tag{4.2}$$

Proof. By Theorem 2.6 the kernel $D_\Theta(z,\omega)$ defined by (2.12) is nonnegative in some neighbourhood U_α of α. Recall that $\mathcal{D}(\Theta)$ denotes the associated reproducing kernel Hilbert space. We show that it provides the state space for the unitary realization.

As in the proof of Theorem 3.1 we consider finite sums over $\omega \in U_\alpha$ of the form

$$D\begin{pmatrix} g_\omega \\ f_\omega \end{pmatrix} = \sum_\omega D_\Theta(\cdot,\omega)\begin{pmatrix} g_\omega \\ f_\omega \end{pmatrix}, \tag{4.3}$$

$$D_a\begin{pmatrix} g_\omega \\ f_\omega \end{pmatrix} = \sum_\omega D_\Theta(\cdot,\omega)\begin{pmatrix} \frac{a(\omega)^*}{\rho_\alpha(\omega)^*}g_\omega \\ \frac{b(\omega)^*}{\delta_\alpha(\omega)^*}f_\omega \end{pmatrix} - D_\Theta(\cdot,\alpha)\begin{pmatrix} 0 \\ \frac{b(\alpha)}{\delta_\alpha(\omega)^*}f_\omega + \frac{b(\alpha)}{\rho_\alpha(\omega)^*}\Theta(\omega)^*g_\omega \end{pmatrix}, \tag{4.4}$$

$$D_b\begin{pmatrix} g_\omega \\ f_\omega \end{pmatrix} = \sum_\omega D_\Theta(\cdot,\omega)\begin{pmatrix} \frac{b(\omega)^*}{\rho_\alpha(\omega)^*}g_\omega \\ \frac{a(\omega)^*}{\delta_\alpha(\omega)^*}f_\omega \end{pmatrix} - D_\Theta(\cdot,\alpha)\begin{pmatrix} 0 \\ \frac{a(\alpha)}{\delta_\alpha(\omega)^*}f_\omega + \frac{a(\alpha)}{\rho_\alpha(\omega)^*}\Theta(\omega)^*g_\omega \end{pmatrix}, \tag{4.5}$$

and

$$D_r\begin{pmatrix} g_\omega \\ f_\omega \end{pmatrix} = \sum_\omega \frac{\tilde\Theta(\omega)^* - \Theta(\alpha)}{\delta_\alpha(\omega)^*}f_\omega + \frac{I_\mathcal{G} - \Theta(\alpha)\Theta(\omega)^*}{\rho_\alpha(\omega)^*}g_\omega, \tag{4.6}$$

where $f_\omega \in \mathcal{F}$, $g_\omega \in \mathcal{G}$. We define for $f \in \mathcal{F}$,

$$Ff = D_\Theta(\cdot,\alpha)\begin{pmatrix} 0 \\ f \end{pmatrix}, \qquad Hf = \Theta(\alpha)f, \tag{4.7}$$

which are clearly bounded operators: $F \in \mathcal{L}(\mathcal{F}, \mathcal{D}(\Theta))$, $H \in \mathcal{L}(\mathcal{F}, \mathcal{G})$. Then the relation $R_i \in (\mathcal{D}(\Theta) \oplus \mathcal{F}) \times (\mathcal{D}(\Theta) \oplus \mathcal{G})$ defined by

$$R_i = \left\{ \left\{ \begin{pmatrix} D_b\begin{pmatrix} g_\omega \\ f_\omega \end{pmatrix} + b(\alpha)Ff \\ f \end{pmatrix}, \begin{pmatrix} D_a\begin{pmatrix} g_\omega \\ f_\omega \end{pmatrix} + a(\alpha)Ff \\ D_r\begin{pmatrix} g_\omega \\ f_\omega \end{pmatrix} + Hf \end{pmatrix} \right\}; \begin{pmatrix} g_\omega \\ f_\omega \end{pmatrix} \in \begin{pmatrix} \mathcal{G} \\ \mathcal{F} \end{pmatrix}, f \in \mathcal{F} \right\}$$

is obviously linear, but also isometric. The proof of the latter is similar to the proof that the relation R is isometric in the previous section and will be omitted.

Since

$$a(\alpha)D_b\begin{pmatrix} g_\omega \\ f_\omega \end{pmatrix} - b(\alpha)D_a\begin{pmatrix} g_\omega \\ f_\omega \end{pmatrix} = D\begin{pmatrix} g_\omega \\ f_\omega \end{pmatrix}, \tag{4.8}$$

we can derive as in the proof of Theorem 3.1 simple estimates which imply that the closure of the relations

$$\left\{ \left\{ D\begin{pmatrix} g_\omega \\ f_\omega \end{pmatrix}, D_b\begin{pmatrix} g_\omega \\ f_\omega \end{pmatrix} \right\}; \begin{pmatrix} g_\omega \\ f_\omega \end{pmatrix} \in \begin{pmatrix} \mathcal{G} \\ \mathcal{F} \end{pmatrix} \right\},$$

$$\left\{ \left\{ D\begin{pmatrix} g_\omega \\ f_\omega \end{pmatrix}, D_a\begin{pmatrix} g_\omega \\ f_\omega \end{pmatrix} \right\}; \begin{pmatrix} g_\omega \\ f_\omega \end{pmatrix} \in \begin{pmatrix} \mathcal{G} \\ \mathcal{F} \end{pmatrix} \right\},$$

22   D. Alpay et al.

$$\left\{\left\{D\begin{pmatrix}g_\omega\\f_\omega\end{pmatrix}, D_r\begin{pmatrix}g_\omega\\f_\omega\end{pmatrix}\right\}; \begin{pmatrix}g_\omega\\f_\omega\end{pmatrix} \in \begin{pmatrix}\mathcal{G}\\\mathcal{F}\end{pmatrix}\right\}$$

are the graphs of bounded operators $A, B \in \mathcal{L}(\mathcal{D}(\Theta))$ and $G \in \mathcal{L}(\mathcal{D}(\Theta), \mathcal{F})$, respectively. Note that, by (4.3)–(4.6), for $f \in \mathcal{F}, g \in \mathcal{G}$ and $\omega \in U_\alpha$,

$$AD_\Theta(\cdot,\omega)\begin{pmatrix}g\\f\end{pmatrix} = D_\Theta(\cdot,\omega)\begin{pmatrix}\frac{a(\omega)^*}{\rho_\alpha(\omega)^*}g\\\frac{b(\omega)^*}{\delta_\alpha(\omega)^*}f\end{pmatrix} - D_\Theta(\cdot,\alpha)\begin{pmatrix}0\\\frac{b(\alpha)}{\delta_\alpha(\omega)^*}f + \frac{b(\alpha)}{\rho_\alpha(\omega)^*}\Theta(\omega)^*g\end{pmatrix}, \quad (4.9)$$

$$BD_\Theta(\cdot,\omega)\begin{pmatrix}g\\f\end{pmatrix} = D_\Theta(\cdot,\omega)\begin{pmatrix}\frac{b(\omega)^*}{\rho_\alpha(\omega)^*}g\\\frac{a(\omega)^*}{\delta_\alpha(\omega)^*}f\end{pmatrix} - D_\Theta(\cdot,\alpha)\begin{pmatrix}0\\\frac{a(\alpha)}{\delta_\alpha(\omega)^*}f + \frac{a(\alpha)}{\rho_\alpha(\omega)^*}\Theta(\omega)^*g\end{pmatrix} \quad (4.10)$$

and

$$GD_\Theta(\cdot,\omega)\begin{pmatrix}g\\f\end{pmatrix} = \frac{\tilde{\Theta}(\omega)^* - \Theta(\alpha)}{\delta_\alpha(\omega)^*}f + \frac{I_\mathcal{G} - \Theta(\alpha)\Theta(\omega)^*}{\rho_\alpha(\omega)^*}g. \quad (4.11)$$

From (4.8)–(4.10) it follows that

$$a(\alpha)A - b(\alpha)B = I_{\mathcal{D}(\Theta)}. \quad (4.12)$$

Since the operators A, B, G, F and H are bounded, the adjoint operators $A^*, B^* \in \mathcal{L}(\mathcal{D}(\Theta))$, $G^* \in \mathcal{L}(\mathcal{G}, \mathcal{D}(\Theta))$, $F^* \in \mathcal{L}(\mathcal{D}(\Theta), \mathcal{F})$ and $H^* \in \mathcal{L}(\mathcal{G}, \mathcal{F})$ exist. They are given by

$$A^*D_\Theta(\cdot,\omega)\begin{pmatrix}g\\f\end{pmatrix} = D_\Theta(\cdot,\omega)\begin{pmatrix}\frac{b(\omega)^*}{\delta_\alpha(\omega)^*}g\\\frac{a(\omega)^*}{\rho_\alpha(\omega)^*}f\end{pmatrix} - D_\Theta(\cdot,\alpha)\begin{pmatrix}\frac{b(\alpha)}{\rho_\alpha(\omega)^*}\tilde{\Theta}(\omega)^*f + \frac{b(\alpha)}{\delta_\alpha(\omega)^*}g\\0\end{pmatrix},$$
$$(4.13)$$

$$B^*D_\Theta(\cdot,\omega)\begin{pmatrix}g\\f\end{pmatrix} = D_\Theta(\cdot,\omega)\begin{pmatrix}\frac{a(\omega)^*}{\delta_\alpha(\omega)^*}g\\\frac{b(\omega)^*}{\rho_\alpha(\omega)^*}f\end{pmatrix} - D_\Theta(\cdot,\alpha)\begin{pmatrix}\frac{a(\alpha)}{\rho_\alpha(\omega)^*}\tilde{\Theta}(\omega)^*f + \frac{a(\alpha)}{\delta_\alpha(\omega)^*}g\\0\end{pmatrix};$$
$$(4.14)$$

$$G^*g = D_\Theta(\cdot,\alpha)\begin{pmatrix}g\\0\end{pmatrix}, \quad (4.15)$$

$$F^*\begin{pmatrix}k\\h\end{pmatrix} = h(\alpha), \quad (4.16)$$

$$H^*g = \Theta(\alpha)^* = \tilde{\Theta}(\alpha)g, \quad (4.17)$$

where $f \in \mathcal{F}$, $g \in \mathcal{G}$ and $\binom{k}{h} \in \mathcal{D}(\Theta)$.

The formula (4.17) for H^* is clear. Let us verify the formulas (4.15) and (4.16) for the adjoints of G and F. By the reproducing kernel property of D_Θ and (4.11), we have for arbitrary $f \in \mathcal{F}$, $g, y \in \mathcal{G}$ and $\binom{k}{h} \in \mathcal{D}(\Theta)$,

$$
\left[G^*g(z), \binom{y}{f} \right]_{\mathcal{G} \oplus \mathcal{F}} = \left[G^*g, \, D_\Theta(\cdot, z) \binom{y}{f} \right]_{\mathcal{D}(\Theta)}
$$
$$
= \left[g, \, \frac{\widetilde{\Theta}(z)^* - \Theta(\alpha)}{\delta_\alpha(z)^*} f + \frac{I - \Theta(\alpha)\Theta(z)^*}{\rho_\alpha(z)^*} y \right]_{\mathcal{G}}
$$
$$
= \left[\frac{\widetilde{\Theta}(z) - \Theta(\alpha)^*}{\delta_\alpha(z)} g, f \right]_{\mathcal{F}} + \left[\frac{I - \Theta(z)\Theta(\alpha)^*}{\rho_\alpha(z)} g, y \right]_{\mathcal{F}}
$$
$$
= \left[\begin{pmatrix} \frac{\widetilde{\Theta}(z) - \Theta(\alpha)^*}{\delta_\alpha(z)} g \\ \frac{I - \Theta(z)\Theta(\alpha)^*}{\rho_\alpha(z)} g \end{pmatrix}, \binom{y}{f} \right]_{\mathcal{G} \oplus \mathcal{F}} = \left[D_\Theta(z, \alpha) \binom{g}{0}, \binom{y}{f} \right]_{\mathcal{G} \oplus \mathcal{F}}.
$$
$$
\tag{4.18}
$$

Similarly,

$$
\left[\binom{k}{h}, Ff \right]_{\mathcal{D}(\Theta)} = \left[\binom{k}{h}, \, D_\Theta(\cdot, \alpha) \binom{0}{f} \right]_{\mathcal{D}(\Theta)}
$$
$$
= \left[\binom{k(\alpha)}{h(\alpha)}, \binom{0}{f} \right]_{\mathcal{G} \oplus \mathcal{F}} = \left[h(\alpha), f \right]_{\mathcal{F}}.
$$

To obtain (4.14) we use the following identity

$$
\begin{pmatrix} \frac{b(z)}{\rho_\alpha(z)} I & 0 \\ 0 & \frac{a(z)}{\delta_\alpha(z)} I \end{pmatrix} D_\Theta(z, \omega) - D_\Theta(z, \omega) \begin{pmatrix} \frac{a(\omega)^*}{\delta_\alpha(\omega)^*} I & 0 \\ 0 & \frac{b(\omega)^*}{\rho_\alpha(\omega)^*} I \end{pmatrix}
$$
$$
= \begin{pmatrix} 0 & \frac{a(\alpha)}{\rho_\alpha(z)} \Theta(z) \\ 0 & \frac{a(\alpha)}{\delta_\alpha(z)} \end{pmatrix} D_\Theta(\alpha, \omega) - D_\Theta(z, \alpha) \begin{pmatrix} \frac{a(\alpha)}{\delta_\alpha(\omega)^*} I & \frac{a(\alpha)}{\rho_\alpha(\omega)^*} \widetilde{\Theta}(\omega)^* \\ 0 & 0 \end{pmatrix},
\tag{4.19}
$$

which holds for all $z, \omega \in U_\alpha$ and can be checked by a direct computation. Using (4.10), the reproducing kernel property of D_Θ and (4.19) we find (with $x \in \mathcal{F}$)

$$
\left[B^* D_\Theta(\cdot, \omega) \binom{g}{f}, D_\Theta(\cdot, z) \binom{y}{x} \right]_{\mathcal{D}(\Theta)}
$$
$$
= \left[D_\Theta(\cdot, \omega) \binom{g}{f}, \, D_\Theta(\cdot, z) \begin{pmatrix} \frac{b(z)^*}{\rho_\alpha(z)^*} y \\ \frac{a(z)^*}{\delta_\alpha(z)^*} x \end{pmatrix} \right.
$$
$$
\left. - D_\Theta(\cdot, \alpha) \begin{pmatrix} 0 \\ \frac{b(\alpha)}{\delta_\alpha(z)^*} x + \frac{b(\alpha)}{\rho_\alpha(z)^*} \Theta(z)^* y \end{pmatrix} \right]_{\mathcal{D}(\Theta)}
$$

$$
= \left[\left(\left(\begin{pmatrix} \frac{b(z)}{\rho_\alpha(z)} I & 0 \\ 0 & \frac{a(z)}{\delta_\alpha(z)} I \end{pmatrix} D_\Theta(z,\omega) \right. \right. \right.
$$

$$
\left. \left. \left. - \begin{pmatrix} 0 & \frac{a(\alpha)}{\delta_\alpha(z)} I \\ 0 & \frac{a(\alpha)}{\rho_\alpha(z)} \Theta(z) \end{pmatrix} D_\Theta(\alpha,\omega) \right) \begin{pmatrix} g \\ f \end{pmatrix}, \begin{pmatrix} y \\ x \end{pmatrix} \right]_{\mathcal{G} \oplus \mathcal{F}}
$$

$$
= \left[\left(D_\Theta(z,\omega) \begin{pmatrix} \frac{a(\omega)^*}{\delta_\alpha(\omega)^*} I & 0 \\ 0 & \frac{b(\omega)^*}{\rho_\alpha(\omega)^*} I \end{pmatrix} \right. \right.
$$

$$
\left. \left. - D_\Theta(z,\alpha) \begin{pmatrix} \frac{a(\alpha)}{\delta_\alpha(\omega)^*} I & \frac{a(\alpha)}{\rho_\alpha(\omega)^*} \tilde{\Theta}(\omega)^* \\ 0 & 0 \end{pmatrix} \right) \begin{pmatrix} g \\ f \end{pmatrix}, \begin{pmatrix} y \\ x \end{pmatrix} \right]_{\mathcal{G} \oplus \mathcal{F}}
$$

$$
= \left[D_\Theta(\cdot,\omega) \begin{pmatrix} \frac{a(\omega)^*}{\delta_\alpha(\omega)^*} g \\ \frac{b(\omega)^*}{\rho_\alpha(\omega)^*} f \end{pmatrix} \right.
$$

$$
\left. - D_\Theta(\cdot,\alpha) \left(\frac{a(\alpha)}{\rho_\alpha(\omega)^*} \tilde{\Theta}(\omega)^* f + \frac{a(\alpha)}{\delta_\alpha(\omega)^*} g \right), D_\Theta(\cdot,z) \begin{pmatrix} y \\ x \end{pmatrix} \right]_{\mathcal{D}(\Theta)},
$$

which implies (4.14). The equality (4.13) follows from (4.12) and (4.14). Note that the closure of the relation R_i can be written as

$$
\bar{R}_i = \left\{ \left\{ \mathcal{A} \begin{pmatrix} h \\ f \end{pmatrix}, \mathcal{B} \begin{pmatrix} h \\ f \end{pmatrix} \right\}; h \in \mathcal{D}(\Theta), f \in \mathcal{F} \right\},
$$

where $\mathcal{A}$ and $\mathcal{B}$ are given by (2.4). Since R_i is isometric, the equality (2.3) holds. We define $\tilde{A}$ and $\tilde{B}$ by (2.6). It follows from (4.7), (4.9)–(4.11) and (4.13)–(4.17) that (2.5) is valid. We omit the calculations. Hence (2.3) and (2.5) are valid simultaneously and the colligation is unitary.

We now show that the colligation is closely connected. By (4.9) and (4.10),

$$
\left(a(\omega)^* A - b(\omega)^* B \right) D_\Theta(\cdot,\omega) \begin{pmatrix} 0 \\ f \end{pmatrix} = D_\Theta(\cdot,\alpha) \begin{pmatrix} 0 \\ f \end{pmatrix}, \qquad f \in \mathcal{F},
$$

which together with (4.7) implies that

$$
\left(a(\omega)^* A - b(\omega)^* B \right)^{-1} F f = D_\Theta(\cdot,\omega) \begin{pmatrix} 0 \\ f \end{pmatrix}, \qquad f \in \mathcal{F}. \tag{4.20}
$$

Similarly, (4.13) and (4.15) imply that

$$
\left(a(\omega)^* A^* - b(\omega)^* B^* \right) D_\Theta(\cdot,\omega) \begin{pmatrix} g \\ 0 \end{pmatrix} = D_\Theta(\cdot,\alpha) \begin{pmatrix} g \\ 0 \end{pmatrix}, \qquad g \in \mathcal{G},
$$

which together with (4.15) gives

$$
\left(a(\omega)^* A^* - b(\omega)^* B^* \right)^{-1} G^* g = D_\Theta(\cdot,\omega) \begin{pmatrix} g \\ 0 \end{pmatrix}, \qquad g \in \mathcal{G}.
$$

Therefore

$$\bigvee_{z,\omega\in U_\alpha} \left\{ \mathrm{ran}\left(\left(a(z)^*A - b(z)^*B\right)^{-1}F\right), \mathrm{ran}\left(\left(a(\omega)^*A - b(\omega)^*B\right)^{-1}G^*\right)\right\}$$

$$= \bigvee_{z,\omega\in U_\alpha} \left\{ \left\{ D_\Theta(\cdot,z)\begin{pmatrix}0\\f\end{pmatrix}, \ f\in\mathcal{F}\right\}, \left\{ D_\Theta(\cdot,\omega)\begin{pmatrix}g\\0\end{pmatrix}, \ g\in\mathcal{G}\right\}\right\} = \mathcal{D}(\Theta),$$

and the colligation is closely connected.

Finally, using the formulas (4.20) and (4.11) we obtain that for all $f \in \mathcal{F}$ and $\omega \in U_\alpha$,

$$\left(H + \delta_\alpha(\omega)^*G\left(a(\omega)^*A - b(\omega)^*B\right)^{-1}F\right)f = \Theta(\alpha)f + \delta_\alpha(\omega)^*GD_\Theta(\cdot,\omega)\begin{pmatrix}0\\f\end{pmatrix}$$

$$= \Theta(\alpha)f + \delta_\alpha(\omega)^*\frac{\tilde\Theta(\omega)^* - \Theta(\alpha)}{\delta_\alpha(\omega)^*}f$$

$$= \tilde\Theta(\omega)^*f.$$

Hence

$$\Theta(\omega) = H + \delta_\alpha(\omega)G\left(a(\omega)A - b(\omega)B\right)^{-1}F = \Theta_\Delta(\omega).$$

To prove (4.2) we begin with the following identity, which is similar to formula (4.19),

$$\begin{pmatrix}\frac{a(z)}{\delta_\alpha(z)}I & 0\\ 0 & \frac{b(z)}{\rho_\alpha(z)}I\end{pmatrix} D_\Theta(z,\omega) - D_\Theta(z,\omega)\begin{pmatrix}\frac{b(\omega)^*}{\rho_\alpha(\omega)^*}I & 0\\ 0 & \frac{a(\omega)^*}{\delta_\alpha(\omega)^*}I\end{pmatrix}$$

$$= \begin{pmatrix}\frac{a(\alpha)}{\delta_\alpha(z)}I & 0\\ \frac{a(\alpha)}{\rho_\alpha(z)}\tilde\Theta(z) & 0\end{pmatrix} D_\Theta(\alpha,\omega) - D_\Theta(z,\alpha)\begin{pmatrix}0 & 0\\ \frac{a(\alpha)}{\rho_\alpha(\omega)^*}\Theta(\omega)^* & \frac{a(\alpha)}{\delta_\alpha(\omega)^*}I\end{pmatrix}, \tag{4.21}$$

which holds for all z, ω in U_α and is readily checked. By (4.8) and (4.21) we obtain that for $h \in \mathcal{D}(\Theta)$ of the form $h = D_\Theta(\cdot,\omega)\begin{pmatrix}g\\f\end{pmatrix}$ with $f \in \mathcal{F}$ and $g \in \mathcal{G}$,

$$(Bh)(z) = D_\Theta(z,\omega)\begin{pmatrix}\frac{b(\omega)^*}{\rho_\alpha(\omega)^*}I & 0\\ 0 & \frac{a(\omega)^*}{\delta_\alpha(\omega)^*}I\end{pmatrix}\begin{pmatrix}g\\f\end{pmatrix}$$

$$- D_\Theta(z,\alpha)\begin{pmatrix}0 & 0\\ \frac{a(\alpha)}{\rho_\alpha(\omega)^*}\Theta(\omega)^* & \frac{a(\alpha)}{\delta_\alpha(\omega)^*}I\end{pmatrix}\begin{pmatrix}g\\f\end{pmatrix}$$

$$= \begin{pmatrix}\frac{a(z)}{\delta_\alpha(z)}I & 0\\ 0 & \frac{b(z)}{\rho_\alpha(z)}I\end{pmatrix} D_\Theta(z,\omega)\begin{pmatrix}g\\f\end{pmatrix} - \begin{pmatrix}\frac{a(\alpha)}{\delta_\alpha(z)}I & 0\\ \frac{a(\alpha)}{\rho_\alpha(z)}\tilde\Theta(z) & 0\end{pmatrix} D_\Theta(\alpha,\omega)\begin{pmatrix}g\\f\end{pmatrix}$$

$$= \begin{pmatrix}\frac{a(z)}{\delta_\alpha(z)}I & 0\\ 0 & \frac{b(z)}{\rho_\alpha(z)}I\end{pmatrix} h(z) - \begin{pmatrix}\frac{a(\alpha)}{\delta_\alpha(z)}I\\ \frac{a(\alpha)}{\rho_\alpha(z)}\tilde\Theta(z)\end{pmatrix} Gh,$$

which proves (4.2) for all h in a total set of $\mathcal{D}(\Theta)$. By linearity and continuity, (4.2) holds for all $h \in \mathcal{D}(\Theta)$. The equality (4.1) follows from (4.2) and (4.12). ∎

References

[ADPS] D. Alpay, A. Dijksma, J. van der Ploeg and H. de Snoo, Holomorphic operators between Krein spaces and the number of squares of associated kernels, To appear in *Operator Theory: Advances and Applications* **59**, 11–28, Birkhäuser Verlag, 1992.

[ADS]   D. Alpay, A. Dijksma and H. de Snoo, Reproducing kernel Pontryagin spaces, operator colligations and operator models, Preprint, 1992.

[AD1]   D. Alpay and H. Dym, On a new class of reproducing kernel spaces and a new generalization of the Iohvidov's laws, To appear in *Linear Algebra and Applications*.

[AD2]   D. Alpay and H. Dym, On a new class of structural reproducing kernel spaces, To appear in *J. Func. Anal. Appl.* **108** (1992).

[AD3]   D. Alpay and H. Dym, Realization and factorization of a family of meromorphic functions, In preparation.

[Ar]   N. Aronszjan, Theory of reproducing kernels, *Trans. Am. Math. Soc.* **68** (1950) 334–404.

[Br]   P. Bruinsma, *Interpolation problems for Schur and Nevanlinna pairs*, Ph.D. Thesis, University of Groningen, 1991.

[dBR1] L. de Branges and J. Rovnyak, Canonical models in quantum scattering theory, in C. Wilcox (ed.), *Perturbation Theory and Its Applications in Quantum Mechanics*, Holt, Rinehart and Winston, New York, 1966.

[dBR2] L. de Branges and J. Rovnyak, *Square Summable Power Series*, Holt, Rinehart and Winston, New York, 1966.

[dBS]   L. de Branges and L.A. Shulman, Perturbation theory of unitary operators, *J. Math. Anal. Appl.* **23** (1968) 294–326.

[Fu]   P. Fuhrmann, *Linear Systems and Operators in Hilbert space*, McGraw-Hill, New York, 1981.

[GK]   I. Gohberg and M. Kaashoek, Block Toeplitz operators with rational symbols, *Operator Theory: Advances and Applications* **35**, 385–440, Birkhhäuser Verlag, 1990.

[GKR] I. Gohberg, M. Kaashoek and A. Ran, Factorization of and extension to
J-unitary rational matrix functions on the unit circle, *Integral Equations
and Operator Theory* **15** (1992) 262–300.

[vdP] J. van der Ploeg, Operator functions and associated reproducing kernel
Pontryagin spaces, Master Thesis, University of Groningen, 1991 .

Daniel Alpay, Vladimir Bolotnikov   **Aad Dijksma, Henk de Snoo**

Department of Mathematics   Rijkuniversiteit Groningen
Ben-Gurion University of the Negev   Vakgroep Wiskunde
P.O. Box 653   Postbus 800
84105 Beer-Sheva   9700 AV Groningen
Israel   The Netherlands

Operator Theory:
Advances and Applications, Vol. 61
© 1993 Birkhäuser Verlag Basel

On the Spectra of Selfadjoint Extensions

Johannes Brasche, Hagen Neidhardt and Joachim Weidmann

For selfadjoint extensions of closed symmetric operators with a gap and infinite deficiency indices the paper summarizes results on spectra which may occur inside the gap. Generalizing a result of Krein it is found that an arbitrary point spectrum can be generated by extensions inside the gap. In particular, a dense point spectrum is possible. Moreover, modulo a discrete spectrum every purely singular or absolutely continuous spectrum can be obtained provided the operator is an significantly symmetric one and the spectrum is a regular set. Finally, certain combinations of all three kinds of spectra are realizable.

1. Introduction

Let H be a closed symmetric operator in the separable Hilbert space $\mathcal{H}$. In accordance with Krein [8] the operator H has a gap (a, b), $-\infty < a < b < +\infty$, if the condition

$$\| (H - \frac{a+b}{2})f \| \geq \frac{b-a}{2} \| f \|, \qquad f \in \mathcal{D}(H), \tag{1.1}$$

is satisfied. By the help of this notion an operator H is called semibounded from below with lower bound b if every interval (a, b), $-\infty < a < b$, is a gap for H. It is not hard to see that this definition coincides with the usual definition $(Hf, f) \geq b\| f \|^2$, $f \in \mathcal{D}(H)$. We note this because all results in the sequel will be formulated for symmetric operators with a gap. However, taking into consideration this remark one can easily extend the results to semibounded operators.

A point $z \in \mathbb{C}$ is called a regular point of a closed symmetric operator H if there is a constant $\nu > 0$ such that

$$\| (H - z)f \| \geq \nu\| f \|, \qquad f \in \mathcal{D}(H). \tag{1.2}$$

It is well-known that all points of the upper or lower complexe plane are regular ones. Moreover, if the operator has a gap, then all points inside the gap are regular ones, too. This fact immediately implies that the deficiency indices $n_+(H) = dim(ker(H^* + i))$ and $n_- = dim(ker(H^* - i))$ are equal, i.e. $n_+(H) = n_-(H)$. Hence, by von Neumann's extension theory [1], [10], [11] every closed symmetric operator with a gap admits selfadjoint extensions. However, nothing is said about the spectral properties of these extensions. For instance, are there selfadjoint extensions with a given spectrum inside the gap?

In his famous paper [8] on selfadjoint extensions of semibounded closed symmetric operators Krein has studied this problem mainly for the case of finite deficiency indices. It was found by him that for each closed symmetric operator with a gap there exists a selfadjoint extension which has no spectrum inside the gap. See also [3]. Moreover, assuming that the deficiency indices are finite Krein has given a complete solution of the spectral extension problem. In this case the spectrum of any selfadjoint extension $\widetilde{H}$ of H is discrete inside the gap and consists of at most $n(H) = n_+(H) = n_-(H)$ eigenvalues counting multiplicites, where $n(H)$ is called the deficiency index in the following. Conversely, giving any sequence of real numbers $\{\lambda_s\}_{s=1}^L$, $\lambda_s \in (a, b)$, and any sequence of integers $\{p_s\}_{s=1}^L$, $p_s \geq 1$, such that the condition $\sum_{s=1}^L p_s \leq n(H)$ is satisfied, then there exists a selfadjoint extension $\widetilde{H}$ of H such that the spectrum of $\widetilde{H}$ inside the gap exactly coincides with $\{\lambda_s\}_{s=1}^L$ and the multiplicity of each eigenvalue λ_s is equal to p_s. This result was extended by Krein to sequences of real regular points. Notice that if even every point of an interval (a, b) is a regular one this does not imply that the interval (a, b) is a gap. If $\{\lambda_s\}_{s=1}^L$ is a sequence of real regular points of the closed symmetric operator H with finite deficiency indices and if $\{p_s\}_{s=1}^L$ is any sequence of integers obeying $\sum_{s=1}^L p_s \leq n(H)$, then there is a selfadjoint extension $\widetilde{H}$ of H such that the points $\{\lambda_s\}_{s=1}^L$ are eigenvalues of $\widetilde{H}$ of multiplicity greater than or equal to p_s.

In the following we are going to extend these results of Krein to closed symmetric operators with infinite deficiency indices. However, in this case it is quite possible that beside the discrete point spectrum other kinds of spectra occur, for instance, arbitrary point spectrum, singular continuous spectrum, absolutely continuous spectrum and combinations of all these spectra. So the problem arises to find for any closed symmetric operator with gap (a, b) and with infinite deficiency indices a selfadjoint extension spectrum with a pure point or a purely singular or absolutely continuous spectrum inside the gap wich coincides with a given closed subset $\mathcal{G}$ of $[a, b]$. In the following we try to solve this problem.

It turns out that the assumption of infinite deficiency indices is sufficient for the generation of an arbitrary point spectrum by extensions. These results can be found in Section 3. Theorem 3 treats the case of one gap while Theorem 4 is related to the situation of many gaps. This last case was very carefully investigated by Derkach and Malamud [9] and we recommend this paper to the reader for more information.

To obtain a prescribed singular or absolutely continuous spectrum (Sections 5 and 6) we need actually a little bit more. To this end we introduce the class of significantly symmetric operators.

Definition 1 *A closed symmetric operator H is called significantly symmetric if there exists a real regular point λ such that*

$$P_{\mathcal{N}_\lambda}^{\mathcal{H}} \mathcal{D}(H) \neq \mathcal{N}_\lambda, \qquad \mathcal{N}_\lambda = ker(H^* - \lambda). \tag{1.3}$$

The operator H is called insignificantly symmetric if there exists a real regular point λ such that

$$P_{\mathcal{N}_\lambda}^{\mathcal{H}} \mathcal{D}(H) = \mathcal{N}_\lambda, \qquad \mathcal{N}_\lambda = ker(H^* - \lambda). \tag{1.4}$$

Here and in the following we denote by $P_{\mathcal{M}}^{\mathcal{H}}$ the orthogonal projection from the Hilbert space $\mathcal{H}$ to the closed subspace $\mathcal{M}$. However, it is an open problem whether the assumption of a significantly symmetric operator is really necessary in order to obtain the results of Section 5 and 6. It would be very interesting to prove these results without this assumption.

Moreover we note that in general it is not clear whether the class of closed symmetric operators with at least one real regular point is divided by this definition into two disjoint subclasses. However, this is indeed so as we will see in Section 4. Furthermore, the class of significantly symmetric operators includes important examples as, for instance, elliptic differential operators on bounded regions.

Furthermore, we need that the set $\mathcal{G}$ is regular.

Definition 2 *The closed subset $\mathcal{G} \subseteq \mathbf{R}^1$ is called a regular one, if $\mathcal{G}$ is the closure of its interior $\mathcal{G}^0$, i.e. $\overline{\mathcal{G}^0} = \mathcal{G}$.*

Of course, every closed interval and every finite union of closed intervals are regular sets. Further every regular subset is a perfect one. However, the converse is not true as can be seen from the Cantor set. For us it is important that every regular set contains a sufficiently large open set.

In the following we are able to show that for every significantly symmetric operator H with gap (a, b) and every regular set $\mathcal{G} \subseteq [a, b]$ there exists a selfadjoint extension $\widetilde{H}$ such that the spectrum $\sigma(\widetilde{H})$ of $\widetilde{H}$ on $(a, b) \setminus \mathcal{G}$ is discrete and $\mathcal{G}$ is purely singular continuous. This is the main result of Section 5. For the absolutely continuous spectrum of $\widetilde{H}$, which is treated in Section 6, the same holds with one exception. On the closed set $\partial \mathcal{G} = \mathcal{G} \setminus \mathcal{G}^0$ may appear a singular continuous spectrum. Finally, a certain combination of the results of Section 3, 5 and 6 is valid.

The results of Section 5 and 6 were obtained by the help of the so-called Krein model which was introduced in [3]. In the next section we give a short description of this model.

2. Krein model

Let H be a closed symmetric operator with gap (a,b). Without loss of generality we can assume that $0 \in (a,b)$. Since $0 \in (a,b)$ the inverse operator $L = H^{-1} : \mathcal{R} \to \mathcal{H}$, $\mathcal{R} = \mathcal{R}(H)$, exists and is a bounded Hermitian operator. In accordance with Proposition 2.2 of [3] the operator L admits the representation

$$Lg = Ag \oplus \frac{1}{\sqrt{-ab}}\Gamma\sqrt{I - bA}\sqrt{I - aA}g, \qquad g \in \mathcal{R}, \tag{2.1}$$

where $A : \mathcal{R} \to \mathcal{R}$ is a bounded selfadjoint operator which obeys $ker(A) = 0$ and

$$\frac{1}{a}I \le A \le \frac{1}{b}I, \tag{2.2}$$

and $\Gamma : \overline{\mathcal{R}(\sqrt{I - bA}\sqrt{I - aA})} \to \mathcal{N}$, $\mathcal{N} = \mathcal{H} \ominus \mathcal{R}$, is a contraction with dense range. Both operators are related by the condition

$$\mathcal{R}(\Gamma^*) \cap \mathcal{R}(A) = \{0\}. \tag{2.3}$$

With respect to the constants a and b the bounded selfadjoint operator A and the contraction Γ are uniquely defined by H and the representation (2.1).

In honour of Krein the representation (2.1) is called the Krein model of the inverse of a symmetric operator with gap. By $\{\mathcal{R}, A, \Gamma\}_{(a,b)}$ we denote the so-called Krein parameter of the representation (2.1). The subspace $\mathcal{N} = \mathcal{H} \ominus \mathcal{R}$ is called the deficiency subspace and its dimension coincides with the deficiency index.

Introducing the orthogonal decomposition

$$\mathcal{N} = \mathcal{L} \oplus \mathcal{Q} \tag{2.4}$$

and the contractions $\Lambda = P_{\mathcal{L}}^{\mathcal{N}}\Gamma$ and $\Theta = P_{\mathcal{Q}}^{\mathcal{N}}\Gamma$, which satisfy the conditions

$$\Lambda^*\Lambda + \Theta^*\Theta \le I, \qquad \text{and} \qquad \mathcal{R}(\Lambda^*) \cap \mathcal{R}(\Theta^*) = \{0\}, \tag{2.5}$$

the representation (2.1) goes over to

$$Lg = Ag \oplus \frac{1}{\sqrt{-ab}}\Lambda\sqrt{I - bA}\sqrt{I - aA}g \oplus \frac{1}{\sqrt{-ab}}\Theta\sqrt{I - bA}\sqrt{I - aA}g, \qquad g \in \mathcal{R}. \tag{2.6}$$

The representations (2.6) and the 5-tuples $\{\mathcal{R}, \mathcal{L}, A, \Lambda, \Theta\}_{(a,b)}$ are called the modified Krein model and the modified Krein parameter, respectively.

In terms of this model inverses of symmetric extensions $\hat{H}$ of H can be described. By Proposition 4.2 of [3] the bounded Hermitian operator $\hat{L} : \mathcal{R} \oplus \mathcal{L} \to \mathcal{H}$ is the inverse of a closed symmetric extension $\hat{H}$ of H for wich zero is a regular point if and only if there exist a bounded selfadjoint operator $G : \mathcal{L} \to \mathcal{L}$ and a bounded operator $K : \mathcal{L} \mapsto \mathcal{Q}$ such that $\hat{L}$ admits the representation

$$\hat{L} = \begin{pmatrix} A & \frac{1}{\sqrt{-ab}}\sqrt{I - bA}\sqrt{I - aA}\Lambda^* \\ \frac{1}{\sqrt{-ab}}\Lambda\sqrt{I - bA}\sqrt{I - aA} & -\Lambda(A - \frac{a+b}{ab})\Lambda^* + G \\ \frac{1}{\sqrt{-ab}}\Theta\sqrt{I - bA}\sqrt{I - aA} & -\Theta(A - \frac{a+b}{ab})\Lambda^* + K \end{pmatrix} : \begin{array}{c} \mathcal{R} \\ \oplus \\ \mathcal{L} \end{array} \mapsto \begin{array}{c} \mathcal{R} \\ \oplus \\ \mathcal{L} \\ \oplus \\ \mathcal{Q} \end{array} \tag{2.7}$$

With respect to the constants a and b and the subspace $\mathcal{L}$ the operators G and K are uniquely determined by $\hat{H}$ and the representation (2.7). The triplet $\{\mathcal{L}, G, K\}$ is called the extension parameter.

The natural question which conditions must be satisfied by the extension parameter in order that the symmetric extension $\hat{H}$ has the same gap as H is answered by Proposition 4.4 and Theorem 4.7 of [3]. It turns out that $\hat{H} = \hat{L}^{-1}$ has the same gap as H if and only if the operators $G : \mathcal{L} \mapsto \mathcal{L}$ and $K : \mathcal{L} \to \mathcal{Q}$ admit the representations

$$G = D_{\Lambda^*} X D_{\Lambda^*} \tag{2.8}$$

and

$$K = \frac{1}{\sqrt{-ab}} Y \sqrt{I - (a + b)X} D_{\Lambda^*} \tag{2.9}$$

where $X : \overline{\mathcal{R}(D_{\Lambda^*})} \mapsto \overline{\mathcal{R}(D_{\Lambda^*})}$ is a selfadjoint operator satisfying

$$\frac{1}{a} \leq X \leq \frac{1}{b} \tag{2.10}$$

and $Y : \overline{\mathcal{R}(\sqrt{I - (a + b)X})} \to \mathcal{Q}$ is a contraction and, additionally, the operator P,

$$P = \begin{pmatrix} \Lambda & D_{\Lambda^*} \frac{\sqrt{-ab}X}{\sqrt{I-(a+b)X}} \\ \Theta & Y \end{pmatrix}, \tag{2.11}$$

is a contraction. We recall that $D_{\Lambda^*} = \sqrt{I - \Lambda\Lambda^*}$. Notice that the condition (2.11) restricts the contraction Y. The concrete restriction is given by Theroem 4.7 of [3].

Applying Proposition 2.2 of [3] to the symmetric extension $\hat{H}$ we see that

$$\hat{A} = \begin{pmatrix} A & \frac{1}{\sqrt{-ab}}\sqrt{I - aA}\sqrt{I - bA}\Lambda^* \\ \Lambda\frac{1}{\sqrt{-ab}}\sqrt{I - bA}\sqrt{I - aA} & -\Lambda(A - \frac{a+b}{ab})\Lambda^* + D_{\Lambda^*} X D_{\Lambda^*} \end{pmatrix} : \begin{array}{c} \mathcal{R} \\ \oplus \\ \mathcal{L} \end{array} \mapsto \begin{array}{c} \mathcal{R} \\ \oplus \\ \mathcal{L} \end{array} \tag{2.12}$$

obeys

$$\frac{1}{a}I \leq \hat{A} \leq \frac{1}{b}I. \tag{2.13}$$

In the following we are interested in extensions $\hat{H} = \hat{L}^{-1}$ where

$$\hat{L} = \begin{pmatrix} A & \frac{1}{\sqrt{-ab}}\sqrt{I-bA}\sqrt{I-aA}\Lambda^* \\ \frac{1}{\sqrt{-ab}}\Lambda\sqrt{I-bA}\sqrt{I-aA} & -\Lambda(A-\frac{a+b}{ab})\Lambda^* + D_{\Lambda^*}XD_{\Lambda^*} \\ \frac{1}{\sqrt{-ab}}\Theta\sqrt{I-bA}\sqrt{I-aA} & 0 \end{pmatrix} : \begin{matrix} \mathcal{R} \\ \oplus \\ \mathcal{L} \end{matrix} \mapsto \begin{matrix} \mathcal{R} \\ \oplus \\ \mathcal{L} \\ \oplus \\ \mathcal{Q} \end{matrix}$$

$$(2.14)$$

and the operator X satisfies (2.10). Introducing the bounded operator $\hat{B}$,

$$\hat{B} = \left(\frac{1}{\sqrt{-ab}}\Theta\sqrt{I-bA}\sqrt{I-aA}, 0 \right) : \hat{\mathcal{R}} = \begin{matrix} \mathcal{R} \\ \oplus \\ \mathcal{L} \end{matrix} \to \mathcal{Q}. \qquad (2.15)$$

the inverse operator $\hat{L} = \hat{H}^{-1}$ has the representation

$$\hat{L}\hat{g} = \hat{A}\hat{g} \oplus \hat{B}\hat{g}, \qquad \hat{g} \in \hat{\mathcal{R}}. \qquad (2.16)$$

Let $\tilde{Q} : \mathcal{Q} \to \mathcal{Q}$ be a bounded or unbounded selfadjoint operator. Then $\tilde{L}^{-1}$,

$$\tilde{L} = \begin{pmatrix} \hat{A} & \hat{B}^* \\ \hat{B} & \tilde{Q} \end{pmatrix} : \begin{matrix} \hat{\mathcal{R}} \\ \oplus \\ \mathcal{Q} \end{matrix} \mapsto \begin{matrix} \hat{\mathcal{R}} \\ \oplus \\ \mathcal{Q} \end{matrix}, \qquad (2.17)$$

makes sense and defines a selfadjoint extension $\widetilde{H} = \tilde{L}^{-1}$ of H. In the following we construct the desired selfadjoint extension by choosing the subspace $\mathcal{L}$ and the selfadjoint operator $\tilde{Q}$ in an appropriate manner.

3. Point spectrum

The aim of this section is to indicate the generalization of Krein's results [8] on the point spectrum of selfadjoint extensions to infinite dimensional deficiency indices. The proofs of the theorems can be found in [3]. We refer to [9] for related results. By $\sigma_p(\cdot)$ we denote the set of eigenvalues of an operator.

Theorem 3 *Let H be a closed symmetric operator with infinite deficiency indices and gap (a,b). Let $\mathcal{E}$ be a countable subset of (a,b) and let $p : \mathcal{E} \mapsto \mathbf{N} \cup \{\aleph_0\}$ be any mapping. Then there exists a selfadjoint extension $\widetilde{H}$ of H with the following properties:*

(i) $\sigma_p(\widetilde{H}) \cap (a,b) = \mathcal{E}$,

(ii) each $\lambda \in \mathcal{E}$ is an eigenvalue of $\widetilde{H}$ with multiplicity equal to $p(\lambda)$,

(iii) $\widetilde{H}$ has a pure point spectrum within (a,b).

The generalization to regular points goes as follows.

Theorem 4 *Let $\mathcal{E}$ be a countable set of regular points of the closed symmetric operator H with infinite deficiency indices. Let $p : \mathcal{E} \to \mathbf{N} \cup \{\aleph_0\}$ be any mapping. Then H has a selfadjoint extension $\widetilde{H}$ such that each $\lambda \in \mathcal{E}$ is an eigenvalue of $\widetilde{H}$ with multiplicity greater than or equal to $p(\lambda)$.*

Obviously, above theorems are straightforward generalizations of the corresponding results of Krein. However, for the case that the closed symmetric operator H commutes with a conjugation C both theorems does not answer the question whether the extension can be chosen such that it commutes with the conjugation too.

Theorem 5 *Let $\mathcal{E}$ be a countable set of regular points of the closed symmetric operator H. Let $p : \mathcal{E} \to \mathbf{N} \cup \aleph$, be any mapping such that $\sum_{\lambda \in \mathcal{E}} p(\lambda) \le n(H)$. Suppose that H is C-real for some conjugation C. Then H has a C-real selfadjoint extension $\widetilde{H}$ such that $\lambda \in \mathcal{E}$ is an eigenvalue of H with multiplicity greater than or equal to $p(\lambda)$.*

Since there exist closed symmetric operators which have a gap and which does not commute with any conjugation the last theorem does not imply the previous ones. In accordance with [4] for the existence of such symmetric operators it is sufficient that the deficiency index is greater than one. If the deficiency index is equal to one, then we always can find a conjugation commuting with the symmetric operator.

Now from Theorem 3 immediately follows that we can generate every given spectrum inside the gap (a, b) provided the deficiency indices are infinite. To see this let $\mathcal{G} \subseteq [a, b]$ be an arbitrary closed set. For every such a set there is a countable subset $\mathcal{E} \subseteq \mathcal{G} \cap (a, b)$ obeying $\overline{\mathcal{E}} = \mathcal{G}$. Taking into account Theorem 3 we obtain a selfadjoint extension $\widetilde{H}$ such that $\sigma_p(\widetilde{H}) \cap (a, b) = \mathcal{E}$ and $\widetilde{H}$ has a pure point spectrum within (a, b). Hence we have found an extension satisfying $\sigma(\widetilde{H}) \cap (a, b) = \mathcal{G}$. Naturally the question arises whether this can be done requiring, however, that the occuring spectrum in (a, b) is not a pure point one but a purely singular or absolutely continuous one. In the following we try to answer this question.

4. Significantly symmetric operators

First of all we want to show that the Definition 1 is independent of the chosen regular point. To this end we introduce the open set $\mathcal{F}$ of all (real or complex) regular points of the symmetric operator H. Since in Definition 1 the existence of a real regular point is assumed the set $\mathcal{F}$ is connected. Let $z, \xi \in \mathcal{F}$. Then by section 100 of [1] we have

$$\lim_{z \to \xi} \| P^{\mathcal{H}}_{\mathcal{N}_z} - P^{\mathcal{H}}_{\mathcal{N}_\xi} \| = 0 \tag{4.1}$$

provided z tends to ξ on a curve in $\mathcal{F}$. Consequently, for every $\xi \in \mathcal{F}$ there is a neighbourhood $\mathcal{O}_\xi \subset \mathcal{F}$ such that

$$\| P^{\mathcal{H}}_{\mathcal{N}_z} - P^{\mathcal{H}}_{\mathcal{N}_\xi} \| < 1, \qquad z \in \mathcal{O}_\xi. \tag{4.2}$$

Lemma 6 *Let $z, \xi \in \mathcal{F}$. If $z \in \mathcal{O}_\xi$, then*

$$\overline{\mathcal{R}(P^{\mathcal{H}}_{\mathcal{N}_z} P^{\mathcal{H}}_{\mathcal{N}_\xi})} = \mathcal{N}_z. \tag{4.3}$$

Proof. Let $P^{\mathcal{H}}_{\mathcal{N}_\xi} P^{\mathcal{H}}_{\mathcal{N}_z} f = 0$, $f \in \mathcal{N}_z$, $\| f \| = 1$. Obviously, we have

$$0 = P^{\mathcal{H}}_{\mathcal{N}_\xi} P^{\mathcal{H}}_{\mathcal{N}_z} f = (P^{\mathcal{H}}_{\mathcal{N}_\xi} - P^{\mathcal{H}}_{\mathcal{N}_z}) P^{\mathcal{H}}_{\mathcal{N}_z} f + f. \tag{4.4}$$

By (4.2) we obtain that

$$1 = \| (P^{\mathcal{H}}_{\mathcal{N}_\xi} - P^{\mathcal{H}}_{\mathcal{N}_z}) P^{\mathcal{H}}_{\mathcal{N}_z} f \| < \| f \| = 1 \tag{4.5}$$

Since this is impossible we find that $ker(P^{\mathcal{H}}_{\mathcal{N}_\xi} P^{\mathcal{H}}_{\mathcal{N}_z}) = \mathcal{H} \ominus \mathcal{N}_z$. Hence (4.3) holds. ∎

Furthermore, we need the following technical lemma.

Lemma 7 *Let z be a regular point of the closed symmetric operator H, i.e., $z \in \mathcal{F}$, and let $\check{H}$ be a restriction of H. If $P^{\mathcal{H}}_{\mathcal{N}_z} \mathcal{D}(H) \neq \mathcal{N}_z$, $\mathcal{N}_z = ker(H^* - \bar{z})$, then $P^{\mathcal{H}}_{\check{\mathcal{N}}_z} \mathcal{D}(\check{H}) \neq \check{\mathcal{N}}_z$, $\check{\mathcal{N}}_z = ker(\check{H}^* - \bar{z})$.*

Proof. Let $P^{\mathcal{H}}_{\check{\mathcal{N}}_z} \mathcal{D}(\check{H}) = \check{\mathcal{N}}_z$. Since $\mathcal{N}_z \subseteq \check{\mathcal{N}}_z$ we find $P^{\mathcal{H}}_{\mathcal{N}_z} \leq P^{\mathcal{H}}_{\check{\mathcal{N}}_z}$. Hence we obtain $P^{\mathcal{H}}_{\mathcal{N}_z} \mathcal{D}(\check{H}) = \mathcal{N}_z$. Using $\mathcal{D}(\check{H}) \subseteq \mathcal{D}(H)$ we get $P^{\mathcal{H}}_{\mathcal{N}_z} \mathcal{D}(H) = \mathcal{N}_z$ which contraticts the assumption. ∎

Crucial will be the following assertion in the sequel.

Lemma 8 *If $B : \mathcal{R} \to \mathcal{N}$ is a bounded operator with dense but not closed range, then there is an infinite dimensional subspace $\mathcal{Q} \subseteq \mathcal{N}$ such that $P^{\mathcal{N}}_{\mathcal{Q}} B$ is a compact operator, i.e., $P^{\mathcal{N}}_{\mathcal{Q}} B \in \mathcal{L}_\infty(\mathcal{R}, \mathcal{Q})$.*

Proof. Let us introduce the polar decomposition $B^* = VT$, $V = sign(B^*) : \mathcal{N} \to \mathcal{R}$, $T = | B^* | = \sqrt{BB^*} : \mathcal{N} \to \mathcal{N}$, of the operator $B^* : \mathcal{N} \to \mathcal{R}$. Since $\mathcal{R}(B)$ is dense but not closed we have $0 \in \sigma_e(T)$ where $\sigma_e(\cdot)$ denotes the essential spectrum. Hence by Theorem VII.12 of [10] there is an orthonormal sequence $\{\varphi_n\}_{n=1}^\infty$, $\| \varphi_n \| = 1$, such that $\| T\varphi_n \| \to 0$ as $n \to \infty$. Obviously, we can assume that

$$\| T\varphi_n \| \leq \frac{c}{n}, \qquad n = 1, 2, ..., \qquad c > 0. \tag{4.6}$$

By Q we denote the subspace which is spanned by $\{\varphi_n\}_{n=1}^\infty$. By (4.6) we have

$$\sum_{n=1}^\infty \| TP_Q^N \varphi_n \|^2 = \sum_{n=1}^\infty \| T\varphi_n \|^2 \le \sum_{n=1}^\infty \frac{c^2}{n^2} < +\infty. \tag{4.7}$$

Hence TP_Q^N is a Hilbert-Schmidt operator. However, this immediately yields that that $P_Q^N B$ is compact. ∎

Next we prove the independence of λ of Definition 1.

Proposition 9 *Let H be a closed symmetric operator. If there is a real regular point λ such that condition (1.3) is satisfied, then (1.3) holds for every (real or complex) regular point.*

Proof. Without loss of generality we assume $\lambda = 0$. Furthermore, we note that the regularity of zero implies the existence of an interval (a,b), $0 \in (a,b)$, which is a gap of H. Thus we can use the Krein model [3]. We introduce the operator $B = \mathcal{R} \to \mathcal{N}$,

$$B = \frac{1}{\sqrt{-ab}} \Gamma \sqrt{I - bA} \sqrt{I - aA}. \tag{4.8}$$

On account of (1.3) we get $\mathcal{R}(B) \ne \mathcal{N}$. Applying Lemma 8 there is an infinite dimensional subspace $Q \subseteq \mathcal{N}$ such that $P_Q^N B \in \mathcal{L}_\infty(\mathcal{R}, \mathcal{N})$. Setting $\mathcal{L} = \mathcal{N} \ominus Q$ and choosing the extension parameter in accordance with (2.14) we obtain a closed symmetric extension $\hat{H}$ of H satisfying

$$P_Q^{\mathcal{H}} \mathcal{D}(\hat{H}) \ne Q. \tag{4.9}$$

This comes from the fact that $\hat{B} : \hat{\mathcal{R}} \to Q$ given by (2.15) is a compact operator and the relation $P_Q^{\mathcal{H}} \mathcal{D}(\hat{H}) = \mathcal{R}(\hat{B})$ holds. Notice that zero is a regular point for $\hat{H}$. So we have found an extension $\hat{H}$ of H preserving the condition (1.3).

Let z be a nonreal number. We have $(\hat{H} - z)\hat{L}g = g - z\hat{L}g$, $g \in \hat{\mathcal{R}}$. Notice that if g runs through $\hat{\mathcal{R}}$, then $g - z\hat{L}g$ runs through $\hat{\mathcal{R}}_z = \mathcal{R}(\hat{H} - z)$. Since $Q_z = \mathcal{H} \ominus \hat{\mathcal{R}}_z$, $Q_z = ker(\hat{H}^* - \bar{z})$, we get that $f \in Q_z$ if and only if $(g - z\hat{L}g, f) = 0$. Using the decomposition $f = \hat{g} \oplus h$, $\hat{g} \in \hat{\mathcal{R}}$, $h \in Q$, and the representation (2.16) we obtain that $\hat{g} \oplus h \in Q_z$ if and only if

$$\hat{g} = \bar{z}(I - \bar{z}\hat{A})^{-1}\hat{B}^* h, \qquad h \in Q. \tag{4.10}$$

Introducing the operator K_z,

$$K_z = \begin{pmatrix} M_z \\ I \end{pmatrix} : Q \to \begin{matrix} \hat{\mathcal{R}} \\ \oplus \\ Q \end{matrix} , \qquad M_z = \bar{z}(I - \bar{z}\hat{A})^{-1}\hat{B}^* : Q \to \hat{\mathcal{R}}, \tag{4.11}$$

we find that $\mathcal{R}(K_z) = \mathcal{Q}_z$. Let $K_z = V_z \mid K_z \mid$ be the polar decomposition of K_z. It is not hard to see that $V_z : \mathcal{Q} \mapsto \mathcal{H}$ admits the representation

$$V_z = \left(\begin{array}{c} M_z(I + M_z^* M_z)^{-\frac{1}{2}} \\ (I + M_z^* M_z)^{-\frac{1}{2}} \end{array} \right) : \mathcal{Q} \to \begin{array}{c} \hat{\mathcal{R}} \\ \oplus \\ \mathcal{Q} \end{array} . \tag{4.12}$$

Hence the projection $P_{\mathcal{Q}_z}^{\mathcal{H}} = V_z V_z^*$ has the representation

$$P_{\mathcal{Q}_z}^{\mathcal{H}} = \left(\begin{array}{cc} M_z(I + M_z^* M_z)^{-1} M_z^* & M_z(I + M_z^* M_z)^{-1} \\ (I + M_z^* M_z)^{-1} M_z^* & (I + M_z^* M_z)^{-1} \end{array} \right) : \begin{array}{c} \hat{\mathcal{R}} \\ \oplus \\ \mathcal{Q} \end{array} \mapsto \begin{array}{c} \hat{\mathcal{R}} \\ \oplus \\ \mathcal{Q} \end{array} . \tag{4.13}$$

Notice that $z = 0$ corresponds to $P_{\mathcal{Q}}^{\mathcal{H}}$.

Since B is compact the operator M_z is compact for every z, $\mathrm{Im}(z) \neq 0$. This immediately yields that

$$P_{\mathcal{Q}_z}^{\mathcal{H}} - P_{\mathcal{Q}}^{\mathcal{H}} \in \mathcal{L}_\infty(\mathcal{H}, \mathcal{H}), \qquad \mathrm{Im}(z) \neq 0. \tag{4.14}$$

Now we choose a selfadjoint extension $\widetilde{H}$ of $\hat{H}$ by setting $\tilde{Q} = 0$ in (2.17). The selfadjoint extension obeys

$$P_{\mathcal{Q}}^{\mathcal{H}} \widetilde{L} = P_{\mathcal{Q}}^{\mathcal{H}} \widetilde{H}^{-1} = \hat{B} P_{\hat{\mathcal{R}}}^{\mathcal{H}} \tag{4.15}$$

which yields

$$P_{\mathcal{Q}}^{\mathcal{H}} \widetilde{L} = P_{\mathcal{Q}}^{\mathcal{H}} \widetilde{H}^{-1} \in \mathcal{L}_\infty(\mathcal{H}, \mathcal{Q}). \tag{4.16}$$

Using the identity

$$(\widetilde{H} - z)^{-1} = \widetilde{H}^{-1}(I + z(\widetilde{H} - z)^{-1}), \qquad \mathrm{Im}(z) \neq 0, \tag{4.17}$$

we immediately get

$$P_{\mathcal{Q}}^{\mathcal{H}}(\widetilde{H} - z)^{-1} \in \mathcal{L}_\infty(\mathcal{H}, \mathcal{Q}), \qquad \mathrm{Im}(z) \neq 0. \tag{4.18}$$

Taking into account (4.14) we find

$$P_{\mathcal{Q}_z}^{\mathcal{H}}(\widetilde{H} - z)^{-1} \in \mathcal{L}_\infty(\mathcal{H}, \mathcal{Q}_z), \qquad \mathrm{Im}(z) \neq 0, \tag{4.19}$$

which yields

$$P_{\mathcal{Q}_z}^{\mathcal{H}} \mathcal{D}(\widetilde{H}) \neq \mathcal{Q}_z, \qquad \mathrm{Im}(z) \neq 0. \tag{4.20}$$

Since $\hat{H}$ is a restriction of $\widetilde{H}$ we obtain

$$P_{\mathcal{Q}_z}^{\mathcal{H}} \mathcal{D}(\hat{H}) \neq \mathcal{Q}_z, \qquad \mathrm{Im}(z) \neq 0. \tag{4.21}$$

Finally, applying Lemma 7 we find

$$P_{\mathcal{N}_z}^{\mathcal{H}}\mathcal{D}(H) \neq \mathcal{N}_z, \qquad \mathrm{Im}(z) \neq 0. \tag{4.22}$$

It remains to prove the assertion for real regular points. To this end we assume that there is a real regular point μ such that $P_{\mathcal{N}_\mu}^{\mathcal{H}}\mathcal{D}(H) = \mathcal{N}_\mu$. Let us introduce the operator $P_{\mathcal{N}_\mu}^{\mathcal{H}}L : \mathcal{R} \to \mathcal{N}_\mu$. Since $P_{\mathcal{N}_\mu}^{\mathcal{H}}\mathcal{D}(H) = \mathcal{N}_\mu$ we have $\mathcal{R}(P_{\mathcal{N}_\mu}^{\mathcal{H}}L) = \mathcal{N}_\mu$. Hence $P_{\mathcal{N}_\mu}^{\mathcal{H}}L : \mathcal{R} \to \mathcal{N}_\mu$ is a semi-Fredholm operator. Consequently, the operator $L^*P_{\mathcal{N}_\mu}^{\mathcal{H}} : \mathcal{N}_\mu \to \mathcal{R}$ is a semi-Fredholm operator, too. Using the representation

$$L^*P_{\mathcal{N}_z}^{\mathcal{H}}P_{\mathcal{N}_\mu}^{\mathcal{H}} = L^*(P_{\mathcal{N}_z}^{\mathcal{H}} - P_{\mathcal{N}_\mu}^{\mathcal{H}})P_{\mathcal{N}_\mu}^{\mathcal{H}} + L^*P_{\mathcal{N}_\mu}^{\mathcal{H}}, \tag{4.23}$$

taking into account (4.1) and applying Theorem 4.22 of [7] we find that there is a neighbourhood $\mathcal{O}_\mu$ such that for $z \in \mathcal{O}_\mu$ the operator $L^*P_{\mathcal{N}_z}^{\mathcal{H}}P_{\mathcal{N}_\mu}^{\mathcal{H}} : \mathcal{N}_\mu \mapsto \mathcal{R}$ is a semi-Fredholm operator, too. In particular, this yields that $\mathcal{R}(L^*P_{\mathcal{N}_z}^{\mathcal{H}}P_{\mathcal{N}_\mu}^{\mathcal{H}})$ is closed. Restricting in a suitable manner the neighbourhood $\mathcal{O}_\mu$ we can assume by Lemma 6 that for $z \in \mathcal{O}_\mu$ we have $\overline{P_{\mathcal{N}_z}^{\mathcal{H}}P_{\mathcal{N}_\mu}^{\mathcal{H}}} = \mathcal{N}_z$. Therefore we obtain

$$\mathcal{R}(L^*P_{\mathcal{N}_z}^{\mathcal{H}}P_{\mathcal{N}_\mu}^{\mathcal{H}}) \subseteq \mathcal{R}(L^*P_{\mathcal{N}_z}^{\mathcal{H}}) \subseteq \overline{\mathcal{R}(L^*P_{\mathcal{N}_z}^{\mathcal{H}}P_{\mathcal{N}_\mu}^{\mathcal{H}})} = \mathcal{R}(L^*P_{\mathcal{N}_z}^{\mathcal{H}}P_{\mathcal{N}_\mu}^{\mathcal{H}}). \tag{4.24}$$

Consequently, for $z \in \mathcal{O}_\mu$ the operator $L^*P_{\mathcal{N}_z}^{\mathcal{H}} : \mathcal{N}_z \to \mathcal{R}$ is a semi-Fredholm operator, too. Hence $P_{\mathcal{N}_z}^{\mathcal{H}}L : \mathcal{R} \to \mathcal{N}_z$ is a semi-Fredholm operator which in particular yields that $\mathcal{R}(P_{\mathcal{N}_z}^{\mathcal{H}}L) = P_{\mathcal{N}_z}^{\mathcal{H}}\mathcal{D}(H)$ is closed for $z \in \mathcal{O}_\mu$. Thus $P_{\mathcal{N}_z}^{\mathcal{H}}\mathcal{D}(H) = \mathcal{N}_z$ for $z \in \mathcal{O}_\mu$. However, choosing a nonreal $z \in \mathcal{O}_\mu$ we find that this is impossible by the first part of the proof. Consequently, the assumption $P_{\mathcal{N}_\mu}^{\mathcal{H}}\mathcal{D}(H) = \mathcal{N}_\mu$ is false. ∎

Notice that Proposition 9 immediately implies that for every real regular point either the condition (1.3) or the condition (1.4) is satisfied. So the the classes of significantly and insignificantly symmetric operators are disjoint.

We note that every closed symmetric operator with finite deficiency indices and at least one regular point obeys the condition (1.4). Hence such an operator is an insignificantly symmetric operator. Let us show that there are insignificantly symmetric operators with infinite deficiency indices.

Example 10 Let H_1 be a closed symmetric operator with gap (a, b) and deficiency indices $(1, 1)$. Setting $H_l = H_1$, $l = 2, 3, \ldots$ and

$$H = \bigoplus_{l=1}^{\infty} H_l \tag{4.25}$$

it is not hard to see that this operator is an insignificantly symmetric one with gap (a, b) and infinite deficiency indices

Further, if H is an insignificantly symmetric operator, then every extension $\hat{H}$ of H which has at least one real regular point is an insignificantly symmetric operator, too. This can be easily proved using the Krein model. If H is a significantly symmetric operator, then every restriction of it is also a significantly symmetric operator. To see this we have to use Lemma 7. So by extensions or restrictions we can not leave the classes of insignificantly or significantly symmetric operators, respectively. Naturally the question arises whether this is possible by the opposite operations.

Lemma 11 *For every significantly symmetric operator H there is an insignificantly symmetric extension with infinite deficiency indices.*

For every insignificantly symmetric operator there is a significantly symmetric restriction.

Proof. Let H be a significantly symmetric operator with regular point λ. Without loss of generality we can assume that $\lambda = 0$. Since H is significantly symmetric the deficiency indices of H are infinite. Consequently, the subspace $\mathcal{N}$ allows an orthogonal decomposition $\mathcal{N} = \mathcal{L} \oplus \mathcal{Q}$ such that $dim(\mathcal{L}) = dim(\mathcal{Q}) = +\infty$. Let $\hat{H}$ be an extension of H which is given by the extension parameter $\{\mathcal{L}, G, K\}$, $K = \theta(A - \frac{a+b}{ab}))\Lambda^* + W$, where $W : \mathcal{L} \to \mathcal{Q}$ obeys $\mathcal{R}(W) = \mathcal{Q}$. We easily see that $P_{\mathcal{Q}}^{\mathcal{H}} \mathcal{D}(\hat{H}) = \mathcal{Q}$.

To prove the second part we note that on account of Theorem 4 of [2] there is a closed symmetric restriction $\check{H}$ of H such that $\mathcal{D}(\check{H}^2) = \{0\}$. Let $\{\check{\mathcal{R}}, \check{A}, \check{\Gamma}\}_{(a,b)}$ be the Krein parameter of $\check{\mathcal{H}}$ and let $\check{B} = \frac{1}{\sqrt{-ab}}\check{\Gamma}\sqrt{I - b\check{A}}\sqrt{I - a\check{A}} : \check{\mathcal{R}} \to \check{\mathcal{N}}$, $\mathcal{R}(\check{H}) = \check{\mathcal{R}}$, $\check{\mathcal{N}} = ker(\check{H}^*)$. Since

$$\mathcal{D}(\check{H}^2) = \check{L}\check{A}ker(\check{B}), \qquad \check{L} = \check{H}^{-1}, \tag{4.26}$$

and $ker(\check{A}) = \{0\}$ we obtain that $\mathcal{D}(\check{H}^2) = \{0\}$ if and only if $ker(\check{B}) = \{0\}$. Moreover, if $P_{\check{\mathcal{N}}}^{\mathcal{H}} \mathcal{D}(\check{H}) = \mathcal{R}(\check{B}) = \check{\mathcal{N}}$, then $\mathcal{R}(\check{B}^*) = \check{\mathcal{R}}$. Hence condition (2.3) $\mathcal{R}(\check{B}^*) \cap \mathcal{R}(\check{A}) = \{0\}$ cannot be satisfied. Consequently, the assumption $P_{\check{\mathcal{N}}}^{\mathcal{H}} \mathcal{D}(\check{H}) = \check{\mathcal{N}}$ is false and $\check{H}$ is a significantly symmetric operator. ∎

By the way we have shown that every closed symmetric operator which has a real regular point and which satisfies $\mathcal{D}(H^2) = \{0\}$ is significantly symmetric. However, this is not the only case.

Lemma 12 *Let H be a closed symmetric operator with infinite deficiency indices. If H possesses a selfadjoint extension with purely discrete spectrum, then H is significantly symmetric.*

Proof. If $\widetilde{H} \supset H$ has a purely discret spectrum, then there is a real point λ such that $\lambda \in \varrho(\widetilde{H})$. Hence λ is a real regular point of H. Notice that without loss of generality we can assume that $\lambda = 0$.

Since $\widetilde{H}$ has a purely discrete spectrum the operator $\widetilde{L} = \widetilde{H}^{-1}$ is compact. Using the Krein model we immediately get that $B = P_{\mathcal{N}}^{\mathcal{H}} L : \mathcal{R} \to \mathcal{N}$ is a compact operator. Since $\mathcal{N}$ is infinite dimensional we find $\mathcal{R}(B) \neq \mathcal{N}$. However, $\mathcal{R}(B) = P_{\mathcal{N}}^{\mathcal{H}} \mathcal{D}(H)$. ∎

The last lemma has an important application to partial differential operators.

Example 13 Let Ω be a bounded open subset of $\mathbf{R}^n$, $n \geq 2$, with smooth boundary and let $\dot{H}$ be given by

$$(\dot{H}f)(x) = -\sum_{i,j=1}^{n} \frac{\partial}{\partial x_i} a_{ij}(x) \frac{\partial}{\partial x_j} f(x) + \sum_{j=1}^{n} \left\{ b_j(x) \frac{\partial}{\partial x_j} f(x) + \frac{\partial}{\partial x_j} \overline{b_j(x)} f(x) \right\} + q(x)f(x),$$

$$(4.27)$$

$$f \in \mathcal{D}(\dot{H}) = C_0^\infty(\Omega),\ a_{ij} = \overline{a_{ji}} \in C^1(\overline{\Omega}),\ b_j \in C^1(\overline{\Omega}),\ q = \overline{q} \in C^1(\overline{\Omega}),$$

$$c^{-1}|\xi|^2 \leq \sum_{i,j=1}^{n} a_{ij}(x)\xi_i\xi_j \leq c|\xi|^2, \qquad x \in \overline{\Omega}. \tag{4.28}$$

Obviously, $\dot{H}$ is a symmetric operator on $\mathcal{H} = L^2(\Omega)$. By H we denote the closure of $\dot{H}$. Since $\dot{H}$ is semibounded from below the operator possesses a real regular point. Since $n \geq 2$ the deficiency indices of H are infinite. Further, by the boundedness of Ω the Friedrichs extension of H has a purely discrete spectrum. Hence by Lemma 12 H is a significantly symmetric operator. Setting $a_{ij} = \delta_{ij}$, $b_j = 0$ we see that the Schrödinger operator defined on bounded regions is significantly symmetric. In particular, this holds for the Laplace operator, i.e. q = 0.

5. Singular continuous spectrum

The aim of this section is to find an extension with a given singular continuous spectrum inside the gap. However, in order to solve this problem we were forced to restrict the class of closed symmetric operators to the subclass of significantly symmetric ones. Without this restriction we were unable to prove the desired result. Moreover, we do not know whether the result holds for arbitrary closed symmetric operators with a gap and infinite deficiency indices. In the following we denote the singular and absolutely continuous spectrum of an operator by $\sigma_{sc}(\cdot)$ and $\sigma_{ac}(\cdot)$, respectively.

Theorem 14 *Let H be a significantly symmetric operator with gap (a, b) and let $\mathcal{G}$ be a regular subset of $\mathbf{R}^1$. Then there is a selfadjoint extension $\widetilde{H}$ of H such that*

(i) $\sigma_e(\widetilde{H}) \cap (a, b) = \mathcal{G} \cap (a, b),$

(ii) $\sigma_p(\widetilde{H}) \cap (a, b) \cap \mathcal{G} = \emptyset,$

(iii) $\sigma_{sc}(\widetilde{H}) \cap (a,b) = \mathcal{G} \cap (a,b),$

(iv) $\sigma_{ac}(\widetilde{H}) \cap (a,b) = \emptyset.$

The proof of this theorem can be find in [5]. It uses the Krein model for closed symmetric operators with a gap which was developed in [3] and a refinement of Lemma 8 which guarantees that the operator $P_{\mathcal{Q}}^{\mathcal{N}} B$ is not only a compact one but an operator which singular numbers decay sufficiently fast. In particular, this yields that $P_{\mathcal{Q}}^{\mathcal{N}} B$ is a nuclear operator. Furthermore, the proof is based on the fact that for every perfect subset $\mathcal{G}$ of $\mathbf{R}^1$ we can find a finite singular continuous measure μ_0 such that $\mathcal{G} = supp(\mu_0)$. Finally, crucial was the simple observation that for every finite Borel measure μ_0 on $\mathbf{R}^1$ we can find another measure μ which is absolutely continuous with respect to μ_0, which has the same support as μ_0 and which obeys

$$\int_{\mathbf{R}^1} \frac{1}{(t-\lambda)^2} d\mu(t) = \infty, \qquad \forall \lambda \in supp(\mu). \tag{5.1}$$

Applying Theorem 14 to the Schrödinger or Laplace operators on bounded regions Ω of $\mathbf{R}^n$, $n \geq 2$, we see that these operators admit selfadjoint extensions with a given nonpositive singular continuous spectrum. Moreover, since the spectrum of the Friedrichs extension of the Schrödinger or Laplce operators, which are defined in accordance with (4.27), is discrete the closure H of $\dot{H}$ has a lot of gaps on the positive real axis. Consequently, there exist Schrödinger and Laplace operators with positive singular continuous spectrum.

6. Absolutely continuous spectrum

The method used to prove Theorem 14 is also applicable to the generation of absolutely continuous spectrum by extensions provided we consider the class of significantly symmetric operators.

Theorem 15 *Let H be a significantly symmetric operator and let $\mathcal{G} \subseteq \mathbf{R}^1$ be the support of a finite absolutely continuous measure, i.e., let $\mathcal{G}$ be a closed subset of $\mathbf{R}^1$ such that for every $\lambda \in \mathcal{G}$ and every neighbourhood $\mathcal{U}_\lambda$ of λ the Lebesgue measure of the set $\mathcal{G} \cap \mathcal{U}_\lambda$ is greater than zero. Then there is a selfadjoint extension $\widetilde{H}$ of H such that $\mathcal{G} \subseteq \sigma_{ac}(\widetilde{H})$. In particular we have $\sigma_{ac}(\widetilde{H}) \cap (a,b) = \mathcal{G} \cap (a,b).$*

Theorem 15 is proved in [6]. The main disadvantage of this theorem is that it does not exclude that besides the absolutely continuous spectrum it occurs point and singular continuous spectrum within the gap. To exclude these possibilities we have to strengthen a little bit the assumptions.

Theorem 16 *Let H be a significantly symmetric operator with gap (a,b) and let $\mathcal{G}$ be a regular subset of $\mathbf{R}^1$. Then there exists a selfadjoint extension $\widetilde{H}$ of H such that*

(i) $\sigma_e(\widetilde{H}) \cap (a,b) = \mathcal{G} \cap (a,b)$,

(ii) $\sigma_p(\widetilde{H}) \cap (a,b) \cap \mathcal{G} = \emptyset$,

(iii) $\sigma_{sc}(\widetilde{H}) \cap (a,b) \subseteq \partial\mathcal{G}, \qquad \partial\mathcal{G} = \mathcal{G} \setminus \mathcal{G}^0$,

(iv) $\sigma_{ac}(\widetilde{H}) \cap (a,b) = \mathcal{G} \cap (a,b)$.

The proof of Theorem 16 can be found in [6]. We note that if the closed set $\partial\mathcal{G}$ is countable, then

$$\sigma_{sc}(\widetilde{H}) \cap (a,b) = \emptyset. \tag{6.1}$$

This comes from the fact that a countable set cannot be the support of a continuous measure.

Applying the last theorem to the Schrödinger or Laplace operators on bounded regions Ω of $\mathbf{R}^n$, $n \geq 2$, we get that there are selfadjoint extensions of them which have a nonpositive absolutely continuous spectrum.

Finally, it is possible to combine Theorem 3, 14 and 16. By $\sigma_c(\cdot)$ we denote the continuous spectrum of an operator.

Theorem 17 *Let H be a significantly symmetric operator with gap (a,b). Let $\mathcal{E} \subset (a,b)$ be a countable set and let $p : \mathcal{E} \to \mathcal{N} \cup \{\aleph_0\}$ be any mapping. Further, let $\mathcal{G}_{sc} \subseteq [a,b]$ and $\mathcal{G}_{ac} \subseteq [a,b]$ be two closed subsets which are closures of disjoint open subsets $\mathcal{O}_{sc} \subseteq (a,b)$ and $\mathcal{O}_{ac} \subseteq (a,b)$, respectively, i.e., $\mathcal{G}_{sc} = \overline{\mathcal{O}_{sc}}$ and $\mathcal{G}_{ac} = \overline{\mathcal{O}_{ac}}$. Then there is a selfadjoint extension $\widetilde{H}$ of H such that*

(i) $\sigma_c(\widetilde{H}) \cap (a,b) = \mathcal{G} \cap (a,b), \qquad \mathcal{G} = \mathcal{G}_{sc} \cup \mathcal{G}_{ac}$,
$\mathcal{E} \setminus \mathcal{G} \subseteq \sigma_p(\widetilde{H}) \cap (a,b)$,
if $\lambda \in \mathcal{E} \setminus \mathcal{G}$, then the multiplicity of λ is greater than or equal to $p(\lambda)$,

(ii) $\sigma_p(\widetilde{H}) \cap (a,b) \cap \mathcal{G} = \mathcal{E} \cap \mathcal{G}$,
if $\lambda \in \mathcal{E} \cap \mathcal{G}$, then the multiplicity of λ is equal to $p(\lambda)$,

(iii) $\mathcal{G}_{sc} \subseteq \overline{\sigma_{sc}(\widetilde{H}) \cap (a,b)} \subseteq \mathcal{G}_{sc} \cup \Delta\mathcal{G}_{ac}, \qquad \Delta\mathcal{G}_{ac} = \mathcal{G}_{ac} \setminus \mathcal{O}_{ac}$,

(iv) $\sigma_{ac}(\widetilde{H}) \cap (a,b) = \mathcal{G}_{ac} \cap (a,b)$.

Again, if $\Delta\mathcal{G}_{ac}$ is a countable set, then property (iii) reduces to

$$\overline{\sigma_{sc}(\widetilde{H}) \cap (a,b)} = \mathcal{G}_{sc}. \tag{6.2}$$

So in this case we can determine the singular continuous spectrum exactly. The last theorem shows that the Schrödinger or Laplace operators defined on a bounded domain in $\mathbf{R}^n$, $n \geq 1$, admit selfadjoint extensions with spectra which are far from the usual ones.

References

[1] Achieser, N.I.; Glasmann, I.M.: *Theorie der linearen Operatoren im Hilbert-Raum.* Akademie-Verlag, Berlin 1975.

[2] Brasche, J.F.; Neidhardt, H.: *Has every symmetric operator a closed symmetric restriction whose square has a trivial domain?* Preprint SFB 288 No. 5, TU Berlin Berlin 1992, to appear in *Acta Scientiarum Mathematicarum.*

[3] Brasche, J.F.; Neidhardt, H.: *Krein's extension theory revised.* Preprint SFB 288 No. 4, TU Berlin, Berlin 1992.

[4] Brasche, J.F.; Neidhardt, H.; Weidmann, J.: On the point spectrum of selfadjoint extensions. to appear

[5] Brasche, J.F.; Neidhardt, H.: On the singular continuous spectrum of selfadjoint extensions, to appear

[6] Brasche, J.F.; Neidhardt, H.: On the absolutely continuous spectrum of selfadjoint extension, to appear

[7] Kato, T.: *Perturbation theory for linear operators,* Springer-Verlag, Berlin-Heidelberg-New York 1966.

[8] Krein, M.G.: Theory of selfadjoint extensions of semi-bounded Hermitian operators and its application. I. *Mat. Sbornik* 20 (1947), No. 3, 431-490 (in Russian).

[9] Derkach, V.A.; Malamud, M.M.: Generalized resolvents and the boundary value problems for Hermitian operators with gaps. *Journal of Functional Analysis* 95 (1991), No. 1, 1-95.

[10] Reed, M.; Simon, B.: *Methods of Modern Mathematical Physics. II. Fourier Analysis, Self-Adjointness.* Academic Press, New York-London 1975.

[11] Weidmann, J.: *Lineare Operatoren in Hilberträumen.* B.G.Teubner Stuttgart, Stuttgart 1976.

Johannes Brasche   **Hagen Neidhardt**
Joachim Weidmann

Johann Wolfgang Goethe-Universität   Technische Universität Berlin
Fachbereich Mathematik   Fachbereich Mathematik MA 7-2
Postfach 11 19 32   Str. des 17. Juni 136
W-6000 Frankfurt am Main 11   W-1000 Berlin 12
BRD   BRD

Operator Theory:
Advances and Applications, Vol. 61

On the Spectrum of Singular Nonselfadjoint Differential Operators

Petru A. Cojuhari

Abstract. In this work there are obtained results on the nonexistence of eigenvalues of operators generated by differential expressions of arbitrary order. It is supposed that the operators act in one of the spaces $L_p(\mathbf{R})$ or $L_p(\mathbf{R_+})$ $(1 \leq p \leq \infty)$. In the semi-axis case the boundary conditions are unimportant. Applications to the Schrödinger operator, to the Dirac operator and to the perturbed Hill operator are considered.

1. Introduction

The knowledge of the nature of the spectrum of an operator is of great importance in the scattering theory, in the eigenfunction expansions problems, in the spectral operator theory itself, etc. While the location of the essential spectrum can be determined rather exactly (due to the Weyl-type theorems on the essential spectrum stability), this is not the case for other components of the spectrum, a point component for instance. The reason is that even for small perturbations there may arise a number of eigenvalues as large as we wish, even though these did not exist before the perturbation. In this connection it would be enough to take into account the Weyl-von Neumann Theorem (see, e.g., [22], chapter X, Theorem 2.1). This brings up an important problem: under what conditions is a set of points in the complex plane free of eigenvalues either of the unperturbed or of the perturbed operator? The work of T.Kato [21] contains one of the first results in this direction. The history of the problem can be found in [33,19]. In the work [21] it is proved that the Schrödinger operator $-\Delta + q(x)$ (acting in the space $L_2(\mathbf{R}^3)$) with a potential $q(x)$ such that

$$|q(x)| \leq c(1 + |x|)^{-\delta} \quad (\delta > 1; q(x) \text{ is a real function}),$$

has no positive eigenvalues. Kato's research is based on some complicated integro-differential inequalities. The works [14,36] should be mentioned for simpler proofs of this statement. The proof given in [36] (where the Schrödinger operator with a complex potential is considered) corresponds to the Aronszajn's proof [4] and Cordes's

Theorem [12] on the unique continuation for solutions of elliptic differential equations of second order.

Similar results for the one-dimensional Schrödinger operator $Ly = -y'' + q(x)y$ have been obtained by other authors [15,30,32], with more detailed references and comments in [31,13]. A number of works give the investigation of the one-dimensional Schrödinger operator along the semi-axis $\mathbf{R}_+ = [0, \infty]$ and acting in the space $L_2(\mathbf{R}_+)$. Some of its spectral properties, in particular the conditions producing the absence of eigenvalues (immersed in a continuouse part of the spectrum), appear to be the same under various boundary conditions, for instance under the boundary conditions $y(0) = 0$ and $y'(0) - \theta y(0) = 0$ (θ is a complex number) [31], Supplement 1, pp.443. In this connection, A. M. Krall's works [23,24,25] should be mentioned as places where the boundary conditions of the type

$$\int_0^\infty K(x)y(x)dx - \beta y(0) + \alpha y'(0) = 0 \quad (K(x) \in L_2(\mathbf{R}_+), |\alpha|^2 + |\beta|^2 > 0)$$

are considered. If $K(x), q(x) \in L_1(\mathbf{R}_+)$ and $\alpha \neq 0$, then the spectral structure is as in the case of the bondary condition $y(0) = 0$.

In this work we obtain further results on the nonexistence of eigenvalues of the operator derived from differential expressions of arbitrary order and assigned along the real axis $\mathbf{R}$ or along the positive semi-axis $\mathbf{R}_+$. The operators studied are supposed to act in one of the spaces $L_p(\mathbf{R})$ or $L_p(\mathbf{R}_+)$ ($1 \leq p < \infty$). In doing so, in the semi-axis cases, the choice of the boundary conditions is of no importance here. More precisely, for differential operators assigned along the semi-axis, the maximal differential operator has no eigenvalues (the term "maximal" is used in the sense of definition of [22], Chapter III, §2). Hence, we will get similar results for the corresponding restrictions of the maximal operator as well.[1]

Some methods of perturbation theory are used to get the results in this paper. In Section 2 there is a general scheme enabling us to define the conditions under which the nonexistence of eigenvalues of the operators acting in a Banach space holds. The scheme given here is in fact a more refined version of the scheme described by the author in [7] (see also [8]).

The abstract scheme of Section 2 is applied to investigate the point spectrum of differential operators of general type. Section 3 deals with the perturbation of differential operators with constant coefficients and Section 4 considers the case when the unperturbed operator is a differential operator with periodic coefficients. There is also a number of examples. On the one hand, we tried to illustrate the peculiarities of applying our general scheme in various cases. On the other hand, in our opinion, the

---

[1] If $A_1 \subset A_2$, it is obvious that $\sigma_p(A_1) \subset \sigma_p(A_2)$, where $\sigma_p(A_j)$ denotes the point spectrum of the operator $A_j (j = 1, 2)$.

other examples are of interest in their own. Primarily these concern the Schrödinger operator, the Dirac operator, the perturbed Hill operator , etc.

It should be noted that the results obtained in this paper can be applied for the discrete case as well. Discrete analogues of differential operator are difference operators. In the paper [27] it is studied the operator L generated in the space l_2 by the difference expression

$$(ly)_j = \frac{1}{2}(y_{j-1} + y_{j+1}) - b_j y_j$$

and by the boundary condition $y_0 = \theta y_1$, with complex numbers $b_j (j = 1, 2, \ldots)$ and θ. In [27] it is shown that if $b_j = O(j^{-1-\epsilon})(\epsilon > 0)$, then the operator L has no eigenvalues in the interval $(-1, 1)$; if $b_j = O(j^{-2-\epsilon})$, then the points $\lambda = \pm 1$ belong to the point spectrum. More general perturbations $B = \|b_{jk}\|_{j,k=1}^{\infty}$ (not necessarily diagonal) have been considered by the author in [8]. The extension of the results in [8] and [27] is given in [7] where the perturbed discrete Wiener-Hopf operators are investigated.

The core of this paper is our report to the International Conference on Operator Theory in Timişoara in June 1992. Some of the results have been stated without any proof in our article [9].

2. The abstract scheme

Let $\mathfrak{B}, \mathfrak{C}, \ldots$ denote Banach spaces. The domain and the range of an operator A are denoted by $\mathfrak{D}(A)$ and $\mathfrak{R}(A)$, respectively.

Let A and B be linear operators in a space $\mathfrak{B}$ such that the following assumptions are fulfilled.

Assumption 1. *The number λ is not an eigenvalue of the operator A, i.e. $\lambda \notin \sigma_p(A)$.*

Assumption 2. *Let B be written as $B = TS$, where S is an operator acting from $\mathfrak{B}$ into $\mathfrak{C}$, and the operator T is acting from $\mathfrak{C}$ into $\mathfrak{D}$, provided that $\mathfrak{D}(S) \supset \mathfrak{D}(A)$.* [2]

Assumption 3. *If $0 \neq \psi \in \mathfrak{C}$ and $T\psi \in \mathfrak{R}(A - \lambda I)$, then there exists a normed space $\mathfrak{N}$ such that $\mathfrak{N} \subset \mathfrak{C}$, $\psi \in \mathfrak{N}$, $S(A - \lambda I)^{-1}T\psi \in \mathfrak{N}$ and the following inequality must be satisfied*

$$|S(A - \lambda I)^{-1}T\psi|_{\mathfrak{N}} < |\psi|_{\mathfrak{N}}, \tag{2.1}$$

$| \cdot |_{\mathfrak{N}}$ *denotes a norm in $\mathfrak{N}$ space.*

---

[2]In this paper, as a rule, the operators S and T will be unbounded. In case of unbounded operators S and T the product $B = TS$ should be understood as an ordinary operator product, i.e. defining $\mathfrak{D}(B) = \{\varphi \in \mathfrak{D}(S) | S\varphi \in \mathfrak{D}(T)\}$.

Statement. *If the Assumptions 1-3 are satisfied, then λ is not an eigenvalue of the operator $A + B$.*

The proof of this statement coincides essentially with that from [7], where the case of bounded operators is discussed.

In applications, in order to choose an appropriate space $\mathfrak{N}$ it is convenient to consider in the space $\mathfrak{C}$ a family of operators $L_\tau (\tau \geq 0)$ with the following property.

Assumption 4. *For each $\tau \geq 0$ the operator L_τ is one-to-one, i.e. $\ker L_\tau = 0$. In addition, $L_0 = I_\mathfrak{C}$ ($I_\mathfrak{C}$ is the unit operator in the space $\mathfrak{C}$).*

On the domain $\mathfrak{D}_\tau = \mathfrak{D}(L_\tau)$ of the operator $L_\tau (\tau \geq 0)$ let us introduce a new norm $|u|_\tau = \|L_\tau u\|_\mathfrak{C}$, $(u \in \mathfrak{D}_\tau)$, where $\|\cdot\|_\mathfrak{C}$ is the norm on the space $\mathfrak{C}$. The norm $|\cdot|_\tau$ turns the linear manifold $\mathfrak{D}_\tau$ into a normed space which in general is not complete. Denote this space by $\mathfrak{C}_\tau$. Clearly, $\mathfrak{C}_0 = \mathfrak{C}$.

It is easy to see that Assumption 3 will be fulfilled if Assumption 4 is combined with the following

Assumption 5. *There exists $\tau \geq 0$ such that if $\psi \in \mathfrak{C}$ and $T\psi \in \mathfrak{R}(A - \lambda I)$, then $\psi \in \mathfrak{C}_\tau$, $S(A - \lambda I)^{-1}T\psi \in \mathfrak{C}_\tau$ and*

$$|S(A - \lambda I)^{-1}T\psi| \leq a|\psi|_\tau \quad (0 < a < 1). \tag{2.2}$$

The fact that the vectors ψ belong to the corresponding $\mathfrak{C}_\tau$ (when $\psi \in \mathfrak{C}$ and $T\psi \in \mathfrak{R}(A - \lambda I)$) follows from the following

Assumption 6. *For each $\psi \in \mathfrak{C}_\tau$ such that $T\psi \in \mathfrak{R}(A - \lambda I)$, the following inequality holds*

$$|S(A - \lambda I)^{-1}T\psi|_\tau \leq c|\psi|_{\tau'} \quad (c = \text{const}; \; \tau > \tau' \geq 0).$$

Here and hereafter c (with indices or without) stands for a nonnegative constant, the precise value of which being nonimportant here. It is possible to assign the same symbol to different constants.

3.   A study of the point spectrum of singular differential operators

3.1   Let us begin by considering the differential operator of the type

$$H = P(D) + \sum_{k=0}^{n-1} q_k(x)D^k, \tag{3.1}$$

where $D = i\, d/dx$ and $P(\xi) = a_n\xi^n + \ldots + a_1\xi + a_0$ is a polynomial over the field of complex numbers $(a_n \neq 0)$ and $q_k(x)(k = 0, 1, \ldots, n - 1; 0 < x < \infty)$ are measurable (generally speaking, complex) functions. The operator H is supposed to act in one of the spaces $L_p(\mathbf{R_+})$ or $L_p(\mathbf{R})$ $(1 \leq p < \infty)$. The domain of the operator H consists of all functions $\varphi \in L_p(\mathbf{R_+})$ (or $L_p(\mathbf{R})$) having absolutely continuous derivatives of the $n - 1$ order on each bounded interval of the positive semi-axis (on the whole real axis, respectively) and the derivative of the n-th order belonging to $L_p(\mathbf{R_+})$ (to $L_p(\mathbf{R})$, respectively).

The operator H will be considered as a perturbation of the operator $A = P(D)$ by the operator $B = \sum_{k=0}^{n-1} q_k(x)D^k$. Information on the spectrum of the operator A can be obtained, for instance from [22], [31],[13]. Here we shall restrict our considerations to a few comments necessary in the hereafter. For the sake of definiteness let the operator A be assigned on the positive semi-axis. The operator A is considered as a polynomial value of $P(\xi)$ of the differential operator $D = i\, d/dx$. Let λ be some number of the complex plane and let ξ_k $(k = 1, \ldots, n)$ be all the roots of the polynomial $P(\xi) - \lambda$. Then

$$A - \lambda I = a_n \prod_{k=1}^{n}(D - \xi_k I). \tag{3.2}$$

It is well-known that the resolvent set of the operator D coincides with the open upper half-plane $\{z \in \mathbf{C} \,|\, \mathrm{Im}\, z > 0\}$, its point spectrum filling the open lower half-plane and all the real axis being in the continuous part of the spectrum. Also, if $\mathrm{Im}\, z > 0$ then

$$(D - zI)^{-1}u(x) = i \int_x^\infty \exp(iz(y - x))u(y)dy. \tag{3.3}$$

At $\mathrm{Im}\, z = 0$ the formula (3.3) is also true, provided that $u \in \mathfrak{R}(D - zI)$.

Now, from the equality (3.2) it easily follows that $\lambda \in \sigma(A)$ if and only if the roots of the polynomial $P(\xi) - \lambda$ are in the closed lower half-plane $\mathrm{Im}\,\xi \leq 0$. Furthermore, the point λ will be an eigenvalue of the operator A if the polynomial $P(\xi) - \lambda$ has some zeros in the open lower half-plane. If the polynomial $P(\xi) - \lambda$ has its zeros only in $\mathrm{Im}\,\xi \geq 0$, the point λ is not an eigenvalue of the operator A. We should take into account that if the operator A is specified on the whole axis (i.e. it acts in $L_p(\mathbf{R})$) its spectrum consists of all points λ on the curve $\lambda = A(\xi)$ $(\xi \in \mathbf{R})$.

3.2 Consider first the case when the operator A is acting in the $L_p(\mathbf{R_+})$ space. Let λ be not an eigenvalue of the operator A. As it was mentioned above, this means that the polynomial $P(\xi) - \lambda$ can be represented

$$P(\xi) - \lambda = \prod_{k=0}^{r}(\xi - \xi_k)^{m_k} P_1(\xi), \tag{3.4}$$

where ξ_k $(k = 0, 1, \ldots, r)$ are real distinct numbers, $m_k \geq 0$ $(k = 0, 1, \ldots, r)$ [3] and $P_1(\xi)$ is a polynomial, the roots of which belong to the upper half-plane $\operatorname{Im}\xi > 0$.

Theorem 3.1 *Let H be a differential operator as in (3.1) and acting in the space $L_p(\mathbf{R}_+)$. Assume the number λ is not an eigenvalue of the operator A, i.e. for the polynomial $P(\xi) - \lambda$, the expansion (3.1) holds, and denote $m = \max\{m_k|\ k = 0, 1, \ldots, r\}$. If the functions $q_k(x)$ are such that*

$$(1 + x)^\delta q_k(x) \in L_\infty(\mathbf{R}_+) \quad (\delta > m; k = 0, 1, \ldots, n - 1),$$

then λ is not an eigenvalue of the operator H.

Proof. Let us verify the conditions of Assumptions 1–3 . The Banach space $\mathfrak{C}$ is the direct sum of n copies of $L_p(\mathbf{R}_+)$:

$$\mathfrak{C} = \bigoplus_{k=1}^{n} L_p(\mathbf{R}_+).$$

The norm in $\mathfrak{C}$ will be defined by the following formula

$$\|\psi\| = \sum_{k=1}^{n} \|\psi_k\|_{L_p(\mathbf{R}_+)} \quad (\psi = \{\psi_k\}_1^n \in \mathfrak{C}).$$

By S we denote the operator acting from $L_p(\mathbf{R}_+)$ in $\mathfrak{C}$ in the following way:

$$S\varphi = \{D^k\varphi\}_{k=0}^{n-1} \quad (\varphi \in \mathfrak{D}(D^{n-1})).$$

Let us consider another operator T acting from $\mathfrak{C}$ into $L_p(\mathbf{R}_+)$ according to the formula:

$$T\{\psi_k\}_1^n = \sum_{k=0}^{n-1} q_k(x)\psi_k(x).$$

It is obvious that $B = TS$ and $\mathfrak{D}(S) \supset \mathfrak{D}(A)$. The operators L_τ from Assumption 4 are defined as follows

$$L_\tau\psi = ((1 + x)^\tau \psi_k)_{k=1}^n \quad (\psi = (\psi_k)_{k=1}^n \in \mathfrak{C})$$

and we proceed to verify Assumptions 5 and 6 (from here Assumption 3 will follow). It should be mentioned that the operator $D^k(A - \lambda I)^{-1}(k = 0, 1, \ldots, n - 1)$ on the domain $\mathfrak{R}(A - \lambda I)$ is a linear combination of operators of the type $(D - \mu I)^{-l}$ with $\operatorname{Im}\mu \geq 0$ (the greatest power of l on $\operatorname{Im}\mu = 0$ being equal to m).

Let T_k be the operator of multiplication by the function $q_k(x)$ (acting in the space $L_p(\mathbf{R}_+)$). Since

$$S(A - \lambda I)^{-1}T = (\sum_{k=0}^{n-1} D^j(A - \lambda I)^{-1}T_k)_{j=0}^{n-1},$$

---

[3] If the polynomial $P(\xi) - \lambda$ has no zeros on the real axis, we assume $m_k = 0$.

we must estimate the norm $|(D - \mu I)^{-l} T_j u|_\tau$ ($\varphi \in L_p(\mathbf{R}_+)$). We shall keep for the norm $\|(1 + x)^\tau \cdot \|_{L_p(\mathbf{R}_+)}$ the same symbol $|\cdot|_\tau$ as for the case of the space $\mathfrak{C}$ and the operators L_τ . We can get the required estimates by means of the following inequalities of Hardy type

$$|u|_\tau \leq c(\tau)|(D - \lambda I)u|_{\tau+\delta} \quad (\tau \geq 0, \delta > 0; u \in \mathfrak{D}(D)), \tag{3.5}$$

where $c(\tau) \to 0$ as $t \to \infty$.

We note that on $\operatorname{Im} \lambda > 0$ the inequalities (3.5) are true for $\delta > 0$ either. Inequalities (3.5) can be obtained from integral inequalities to be found in [18], p.294. They can also be easily obtained by means of Lemma 1 and Lemma 2 in [10].

In view of the inequalities (3.5) we obtain

$$|(D - \mu I)^{-l} T_j u|_\tau \leq c(\tau)|T_j u|_{\tau+\delta},$$

where $\delta > l$ if $\operatorname{Im} \mu = 0$ and $\delta > 0$ if $\operatorname{Im} \mu > 0$.

Since

$$(1 + x)^\delta q(x) \in L_\infty(\mathbf{R}_+) \quad (\delta > l; l = 0, 1, \ldots, m),$$

holds, then

$$|T_j u|_{\tau+\delta} \leq c|u|_\tau$$

and Assumption 5 is verified. Assumption 6 follows from (3.5) as well. $\blacksquare$

It is clear from the above proof that in case that $\xi = 0$ is a root of the polynomial $P(\xi) - \lambda$ (let, for instance, $\xi_0 = 0$ with $m_0 > 0$) Theorem 3.1 can be refined to some extent. More exactly, the following statement comes into being.

Theorem 3.2 *Let H be an operator of type (3.1) and acting in the space $L_p(\mathbf{R}_+)$. Assume that the operator B is defined by*

$$B = \sum_{k=k_0}^{n-1} q(x)D^k,$$

the polyomial $P(\xi) - \lambda$ has the representation (3.4) and denote

$$m = \max\{m_0 - k_0, m_1, \ldots, m_r\}.$$

If the functions $q_k(x)$ are such that

$$(1 + x)^\delta q_k(x) \in L_\infty(\mathbf{R}_+) \quad (\delta > m; k = k_0, \ldots, n - 1),$$

then the number λ is not an eigenvalue of the operator H.

Remark 3.3 For the selfadjoint differential operator H (acting in $L_2(\mathbf{R}_+)$) similar conditions on the functions $q_k(x)$ assure the finiteness of the discrete spectrum of the operator H (see [16], [5], [11]).

Remark 3.4 If the polynomial $P(\xi) - \lambda$ has some zeros in the lower half-plane, the Theorem 3.1 is false, as it will be seen in Section 4.

3.3 Now let the operator H be defined along the whole axis. If the polynomial $P(\xi) - \lambda$ has at least one nonreal root, the number λ can appear as an eigenvalue of the operator H even if the coefficients $q_k(x)(k = 0, 1, \ldots, n - 1)$ tend fast enough to infinity. In this connection, it should be noted that for the case of the space $L_p(\mathbf{R})$, the inequalities of Hardy type (3.5) are true only at $\operatorname{Im}\lambda = 0$ (and fail for $\operatorname{Im}\lambda \neq 0$). For the case of the Hilbert space $L_2(\mathbf{R})$ the inequalities of the type (3.5) with ponders $|x|^\tau(\tau \geq 0)$ are given in [1] and for more general results see [3].

However the following holds.

Theorem 3.5 *Let the operator H act in the space $L_p(\mathbf{R})$ and let all the roots of the polynomial $P(\xi) - \lambda$ be real, m being their greatest multiplicity. If the functions $q_k(x)$ are such that*

$$(1 + |x|)^\delta q_k(x) \in L_\infty(\mathbf{R}) \quad (\delta > m; k = 0, 1, \ldots, n - 1),$$

then the number λ is not an eigenvalue of the operator H.

The Theorem 3.2, with corresponding changes, can also be transferred for the differential operator H preassigned along the whole axis.

4. Investigation of the spectrum of the perturbed differential operators with periodic coefficients

4.1 Let us consider the differential operator

$$H = \sum_{k=0}^{n} h_k(t) \left(\frac{d}{dt}\right)^k,$$

where $h_k(t) = a_k(t) + q_k(t)$, $a_k(t)$ $(k = 0, 1, \ldots, n; a_n(t) \neq 0)$ are periodic functions (with period T) and $q_k(t)(k = 0, 1, \ldots, n; q_n(t) \equiv 0)$ are functions vanishing for t tending to infinity. Assume that $a_k(t)(k = 0, 1, \ldots, n)$ are as smooth as required. Consider the operator H as a perturbation of the operator $A = \sum_{k=0}^{n} a_k(t)(\frac{d}{dt})^k$ subject to the differential operator $B = \sum_{k=0}^{n-1} q_k(t)(\frac{d}{dt})^k$.

The spectrum of the operator A has been well studied [34,28]. In [34] the operator A is studied in the space $L_2(\mathbf{R})$ and in [28] it is studied in the space $L_p(\mathbf{R})(1 \leq p \leq \infty)$.

The operator A has a purely continuous spectrum and coincides with a set (the so-called zones of relative stability) of those values λ for which the equation $Au = \lambda u$ has a non-trivial solution, bounded on the whole axis. It is also noted in [34] that for the operator A of the second order acting in the space $L_2(\mathbf{R}_+)$ (under some boundary conditions at zero) the continuous spectrum nature is preserved.

The conditions will be further set up for nonexistence of eigenvalues of the operator H. The operator H is assumed to act in one of the the Banach spaces $L_p(\mathbf{R})$ or $L_p(\mathbf{R}_+)$ $(1 \le p \le \infty)$.

To apply the scheme from Section 2 we have to describe the operator $(A - \lambda I)^{-1}$. For this purpose we will apply the Floquet theory (see e.g. [17,37]).

Let λ be some complx number and consider the spectral problem

$$Hu = \lambda u. \tag{4.1}$$

Equation (4.1) should be written in a vector form

$$\frac{dx}{dt} = A(t,\lambda)x + B(t)x \tag{4.2}$$

where

$$A(t,\lambda) = \begin{pmatrix} 0 & 1 & \ldots & 0 \\ 0 & 0 & \ldots & 0 \\ \vdots & \vdots & \ldots & \vdots \\ 0 & 0 & \ldots & 1 \\ a_n^{-1}(\lambda - a_0) & -a_n^{-1}a_1 & \ldots & -a_n^{-1}a_{n-1} \end{pmatrix},$$

$$B(t) = \begin{pmatrix} 0 & 0 & \ldots & 0 \\ 0 & 0 & \ldots & 0 \\ \vdots & \vdots & \vdots & \vdots \\ 0 & 0 & \ldots & 0 \\ -a_n^{-1}b_n & -a_n^{-1}b_1 & \ldots & -a_n^{-1}b_{n-1} \end{pmatrix}, \quad x = \begin{pmatrix} u \\ u' \\ \vdots \\ u^{(n-1)} \end{pmatrix}.$$

Consider the equation

$$\frac{dx}{dt} = A(t,\lambda)x(t). \tag{4.3}$$

Denote by means of $U(t)$ the matrix satisfying the system

$$\frac{dU(t)}{dt} = A(t,\lambda)U(t), \quad U(0) = E_n.$$

(E_n is unit matrix of the n-th order). The matrix $U(T)$ is called the monodromy matrix of the equation (4.3). The eigenvalues of the matrix $U(T)$ denoted by $\rho_1, \rho_2, \ldots$ $\ldots, \rho_n$, i.e. the roots of the equation $\det(U(T) - \rho E_n) = 0$, are called the multiplicators of the equation (4.3) .

Consider first the case when the operator H acts in the space $L_p(\mathbf{R}_+)$ (with boundary conditions of general type).

Theorem 4.1 *Let the operator H act in the space $L_p(\mathbf{R}_+)$ $(1 \le p < \infty)$, the multiplicators $\rho_j = \rho_j(\lambda)$ of the equation (4.3) be such that $|\rho_j| \ge 1$ $(j = 1, \ldots, r)$ and m is the maximum of orders of the Jordan Canonical Form blocks complying with unimodular multiplicators ρ_j, $|\rho_j| = 1$. If the functions $q_k(t)$ are such that*

$$(1 + t)^\delta q_k(t) \in L_\infty(\mathbf{R}_+) \quad (\delta > m; k = 0, 1, \ldots, n - 1),$$

then the number λ is not an eigenvalue of the operator H.

Proof. According to Floquet's theory let us assume that the matrix $U(t)$ has the representation $U(t) = F(t) \exp(t\Gamma)$, where $\Gamma = \frac{1}{T} \ln U(T)$ (i.e. one of the solutions of the matrix equation $\exp(TY) = U(T)$) , and $F(t)$ is a nondegenerate differentiated matrix function of period T.

Applying the change of variables $x = F(t)y$ to the equation (4.2) we obtain

$$\frac{dy}{dt} = \Gamma y + F(T)^{-1}B(t)x$$

whence it follows that

$$x(t) = -F(t) \int_t^\infty \exp(\Gamma(t - s))F(s)^{-1}B(s)x(s)ds. \tag{4.4}$$

The matrix structure will be apparent if Γ is put into the Jordan Canonical Form, i.e. it is represented in terms of $\Gamma = S\mathcal{I}S^{-1}$, where

$$\mathcal{I} = \operatorname{diag} \|\mathcal{I}(1), \ldots, \mathcal{I}(l)\|,$$

and $\mathcal{I}(j)(j = 1, \ldots, l)$ are the Jordan Canonical Form blocks

$$\mathcal{I}(j) = \begin{pmatrix} \lambda_j & 1 & 0 & \cdots & 0 \\ 0 & \lambda_j & 1 & \cdots & 0 \\ \vdots & \vdots & \vdots & \vdots & \vdots \\ 0 & 0 & 0 & \cdots & \lambda_j \end{pmatrix}.$$

We should point out that

$$\exp(t\Gamma) = S \exp(t\mathcal{I})S^{-1}, \quad \exp(t\mathcal{I}) = \operatorname{diag} \| \exp \mathcal{I}(1)t, \ldots, \exp \mathcal{I}(l)t \| \tag{4.5}$$

and in addition

$$\exp(\mathcal{I}(j)t) = \exp(t\lambda_j) \begin{pmatrix} 1 & t & \cdots & \frac{t^{p_j-1}}{(p_j-1)!} \\ 0 & 1 & \cdots & \frac{t^{p_j-2}}{(p_j-2)!} \\ \vdots & \vdots & \vdots & \vdots \\ 0 & 0 & \cdots & 1 \end{pmatrix} \tag{4.6}$$

where p_j is the order of the block $\mathcal{I}(j)$ $(j = 1, \ldots, l)$.

The multiplicators $\rho_j = \exp(\lambda_j T)$ $(j = 1, \ldots, l)$ correspond to the eigenvalues λ_j of the matrix Γ and so for $|\rho_j| \geq 1$ where $\operatorname{Re} \lambda_j \geq 0$.

Furthermore, let us introduce the operators

$$(L_\tau x)(t) = (1 + t)^\tau x(t) \quad (\tau \geq 0).$$

We also introduce a new norm $| \cdot |_\tau = \|L_\tau \cdot \|$ on the domain $\mathfrak{D}_\tau$ of the operator L_τ and evaluate the solution of the equation (4.2) with respect to this norm. It should be noted that the vector components from the right hand side of the equation (4.4) are the sums of the following quantities (see formulae (4.5) and (4.6)):

$$(K_j x_k)(t) = q(t) \int_t^\infty (t - s)^{l_j - 1} \exp(\lambda_j(t - s)) p(s) b_{n-k+1}(s) x_k(s) ds,$$

where $p(t)$ and $q(t)$ are continuous periodic functions of period T and l_j is one of the values $1, \ldots, p_j$ $(j = 1, \ldots, l)$.

From the equation (4.4) and recalling the conditions set for the coefficients $q_k(x)$ we will get Assumptions 5 and 6 as in proof of the Theorem 3.1. $\blacksquare$

Let now the operator H be assigned along the whole axis. It can be shown that in this case the number λ will be an eigenvalue even if $q_k(t)$ decreases sufficiently fast at infinity and if at least one of the multiplicators is not unimodular. In this respect we have the following result.

Theorem 4.2 *Let the operator H act in the space $L_p(\mathbf{R})$ where all the multiplicators $\rho_j = \rho_j(\lambda)$ of the equation (4.3) are unimodular and let m be the maximum of Jordan Canonical Form block orders corresponding to the multiplicators ρ_j $(j = 1, \ldots, r)$. If the functions $q_k(t)$ are such that*

$$(1 + |t|)^\delta q_k(t) \in L_\infty(\mathbf{R}^n) \quad (\delta > m; k = 0, 1, \ldots, n - 1),$$

then the number λ is not an eigenvalue of the operator H.

4.2 The spectral problem (4.1) has been reduced to studying the corresponding system of the first order differential equations (4.2) and (4.3). Moreover, the dimension of the systems associated with matrices $A(T, \lambda)$ and $B(t)$ should not be taken into account. In this connection Theorems 4.1 and 4.2 are true (with obvious changes) also for differential operators with matrix coefficients.

It should be only taken into account that under the conditions of Theorems 4.1 and 4.2 the coefficients $a_k(t)(k = 0, 1, \ldots, n)$ are periodic and of size $n \times n$. The operator H is considered in the space $L_p(\mathbf{R}_+; \mathbf{C}^n)$ or $L_p(\mathbf{R}; \mathbf{C}^n)$.

58   P.A. Cojuhari

Moreover, the conditions set for the coefficients $q_k(t)$, for example in Theorem 4.1, should be replaced as follows:

$$(1 + t)^\delta |q_k(t)| \in L_\infty(\mathbf{R}_+) \ (\delta > m; k = 0, 1, \ldots, n - 1),$$

where $|\cdot|$ is the operator matrix norm on $\mathbf{C}^n$.

5. Examples. Schrödinger operators, Dirac operators and the perturbed Hill operators

5.1. In the space $L_p(\mathbf{R}_+)$ consider the differential operator

$$H = D^2 + q_1(x)D + q_0(x),$$

where $q_k(x)(k = 0, 1)$ are measurable and complex valued functions on the semiaxis $\mathbf{R}_+$. According to our notation in this case we have $P(\xi) = \xi^2$. The roots of the polynomial $P(\xi) - \lambda$ for $\lambda > 0$ are simple and belong to the real axis. Therefore it is possible to formulate the following

Statement. *If*
$$(1 + x)^\delta q_k(x) \in L_\infty(\mathbf{R}_+) \quad (\delta > 1; k = 0, 1),$$
then $(0, +\infty) \cap \sigma_p(H) = \emptyset$.

Note that for $\lambda = 0$ there is a multiple zero, since $P(\xi) = \xi^2$, and therefore the following result holds.

Statement. *If*
$$(1 + x)^\delta q_k(x) \in L_\infty(\mathbf{R}_+) \quad (\delta > 2; k = 0, 1),$$
then $\lambda = 0$ *is not an eigenvalue of the operator* H *and hence* $\mathbf{R}_+ \cap \sigma_p(h) = \emptyset$.

5.2. Let us consider the operator

$$H = D^2 + a_1 D + q_1(x)D + q_0(x), \tag{5.1}$$

where $a_1 \in \mathbf{C}$, $\mathrm{Im}\, a_1 < 0$, and $q_k(x)(k = 0, 1)$ are measurable (complex valued) functions. These operators occur in various specific problems in physics. For instance, in [6] it was shown that the investigation of the Schrödinger operator in the one-dimensional problem of scattering of the Brown particle is reduced to the spectral analysis of the family of nonselfadjoint operators of type (5.1) with

$$q_1(x) \equiv 0, \ a_1 = -2i\varepsilon\frac{1 + i\kappa}{1 - i\kappa}, \ q_0(x) = \frac{q(x)}{1 - i\kappa},$$

where the parameters ε, κ are positive.

We consider the operator H acting in the space $L_p(\mathbf{R}_+)$. Then $P(\xi) = \xi^2 + a_1\xi$. It is easy to note that in the representation $\omega = P(z)$ the real axis $\operatorname{Im} z = 0$ is reflected into the parabola $\omega = x^2 + a_1 x$, the lower half-plane $\operatorname{Im} z < 0$ being reflected into the inside of this parabola. Denote the above parabola with Γ and the domain consisting of all the points lying inside the parabola Γ with Π. If $\lambda \in \Pi$, the roots of the polynomial $P(\xi) - \lambda$ belong to the upper half-plane $\operatorname{Im} z > 0$. Therefore , for the values λ the following result holds.

Statement. *If*

$$(1 + x)^\delta q_k(x) \in L_\infty(\mathbf{R}_+) \quad (\delta > 0; k = 0, 1),$$

then $\Pi \cap \sigma_p(H) = \emptyset$.

Let $\lambda \in \Pi$. Then $\lambda = x^2 + a_1 x$ $(x \in \mathbf{R})$ hence $p(\xi) - \lambda = (\xi - x)(\xi + x + a_1)$, i.e. the polynomial $P(\xi) - \lambda$ has a single real root $\xi_1 = x$ and a complex root $\xi_2 = -x - a_1$ in the upper half-plane. Therefore, the following result holds.

Statement. *If*

$$(1 + x)^\delta q_k(x) \in L_\infty(\mathbf{R}_+) \quad (\delta > 1; k = 0, 1),$$

then there exist no eigenvalues of the operator H *on the curve* Γ, *i.e.* $\sigma_p(H) \cap \Gamma = \emptyset$.

5.3. Let H be the operator

$$H = D^4 + \sum_{k=0}^{3} q_k(x) D^k.$$

If

$$(1 + |x|)^\delta q_k(x) \in L_\infty(\mathbf{R}) \quad (\delta > 4; k = 1, 2, 3)$$

then $0 \in \sigma_p(H)$. If, for instance, $q_0(x) = q_1(x) = q_2(x) \equiv 0$ and

$$(1 + |x|)^\delta q_3(x) \in L_\infty(\mathbf{R}) \quad (\delta > 1)$$

then $0 \notin \sigma_p(H)$.

5.4. Let

$$H = D^4 + (q_3(x) - 2i)D^3 + (q_2(x) - 1)D^2 + q_1(x)D + q_0(x) + 2,$$

be considered as an operator acting in $L_p(\mathbf{R}_+)$.

In this case

$$P(\xi) - 2 = \xi^4 - 2i\xi^3 - \xi^2 = \xi^2(\xi - i)^2,$$

therefore, the following result holds

Statement. *If*

$$(1 + x)^\delta q_k(x) \in L_\infty(\mathbf{R}_+) \quad (\delta > 2; k = 0, 1, 2, 3),$$

then the point $\lambda = 2$ *is not eigenvalue of the operator* H.

At $q_0(x) = 0$ this result is refined as follows: $\lambda = 2 \notin \sigma_p(H)$ if

$$(1 + x)^\delta q_k(x) \in L_\infty(\mathbf{R}_+) \quad (\delta > 1; k = 1, 2, 3).$$

5.5. Let H be the differential operator

$$H = -\frac{d^2}{dt^2} + q_1(t)\frac{d}{dt} + p(t) + q_2(t),$$

where $p(t + 1) = p(t)$, $q_k(t)(k = 1, 2)$ are measurable complex valued functions. The unperturbed operator is the Hill operator ([16], pp.281)

$$Au = -\frac{d^2u}{dt^2} + p(t)u.$$

As it is known (see [34] or [16]) the multiplicators corresponding to the inside of the point spectrum of the operator A are simple and of modulus 1. Therefore, the following result can be formulated.

Statement. *If*

$$(1 + t)^\delta q_k(t) \in L_\infty(\mathbf{R}_+) \quad (\delta > 1; k = 1, 2) \tag{5.2}$$

then any inner point of the continuous spectrum of the operator H *is not eigenvalue.*

If λ is one of the extreme points of the arcs of the continuous spectrum of the operator H, then there exists only one multiplicator equal to 1 or -1 [34] (or [16], pp.283). This multiplicator is two-fold. Therefore, from Theorem 4.1 we obtain the following statement.

Statement. *Provided that*

$$(1 + t)^\delta q_k(t) \in L_\infty(\mathbf{R}_+) \quad (\delta > 2; k = 1, 2) \tag{5.3}$$

the extreme points of the continuous spectrum of the operator H *cannot be eigenvalues.*

Similar statements hold for the operator H considered along the whole axis $\mathbf{R}$.

To compare the results obtained let us refer to the papers [35] and [20]. In [35] if has been proved that under the condition

$$\int_{-\infty}^{\infty} (1 + |t|)|q_2(t)|dt < \infty$$

for the case of selfadjoint operator H and $q_1(t) = 0$, the continuous spectrum of the operator H has no eigenvalues.

In the paper [20] similar results have been obtained for the non-selfadjoint operator H. However, instead of conditions (5.2) and (5.3) (for $q_1(t) \equiv 0$ and $p = 2$) the following conditions are considered

$$\int_{-\infty}^{\infty} (1 + |t|)|q_2(t)|dt < \infty \quad \text{and} \quad \int_{-\infty}^{\infty} (1 + t^2)|q_2(t)|dt < \infty.$$

5.6 In space the $L_p(\mathbf{R}, \mathbf{C}^2)$ let us consider the one-dimensional Dirac operator

$$Hu = A\frac{du}{dt} + Q(t)u,$$

where

$$A = \begin{pmatrix} 0 & -1 \\ 1 & 0 \end{pmatrix},$$

and $Q(t) = \|q_{jk}(t)\|_{j,k=1}^2$, $q_{jk}(t)(j,k = 1,2)$ are measurable complex valued functions on the whole axis $\mathbf{R}$.

Statement *If*

$$(1 + |t|)^\delta q_k(t) \in L_\infty(\mathbf{R}) \quad (\delta > 1; j, k = 1, 2),$$

then no eigenvalues will be superimposed on the continuous spectrum (which coincides with the whole axis $\mathbf{R}$) of the operator H.

Thus, according to the results from the paper [29] we can draw the conclusion that the conditions obtained are precise to some extent. The paper [29] deals in particular with the one-dimensional Dirac operator with a scalar potential $q(t)$, i.e. $q_{11}(t) = q_{22} = 0$ and $q_{12}(t) = q_{21}(t) = q(t)$ ($q(t)$ is real function) and it is shown how to construct a potential $q(t)$ dicreasing a bit slower than the Coulomb potential (i.e. $q(t)$it is decreasing like t^{-1}) and the point spectrum of the Dirac operator densely fills the semiaxis $\lambda \geq 0$.

Taking into account the results in the papers [29,2], similar comments can be made for the Schrödinger operator also.

References

[1] S. Agmon: Spectral properties of Schrödinger operators and scattering theory, *Ann. Norm. Sup. Pisa* C 1, Sci II 2 (1975) 151-218.

[2] S. Albeverio: On bound states in the continuum of N-body systems and the Virial theorem, *Ann. Phys*, **71**(1972), 167-276.

[3] W. O. Amhrein, A. M. Berthier, V. Georgescu: Hardy type inequalities for abstract differential operators, Preprint UGVA–DPT 1985/10–479.

[4] N. Aronszajn: A unique continuation theorem for solutions of elliptic partial differential equations or inequalities of second order. *J. Math. Pures Appl.* **36**(1957), 235-249.

[5] M. Sh. Birman: On the spectrum of singular boundary problems [Russian], *Mat. Sb.*, **55**(1961), 125-174.

[6] S. E. Cheremshantsev: Spectral Analysis of nonselfadjoint differential operators arising in the one–dimensional scattering problem of the Brown particle [Russian], *Mat. Sb.*, **129(171)**(1986), 358–377.

[7] P. A. Cojuhari: Nonexistence of the eigenvalues of the perturbed discrete Wiener–Hopf operator [Russian], *Bull. Acad. de Ştiinţe R.S.S. Moldov.*, **2**(1990), 15–21.

[8] P. A. Cojuhari: Nonexistence of the eigenvalues of the Operators close to the operators associated with the infinite–dimensional Jacobi Matrices [Russian], *Bull. Acad. de Ştiinţe R.S.S. Moldova*, **2**(1990), 15–21.

[9] P. A. Cojuhari: On the point spectrum of nonselfadjoint singular differential operators [Russian], *Funkts. Analiz i Prilojen.*, **25**(1991), 78-80.

[10] P. A. Cojuhari: On the point spectrum of the perturbed integral Wiener–Hopf operator, *Mat. Zametki*, **51**:1(1990),102-113.

[11] P. A. Cojuhari: On the finite spectrum of ordinary differential operators [Russian], *Mat. Issled.*, **77**(1984),,86-97.

[12] H. O. Cordes: Über die eindeutige Bestimmtheit der Lösungen elliptischer Differentialgleichungen durch Anfangsvorgaben, *Nachr. Akad. Wiss. Götingen Math.-Phys. K 1, II a*, **11**, 1956, 239-259.

[13] N. Dunford and J. T. Schwartz: *Linear Operators, Part.2: Spectral Theory, Self-Adjoint Operators in Hilbert Space.* Interscience Publishers Inc., New-York–London 1964.

[14] D. M. Eidus: Principle of limit amplitude [Russian], *Uspehi Mat. Nauk.***24**:3(1969), 91-156.

[15] I. M. Gelfand: On the spectrum of nonselfadjoint differential operators [Russian], *Uspehi Mat. Nauk.*, **7**(1952), 183-184.

[16] I. M. Glazman: *Direct Methodsof Qualitative Spectral Analysis* [Russian], Fizmatgiz, Moscow 1963.

[17] P. Hartman: *Ordinary Differential Equation*, New-York–London–Sydney, 1964.

[18] G. H. Hardy, I .E. Littlewood, G. Polya:*Inequalities*, Cambridge Univ. Press, London and New-York, 1952.

[19] L. Hörmander: *The Analysis of Linear Partial Differential Operators, Vol. II Differential Operators with Constant Coefficients*, Springer–Verlag, Berlin–Heidelberg–New-York–Tokyo 1983.

[20] V. A. Jeludev: On perturbations of the spectrum of the Schrödinger operator with periodic complex potential [Russian], *Probl. Mat. Fiziki*, LGU, **3**(1968), 49–58.

[21] T. Kato: Growth properties of solutions of the reduced wave equations with a variable coefficient, *Comm. Pure and Appl. Math.*, **12**(1956), 403-425.

[22] T. Kato: *Perturbation theory for linear operators*, Springer–Verlag, Berlin–Heidelberg–New-York 1966.

[23] M. A. Krall: On singular nonselfadjoint differntial operators of second order [Russian], *Dokl. Akad. Nauk SSSR*, **165**(1965), 1235-127.

[24] M. A. Krall: The adjoint of a differential operator with integral boundary condition, *Proc. Amer. Math. Soc.* bf 16: 4(1965), 738-742.

[25] M. A. Krall: Of the eigenvalues of a singular nonselfadjoint differential operator of second order, *Bull. Inst. Politechn. Iaşi*, **12**, 3-4 (1966), 24-32.

[26] B. M. Levitan and I. S. Sargsyan: *Sturm-Liouville and Dirac Opreators* [Russian], Nauka, Moscow 1988.

[27] V. E. Lyantse: Nonselfadjoint difference operators [Russian], *Dokl. Akad. Nauk SSSR*, **173:6**(1965), 1260–1263.

[28] D. C. McGarvey: Operators commuting with translation by one II. Differential operators with periodic coefficients in $L_p(-\infty, \infty)$, *J. of Math. Analysis and Appl.* 1965, **11**, 187-234.

[29] S. N. Naboko: Schrödinger operators with decaying potential and dense point spectrum [Russian], *Dokl. Akad. Nauk SSSR*, **276:6**(1984), 1312–1315.

[30] M. A. Naimark: Investigation of the spectrum and eigenfunctions expansion of second order nonselfadjoint differential operators in the plane [Russian], *Trud. Mosk. Mat. Obsch.*, **3**(1954), 181-270.

[31] M. A. Naimark: *Linear Differential Operators* [Russian], Nauka, Moscow 1969.

[32] B. S. Pavlov: On the nonselfadjoint operator $-y'' + p(x)y$ in the plane [Russian], *Dokl. Akad. Nauk SSSR*, **141**(1961), 807-810.

[33] M. Reed and B. Simon: *Methods of Modern Mathematical Physics. **IV**: Analysis of operators*, Academic Press, London 1978.

[34] F. S. Rofe-Beketov: On the spectrum of the nonselfadjoint differential operators with periodic coefficients [Russian], *Dokl. Akad. Nauk SSSR*, **152:6**(1963), 1312–1315.

[35] F. S. Rofe-Beketov: A test for the finiteness of the number of discrete levels introduced into the gaps of a continuous spectrum by perturbations of a periodic potential [Russian], *Dokl. Akad. Nauk SSSR*, **156:3**(1964), 55–58.

[36] A. F. Vakulenko: The multidimensional Hardy inequality and absence of the positive eigenvalues of the Schrödinger operator with complex potential [Russian], in *Boundary problems of the mathematical physics and related questions of the theory of functions*, LOMI, **127**(1989), 33-34.

[37] V. A. Yakubovich and V. M. Strajinski: *Linear differential equations with periodical coefficients and applications* [Russian], Nauka, Moscow 1972.

Petru A. Cojuhari

Facultatea de Matematică şi Cibernetică
Universitatea de Stat din Moldova
Str. A. Mateevici 60
277003 Chişinău
Moldova

Operator Theory:
Advances and Applications, Vol. 61
© 1993 Birkhäuser Verlag Basel

The Commutant Lifting Theorem for Contractions on Kreĭn Spaces

Aad Dijksma, Michael Dritschel[*] Stefania Marcantognini[†]
and Henk de Snoo

Abstract. A proof of the commutant lifting theorem for contractions on Kreĭn spaces
is given. This is done by associating to the data a suitable isometry V so that a solution
of the lifting problem is obtained directly from a unitary Hilbert space extension of
V. Furthermore, a bijective correspondence between the solutions and the family of
all minimal unitary Hilbert space extensions of V is established. In the Hilbert space
case the method is due to R. Arocena.

0. Introduction and preliminaries

Let $\mathcal{H}_1$, $\mathcal{H}_2$ be Kreĭn spaces, $T_1 \in \mathbf{L}(\mathcal{H}_1)$, $T_2 \in \mathbf{L}(\mathcal{H}_2)$ contractions with minimal
isometric dilations $(W_1, \mathcal{G}_1)$, $(W_2, \mathcal{G}_2)$, respectively. Denote by P_1, P_2 the orthogonal
projections from $\mathcal{G}_1$ onto $\mathcal{H}_1$ and from $\mathcal{G}_2$ onto $\mathcal{H}_2$, respectively. Then for a contraction
$A \in \mathbf{L}(\mathcal{H}_1, \mathcal{H}_2)$ satisfying $AT_1 = T_2A$, we denote by $\mathbf{LIF}(A)$ the set of all liftings of
A, that is,

$$\mathbf{LIF}(A) = \left\{ \tilde{A} \in \mathbf{L}(\mathcal{G}_1, \mathcal{G}_2) : \tilde{A} \text{ is a contraction, } \tilde{A}W_1 = W_2\tilde{A} \text{ and } P_2\tilde{A} = AP_1 \right\}.$$

The commutant lifting theorem states that $\mathbf{LIF}(A)$ is nonempty. When $\mathcal{H}_1$ and
$\mathcal{H}_2$ are Hilbert spaces, this is a well known theorem due to Sz.-Nagy and Foiaş [NF].
It was motivated by problems in classical interpolation theory considered by D. Sarason [SA1]. For other proofs in the Hilbert space setting , including one using the
method of matrix completions, the reader is refered to [FF]. A proof not contained in
[FF] was given by R. Arocena [AR]. Arocena constructs from the triple $\{T_1, T_2, A\}$ an
isometric operator V in a Hilbert space with the property that the contractive liftings
of the intertwining operator A can be obtained directly from the unitary extensions
of V. What is more, he proved that there exists a bijection between $\mathbf{LIF}(A)$ and

---

[*]The second author was supported in part by a grant from the National Science Foundation.
[†]The third author was supported by the "Gran Mariscal de Ayacucho" Foundation (F.G.M.A.,
Venezuela) and the Netherlands Organization for Scientific Research (N.W.O., the Netherlands).

the family of all minimal unitary extensions of V. His approach in addition leads to a Schur type description of the set $\mathbf{LIF}(A)$; see also [MO]. For a lucid treatment of Arocena's method we refer to [SA2].

The commutant lifting theorem is an abstract pattern for different interpolation problems (cf. [SA1]). The method of unitary extensions of a certain isometric operator to solve interpolation problems was used by M. Naimark, M.G. Kreĭn , H. Langer and others. For a discussion of these in connection with commutant lifting, see [C].

In this paper we consider the commutant lifting theorem in the general Kreĭn space setting. The first proof that in this case the set $\mathbf{LIF}(A)$ is nonempty was given in M. Dritschel's doctoral dissertation (University of Virginia, May, 1989); see the self-contained treatment in [DR] by M. Dritschel and J. Rovnyak. His proof is an adaptation of the method of matrix completions mentioned above. In this paper we present a proof using Arocena's method. In adapting this method, we provide a connection between operator-valued analytic functions and the commutant lifting theorem in the indefinite setting. Thus we give not only a new proof of the theorem that $\mathbf{LIF}(A)$ is nonempty, but also information about the problem of describing the set $\mathbf{LIF}(A)$.

An exhaustive description of $\mathbf{LIF}(A)$ in the Hilbert space case was given by the Romanian school of operator theory (Gr. Arsene, Z. Ceauşescu and C. Foiaş). Recent papers, for instance [AR] and [MO], show that the problem of describing $\mathbf{LIF}(A)$ also admits different and interesting approaches. Although all these characterizations of $\mathbf{LIF}(A)$ for contractions in Kreĭn spaces are the same as in Hilbert spaces, their proofs are more involved.

Familiarity with operator theory on Kreĭn spaces is presumed and we refer the reader to [AN], [AI] and [BO] for more details. Most of the notation adopted here is the same as in [DR] and proofs of many of the statements given in the introduction may be found there as well. For the sake of a selfcontained treatment, we recall in this section some elementary notions from the theory of Kreĭn spaces and operators on them. All notions are to be taken in their Kreĭn space versions unless otherwise noted. We use $\mathbf{N}$ for the set of nonnegative integers and $\mathbf{Z}$ for the set of all integers.

A *Kreĭn space* is a linear space $\mathcal{K}$ equipped with an inner product (a hermitian sesquilinear form) $\langle \cdot, \cdot \rangle_{\mathcal{K}}$ such that there exist two subspaces $\mathcal{K}_+$ and $\mathcal{K}_-$ with the following properties:

(1) $\mathcal{K}$ is the direct algebraic sum of $\mathcal{K}_+$ and $\mathcal{K}_-$,

(2) $\langle \mathcal{K}_+, \mathcal{K}_- \rangle_{\mathcal{K}} = \{0\}$,

(3) $(\mathcal{K}_+, \langle \cdot, \cdot \rangle_{\mathcal{K}})$ and $(\mathcal{K}_-, -\langle \cdot, \cdot \rangle_{\mathcal{K}})$ are Hilbert spaces.

We denote orthogonal sums and differences with respect to $\langle \cdot, \cdot \rangle_{\mathcal{K}}$ by $\oplus$ and $\ominus$, respectively, and, more generally, the standard Hilbert space notation is carried over to Kreĭn spaces.

A *fundamental decomposition* of a Kreĭn space $\mathcal{K}$ is an orthogonal sum representation $\mathcal{K} = \mathcal{K}_+ \oplus \mathcal{K}_-$, where $\mathcal{K}_\pm$ are subspaces as those in the above definition. In general, fundamental decompositions are not unique.

A fundamental decomposition $\mathcal{K} = \mathcal{K}_+ \oplus \mathcal{K}_-$ induces a Hilbert space inner product. Namely, if $P_\pm$ are the orthogonal projections from $\mathcal{K}$ onto $\mathcal{K}_\pm$ and $J = P_+ - P_-$, then the Hilbert space inner product of $x, y \in \mathcal{K}$ is given by $\langle Jx, y \rangle_\mathcal{K}$. The operator J is called a *signature operator* or *fundamental symmetry*.

As with fundamental decompositions, a signature operator J is not necessarily unique. Nevertheless, the norms on the Hilbert spaces associated with any signature operator are equivalent and hence generate the same topology. All topological notions on a Kreĭn space are to be understood with respect to that norm topology.

We write $\mathcal{K}_J$ for $\mathcal{K}$ viewed as a Hilbert space relative to the given fundamental decomposition $\mathcal{K} = \mathcal{K}_+ \oplus \mathcal{K}_-$. The symbol $|\mathcal{K}_-|$ shall be used to denote the Hilbert space $(\mathcal{K}_-, -\langle \cdot, \cdot \rangle_\mathcal{K})$. In this notation, $\mathcal{K}_J = \mathcal{K}_+ \oplus |\mathcal{K}_-|$ and $\langle x, y \rangle_{\mathcal{K}_J} = \langle Jx, y \rangle_\mathcal{K}$, $x, y \in \mathcal{K}$.

The *negative* and *positive indices* of the Kreĭn space $\mathcal{K}$

$$\mathrm{ind}_\pm(\mathcal{K}) = \dim(\mathcal{K}_\pm)$$

are independent of the chosen fundamental decomposition $\mathcal{K} = \mathcal{K}_+ \oplus \mathcal{K}_-$. Here "dim" means the algebraic dimension.

If $\{\mathcal{M}_i\}_{i \in I}$ is a family of subsets of a Kreĭn space $\mathcal{K}$, the symbol $LS_{i \in I}\{\mathcal{M}_i\}$ indicates the linear span, while $CLS_{i \in I}\{\mathcal{M}_i\}$ denotes the closed linear span of all these sets.

A vector f of a Kreĭn space $\mathcal{K}$ is *negative* if $\langle f, f \rangle_\mathcal{K} \leq 0$. A subspace of $\mathcal{K}$ is negative if it is composed of negative vectors. It is maximal negative if it is not properly contained in a larger negative subspace. Fix a fundamental symmetry J on $\mathcal{K}$ and let $\|\cdot\|$ denote the associated Hilbert space norm on $\mathcal{K}_J$. A subspace $\mathcal{M}$ is uniformly negative if there is a positive constant ϵ such that for all f in $\mathcal{M}$, $\langle f, f \rangle_\mathcal{K} \leq -\epsilon \|f\|^2$. It is maximal uniformly negative if is not contained in any larger uniformly negative subspace. The same notions with "negative" replaced by "positive" may likewise be defined. Maximal subspaces are necessarily closed. The orthogonal complement of a maximal (uniformly) negative subspace is maximal (uniformly) positive.

By a *regular subspace* of a Kreĭn space $\mathcal{K}$ we mean a closed subspace $\mathcal{M}$ of $\mathcal{K}$ which is a Kreĭn space in the inner product inherited from $\mathcal{K}$. As for closed subspaces of a Hilbert space, a form of the projection theorem holds for regular subspaces of a Kreĭn space. If $\mathcal{M}$ is a regular subspace of $\mathcal{K}$ we write $P_\mathcal{M}^\mathcal{K}$ to indicate the orthogonal projection from $\mathcal{K}$ onto $\mathcal{M}$. Of the closed negative subspaces of a Kreĭn space, only the uniformly negative ones are regular; and likewise for the positive subspaces.

By $\mathbf{L}(\mathcal{H}, \mathcal{K})$ we denote the set of all everywhere defined and bounded linear operators from a Kreĭn space $\mathcal{H}$ to a Kreĭn space $\mathcal{K}$. We use $\mathbf{L}(\mathcal{H})$ for $\mathbf{L}(\mathcal{H}, \mathcal{H})$.

An operator $T \in \mathbf{L}(\mathcal{H}, \mathcal{K})$ is a *contraction* if for all $h \in \mathcal{H}$,

$$\langle Th, Th \rangle_{\mathcal{K}} \leq \langle h, h \rangle_{\mathcal{H}}.$$

If both T and its adjoint T^* are contractions, then T is said to be a *bicontraction*. Contractions have the property that they map closed uniformly negative subspaces to closed uniformly negative subspaces. Bicontractions may be distinguished by the fact that they map some (and indeed every) maximal uniformly negative subspace to a maximal uniformly negative subspace.

An operator in $\mathbf{L}(\mathcal{H}, \mathcal{K})$ is an isometry if it preserves the Kreĭn space scalar product. The range of an isometry is regular, and an isometry is also a bicontraction if and only if the range contains a maximal uniformly negative subspace or, equivalently, the orthogonal complement of the range is a Hilbert space.

A Kreĭn space $\mathcal{H}$ and a Kreĭn space $\mathcal{K}$ are said to be *weakly isomorphic* if there exists a densely defined linear mapping from $\mathcal{H}$ to a dense subspace of $\mathcal{K}$ that preserves the scalar product (that is, is isometric). Such a mapping is called a *weak isomorphism*. A weak isomorphism need not necessarily have a continuous extension to all of $\mathcal{H}$ but it can be extended to an isomorphism from $\mathcal{H}$ onto $\mathcal{K}$ if either its domain or its range contains a maximal uniformly definite subspace. Weakly isomorphic spaces have the same positive and negative indices.

In the formulas the expressions dom , ran and ker are the abbreviations for domain, range and kernel, respectively.

The following theorems concerning dilations are a special case of [AI], Chapter 5, Theorem 3.4. In the sequel we use the explicit formulations and constructions given by Dritschel and Rovnyak in [DR], Theorems 3.1.2 and 3.1.6. To construct their concrete examples of minimal isometric and unitary dilations of a given bounded linear operator T on a Kreĭn space $\mathcal{H}$, Dritschel and Rovnyak introduce the so called defect and Julia operators associated to T.

By a *defect operator* for $T \in \mathbf{L}(\mathcal{H}, \mathcal{K})$, $\mathcal{H}$ and $\mathcal{K}$ Kreĭn spaces, we mean an operator $\tilde{D} \in \mathbf{L}(\tilde{\mathcal{D}}, \mathcal{H})$, where $\tilde{\mathcal{D}}$ is a Kreĭn space, such that $\tilde{D}$ has zero kernel (or, equivalently, $\tilde{D}^*$ has dense range) and $1 - T^*T = \tilde{D}\tilde{D}^*$. The space $\tilde{\mathcal{D}}$ is called the *defect space*; it is a Hilbert space if and only if T is a contraction. We likewise get a defect operator D and a defect space $\mathcal{D}$ for T^* by factoring $1 - TT^* = DD^*$, where $D \in \mathbf{L}(\mathcal{D}, \mathcal{K})$ has zero kernel and $\mathcal{D}$ is a Kreĭn space. To indicate the dependence on T we sometimes write $\tilde{D}_T$, $\tilde{\mathcal{D}}_T$ and D_T, $\mathcal{D}_T$ instead of $\tilde{D}$, $\tilde{\mathcal{D}}$ and D, $\mathcal{D}$. By [D], Theorem 2.17, ran $\tilde{D}^*$ contains a maximal uniformly negative subspace of $\tilde{\mathcal{D}}$ if and only if ran D^* contains a maximal uniformly negative subspace of $\mathcal{D}$.

By a *Julia operator* for $T \in \mathbf{L}(\mathcal{H}, \mathcal{K})$, $\mathcal{H}$ and $\mathcal{K}$ Kreĭn spaces, we mean a unitary operator having the form

$$\begin{pmatrix} T & D \\ \tilde{D}^* & L \end{pmatrix} \in \mathbf{L}(\mathcal{H} \oplus \mathcal{D}, \mathcal{K} \oplus \tilde{\mathcal{D}}), \tag{0.1}$$

where $\tilde{D} \in \mathbf{L}(\tilde{\mathcal{D}}, \mathcal{H})$ and $D \in \mathbf{L}(\mathcal{D}, \mathcal{K})$ are defect operators for T and T^*, respectively, and $L \in \mathbf{L}(\mathcal{D}, \tilde{\mathcal{D}})$. Julia operators always exist.

Let $T \in \mathbf{L}(\mathcal{H})$. A (strong) *isometric dilation* of T is given by a pair $(W, \mathcal{G})$, where $\mathcal{G}$ is a Kreĭn space containing a regular subspace $\mathcal{K}$ isomorphic to $\mathcal{H}$ and $W \in \mathbf{L}(\mathcal{G})$ is an isometry, such that

$$T^n = \tau^{-1} P_{\mathcal{K}}^{\mathcal{G}} W^n \tau | \mathcal{H}, \qquad n \in \mathbf{N},$$

where $\tau : \mathcal{H} \to \mathcal{K}$ is an *isomorphism* from $\mathcal{H}$ onto $\mathcal{K}$. An isometric dilation $(W, \mathcal{G})$ of T is *minimal* if

$$\mathcal{G} = \underset{n \in \mathbf{N}}{CLS} \left\{ W^n \tau \mathcal{H} \right\}.$$

Two minimal isometric dilations $(W, \mathcal{G})$ and $(W', \mathcal{G}')$ of T are said to be isomorphic if there exists a unitary operator ϕ from $\mathcal{G}$ onto $\mathcal{G}'$ such that $\phi\tau = \tau'$ and $\phi W = W'\phi$.

We remark that any given $T \in \mathbf{L}(\mathcal{H})$ has at least one minimal isometric dilation and that any two minimal isometric dilations of T are weakly isomorphic. In Theorem A below, we present the minimal isometric dilation of T, which we term the *canonical* minimal isometric dilation. In the particular case that T is contractive, any two minimal isometric dilations of T are isomorphic and hence indistinguishable from the canonical minimal isometric dilation.

Theorem A. *Let $T \in \mathbf{L}(\mathcal{H})$, $\mathcal{H}$ a Kreĭn space, and let $\tilde{D} \in \mathbf{L}(\tilde{\mathcal{D}}, \mathcal{H})$ be a defect operator for T. Define the Kreĭn space $\mathcal{G} = \mathcal{H} \oplus \tilde{\mathcal{D}} \oplus \tilde{\mathcal{D}} \oplus \ldots$ and the operator matrix*

$$W = \begin{pmatrix} T & 0 & 0 & 0 & \cdots \\ \tilde{D}^* & 0 & 0 & 0 & \cdots \\ 0 & 1 & 0 & 0 & \cdots \\ 0 & 0 & 1 & 0 & \cdots \\ & & \cdots & & \end{pmatrix}.$$

Then $(W, \mathcal{G})$ is a minimal isometric dilation of T satisfying the following:

(a) W is a lifting of T, in the sense that $P_{\mathcal{H}}^{\mathcal{G}} W = T P_{\mathcal{H}}^{\mathcal{G}}$,

(b) $\mathcal{H}$ is invariant under W^* and $W^* | \mathcal{H} = T^*$.

In what follows we also consider unitary dilations of T_1 and T_2. A pair $(U, \tilde{\mathcal{H}})$ is said to be a (strong) *unitary dilation* of $T \in \mathbf{L}(\mathcal{H})$ if $\tilde{\mathcal{H}}$ is a Kreĭn space that contains a regular subspace $\mathcal{K}$ isomorphic to $\mathcal{H}$ and $U \in \mathbf{L}(\tilde{\mathcal{H}})$ is a unitary operator, such that for all $n \in \mathbf{N}$,

$$T^n = \tau^{-1} P_{\mathcal{K}}^{\tilde{\mathcal{H}}} U^n \tau | \mathcal{H} \qquad \text{and} \qquad (T^*)^n = \tau^{-1} P_{\mathcal{K}}^{\tilde{\mathcal{H}}} U^{-n} \tau | \mathcal{H},$$

where $\tau : \mathcal{H} \to \mathcal{K}$ is an isomorphism from $\mathcal{H}$ onto $\mathcal{K}$. A unitary dilation $(U, \tilde{\mathcal{H}})$ of T is *minimal* if

$$\tilde{\mathcal{H}} = \underset{n \in \mathbb{Z}}{CLS} \left\{ U^n \tau \mathcal{H} \right\}.$$

Two minimal unitary dilations $(U, \tilde{\mathcal{H}})$ and $(U', \tilde{\mathcal{H}}')$ of T are always weakly isomorphic, and they are said to be isomorphic if there exists a unitary operator ϕ from $\tilde{\mathcal{H}}$ onto $\tilde{\mathcal{H}}'$ such that $\phi\tau = \tau'$ and $\phi U = U'\phi$. The existence of a minimal unitary dilation of an operator $T \in \mathbf{L}(\mathcal{H})$ is guaranteed by the theorem stated next. For bicontractions, the minimal unitary dilation is unique up to isomorphism.

Theorem B. *Let $\mathcal{H}$ be a Kreĭn space, let $T \in \mathbf{L}(\mathcal{H})$ have Julia operator of the form (0.1), and put $\tilde{\mathcal{H}} = \ldots \oplus \mathcal{D} \oplus \mathcal{D} \oplus \mathcal{H} \oplus \tilde{\mathcal{D}} \oplus \tilde{\mathcal{D}} \oplus \ldots$. Then the operator matrix*

$$
U = \begin{pmatrix}
 & & & \cdots & & & & \\
\cdots & 1 & 0 & 0 & 0 & 0 & 0 & 0 & \cdots \\
\cdots & 0 & 1 & 0 & 0 & 0 & 0 & 0 & \cdots \\
\cdots & 0 & 0 & D & \boxed{T} & 0 & 0 & 0 & \cdots \\
\cdots & 0 & 0 & L & \tilde{D}^* & 0 & 0 & 0 & \cdots \\
\cdots & 0 & 0 & 0 & 0 & 1 & 0 & 0 & \cdots \\
\cdots & 0 & 0 & 0 & 0 & 0 & 1 & 0 & \cdots \\
 & & & \cdots & & & &
\end{pmatrix}
$$

acts as an everywhere defined continuous operator on the Kreĭn space $\tilde{\mathcal{H}}$, and $(U, \tilde{\mathcal{H}})$ is a minimal unitary dilation of T. Furthermore, $(U, \tilde{\mathcal{H}})$ has the following properties:

(a) if $(W, \mathcal{G})$ is the canonical minimal isometric dilation of T then $\mathcal{G}$ is a regular subspace of $\tilde{\mathcal{H}}$, $W = U|\mathcal{G}$ and $(U^{-1}, \tilde{\mathcal{H}})$ is a canonical minimal isometric dilation of W^*,

(b) if $(V, \mathcal{F})$ denotes the canonical minimal isometric dilation of T^* then $\mathcal{F}$ is a regular subspace of $\tilde{\mathcal{H}}$, $V = U^{-1}|\mathcal{F}$ and $(U, \tilde{\mathcal{H}})$ is a canonical minimal isometric dilation of V^*.

As in the isometric case, the minimal unitary dilation defined in Theorem B will be called the *canonical* minimal unitary dilation.

1. The isometry V

In this section we assume that $\mathcal{H}_1$ and $\mathcal{H}_2$ are Kreĭn spaces, $T_1 \in \mathbf{L}(\mathcal{H}_1)$ and $T_2 \in \mathbf{L}(\mathcal{H}_2)$ are contractions and $A \in \mathbf{L}(\mathcal{H}_1, \mathcal{H}_2)$ satisfies $AT_1 = T_2A$. Also we fix defect operators $\tilde{D}_A, D_A$, with corresponding defect spaces $\tilde{\mathcal{D}}_A, \mathcal{D}_A$.

In the following we assume that fundamental symmetries J_i on $\mathcal{H}_i$ are chosen, and we denote by $\mathcal{H}_{J_i}$ the associated Hilbert spaces, $i = 1, 2$. Then $\tilde{J} = \begin{pmatrix} J_1 & 0 \\ 0 & J_2 \end{pmatrix}$

is a fundamental symmetry on $\mathcal{H}_1 \oplus \mathcal{H}_2$. We define the selfadjoint operator $B \in \mathbf{L}(\mathcal{H}_1 \oplus \mathcal{H}_2)$ as the 2×2 block matrix

$$B = \begin{pmatrix} 1 & A^* \\ A & 1 \end{pmatrix}.$$

We denote by $\overline{\operatorname{ran} \tilde{J}B}$ the closure of $\operatorname{ran} \tilde{J}B$ in $\mathcal{H}_1 \oplus \mathcal{H}_2$:

$$\overline{\operatorname{ran} \tilde{J}B} = (\mathcal{H}_{J_1} \oplus \mathcal{H}_{J_2}) \ominus \ker B.$$

For the proof of the following lemma, the reader is referred to [CG], Sections 2.3 and 2.4.

Lemma 1.1 *There exists a Kreĭn space $(\mathcal{G}, \langle \cdot, \cdot \rangle_{\mathcal{G}})$ such that $\overline{\operatorname{ran} \tilde{J}B}$ is dense in $\mathcal{G}$ and*

$$\langle x, y \rangle_{\mathcal{G}} = \langle Bx, y \rangle_{\mathcal{H}_1 \oplus \mathcal{H}_2}, \quad \langle x, y \rangle_{\mathcal{G}_J} = \left\langle |\tilde{J}B| x, y \right\rangle_{\mathcal{H}_{J_1} \oplus \mathcal{H}_{J_2}}, \qquad x, y \in \overline{\operatorname{ran} \tilde{J}B},$$

where $J = \operatorname{sgn}(\tilde{J}B)$ on $\overline{\operatorname{ran} \tilde{J}B}$.

Let Q be the projection from $\mathcal{H}_{J_1} \oplus \mathcal{H}_{J_2}$ onto $\overline{\operatorname{ran} \tilde{J}B}$ and define the operators $\tau_1 : \mathcal{H}_1 \to \mathcal{G}$ and $\tau_2 : \mathcal{H}_2 \to \mathcal{G}$ by

$$\tau_1 h_1 = Q\{h_1, 0\} \quad \text{and} \quad \tau_2 h_2 = Q\{0, h_2\}, \qquad h_i \in \mathcal{H}_i, \ i = 1, 2.$$

Lemma 1.2 *The operators τ_1 and τ_2 have the following properties:*

(a) τ_1 and τ_2 are continuous isometries,
(b) $\mathcal{G} = \operatorname{CLS}\{\tau_1 \mathcal{H}_1, \tau_2 \mathcal{H}_2\}$,
(c) for all $h_1 \in \mathcal{H}_1$, $h_2 \in \mathcal{H}_2$, $\langle \tau_1 h_1, \tau_2 h_2 \rangle_{\mathcal{G}} = \langle A h_1, h_2 \rangle_{\mathcal{H}_2}$,
(d) $P_{\tau_2 \mathcal{H}_2}^{\mathcal{G}} | \tau_1 \mathcal{H}_1 = \tau_2 A \tau_1^{-1} | \tau_1 \mathcal{H}_1$ and $P_{\tau_1 \mathcal{H}_1}^{\mathcal{G}} | \tau_2 \mathcal{H}_2 = \tau_1 A^* \tau_2^{-1} | \tau_2 \mathcal{H}_2$,
(e) $(1 - P_{\tau_2 \mathcal{H}_2}^{\mathcal{G}}) \tau_1 \mathcal{H}_1$ is dense in $\mathcal{G} \ominus \tau_2 \mathcal{H}_2$ and $(1 - P_{\tau_1 \mathcal{H}_1}^{\mathcal{G}}) \tau_2 \mathcal{H}_2$ is dense in $\mathcal{G} \ominus \tau_1 \mathcal{H}_1$.

Proof. Since τ_i maps $\mathcal{H}_i$ into $\overline{\operatorname{ran} \tilde{J}B}$, from the previous lemma we have that for any $h_1 \in \mathcal{H}_1$ and any $h_2 \in \mathcal{H}_2$,

$$\begin{aligned} \langle \tau_1 h_1, \tau_2 h_2 \rangle_{\mathcal{G}} &= \langle BQ\{h_1, 0\}, Q\{0, h_2\} \rangle_{\mathcal{H}_1 \oplus \mathcal{H}_2} \\ &= \langle B\{h_1, 0\}, \{0, h_2\} \rangle_{\mathcal{H}_1 \oplus \mathcal{H}_2} = \langle A h_1, h_2 \rangle_{\mathcal{H}_2}, \end{aligned}$$

yielding (c).

Similarly we obtain

$$\langle \tau_1 h_1, \tau_1 k_1 \rangle_{\mathcal{G}} = \langle B\{h_1, 0\}, \{k_1, 0\} \rangle_{\mathcal{H}_1 \oplus \mathcal{H}_2} = \langle h_1, k_1 \rangle_{\mathcal{H}_1}$$

72   A. Dijksma et al.

and

$$\langle \tau_1 h_1, \tau_1 h_1 \rangle_{\mathcal{G}_J} = \left\langle |\tilde{J}B| Q\{h_1, 0\}, Q\{h_1, 0\} \right\rangle_{\mathcal{H}_{J_1} \oplus \mathcal{H}_{J_2}}$$
$$= \left\langle |\tilde{J}B| \{h_1, 0\}, \{h_1, 0\} \right\rangle_{\mathcal{H}_{J_1} \oplus \mathcal{H}_{J_2}} \leq \| |\tilde{J}B| \| \langle h_1, h_1 \rangle_{\mathcal{H}_{J_1}}$$

for $h_1, k_1 \in \mathcal{H}_1$. Likewise, for $h_2, k_2 \in \mathcal{H}_2$,

$$\langle \tau_2 h_2, \tau_2 k_2 \rangle_{\mathcal{G}} = \langle h_2, k_2 \rangle_{\mathcal{H}} \quad \text{and} \quad \langle \tau_2 h_2, \tau_2 h_2 \rangle_{\mathcal{G}_J} \leq \| |\tilde{J}B| \| \langle h_2, h_2 \rangle_{\mathcal{H}_{J_2}}.$$

Hence it follows that both τ_1 and τ_2 are continuous isometries, giving (a).

Since $\overline{\operatorname{ran} \tilde{J} B} = \operatorname{ran} Q = LS\{\tau_1 \mathcal{H}_1, \tau_2 \mathcal{H}_2\}$ and $\overline{\operatorname{ran} \tilde{J}B}$ is dense in $\mathcal{G}$, (b) is also true.

By (a) the ranges of τ_1 and τ_2 are regular subspaces of $\mathcal{G}$. So (d) follows immediately from (c).

Clearly the range of $1 - P^{\mathcal{G}}_{\tau_2 \mathcal{H}_2}$ is contained in $\mathcal{G} \ominus \tau_2 \mathcal{H}_2$. To see that it is dense, suppose that $g \in \mathcal{G} \ominus \tau_2 \mathcal{H}_2$ satisfies $\left\langle (1 - P^{\mathcal{G}}_{\tau_2 \mathcal{H}_2}) \tau_1 h_1, g \right\rangle_{\mathcal{G}} = 0$ for all $h_1 \in \mathcal{H}_1$. Then $\langle \tau_1 h_1, g \rangle_{\mathcal{G}} = 0$ for all $h_1 \in \mathcal{H}_1$. It follows from (b) that $g = 0$. The last part of (e) can be proved in a similar manner. ∎

Since the ranges of $\tau_1 \mathcal{H}_1$ and $\tau_2 \mathcal{H}_2$ are regular subspaces of $\mathcal{G}$, we get the following from Lemma 1.2.

Corollary 1.3 *$\mathcal{G} \ominus \tau_2 \mathcal{H}_2$ is weakly isomorphic to $\tilde{D}_A$ and $\mathcal{G} \ominus \tau_1 \mathcal{H}_1$ is weakly isomorphic to $\mathcal{D}_A$.*

Proof. By Lemma 1.2 (e), $(1 - P^{\mathcal{G}}_{\tau_2 \mathcal{H}_2}) \tau_1 \mathcal{H}_1$ is dense in $\mathcal{G} \ominus \tau_2 \mathcal{H}_2$. Since $\tilde{D}_A$ is a defect operator, the range of $\tilde{D}_A^*$ is dense in $\tilde{D}_A$. Using Lemma 1.2 (c), we find that the mapping $\tilde{\varphi}_A : (1 - P^{\mathcal{G}}_{\tau_2 \mathcal{H}_2}) \tau_1 \mathcal{H}_1 \to \operatorname{ran} \tilde{D}_A^*$ defined by

$$\tilde{\varphi}_A(\tau_1 h_1 - \tau_2 A h_1) = \tilde{D}_A^* h_1$$

is a weak isomorphism. The second part can be proved in a similar way. ∎

Corollary 1.4 *The following equalities hold:*

(a) $\operatorname{ind}_\pm(\mathcal{G} \ominus \tau_2 \mathcal{H}_2) = \operatorname{ind}_\pm(\tilde{D}_A)$,

(b) $\operatorname{ind}_\pm(\mathcal{G} \ominus \tau_1 \mathcal{H}_1) = \operatorname{ind}_\pm(\mathcal{D}_A)$.

Corollary 1.5 *If either $\operatorname{ran} \tilde{D}_A^*$ or $\operatorname{ran} D_A^*$ contains a maximal uniformly negative subspace then $\mathcal{G} \ominus \tau_2 \mathcal{H}_2$ is isomorphic to $\tilde{D}_A$ and $\mathcal{G} \ominus \tau_1 \mathcal{H}_1$ is isomorphic to $\mathcal{D}_A$.*

Proof. According to [D], Theorem 2.17, $\operatorname{ran} \tilde{D}_A^*$ contains a maximal uniformly negative subspace of $\tilde{\mathcal{D}}_A$ if and only if $\operatorname{ran} D_A^*$ contains a maximal uniformly negative subspace of $\mathcal{D}_A$. The result now follows from Corollary 1.3, since a weak isomorphism is continuous if either the range or the domain contains a maximal uniformly negative subspace. ∎

Corollary 1.6 *The isometry τ_2 is bicontractive if and only if A is contractive. Analogously, the isometry τ_1 is bicontractive if and only if A^* is contractive.*

Proof. The isometry τ_2 is a bicontraction exactly when the orthogonal complement of its range, that is, $\mathcal{G} \ominus \tau_2 \mathcal{H}_2$, is a Hilbert space. By Corollary 1.4, this corresponds to $\tilde{\mathcal{D}}_A$ being a Hilbert space or, equivalently, to A being a contraction. The other assertion can be proved likewise. ∎

In what follows we denote by $(V_2, \mathcal{F}_2)$ the canonical minimal isometric dilation of T_2^* and we consider the Kreĭn space

$$\mathcal{H} = \mathcal{G} \oplus (\mathcal{G}_1 \ominus \mathcal{H}_1) \oplus (\mathcal{F}_2 \ominus \mathcal{H}_2).$$

The elements of $\mathcal{H}$ will be represented as triples of the form $\{g, g_1, f_2\}$, where $g \in \mathcal{G}$, $g_1 \in \mathcal{G}_1 \ominus \mathcal{H}_1$ and $f_2 \in \mathcal{F}_2 \ominus \mathcal{H}_2$. Furthermore, we define the linear operators $\sigma_1 : \mathcal{G}_1 \to \mathcal{H}$ and $\rho_2 : \mathcal{F}_2 \to \mathcal{H}$ by

$$\sigma_1 | \mathcal{H}_1 := \tau_1, \qquad \sigma_1 | \mathcal{G}_1 \ominus \mathcal{H}_1 := 1,$$

and

$$\rho_2 | \mathcal{H}_2 := \tau_2, \qquad \rho_2 | \mathcal{F}_2 \ominus \mathcal{H}_2 := 1.$$

Recall from Lemma 1.2 that $\mathcal{G} = CLS\,\{\tau_1 \mathcal{H}_1, \tau_2 \mathcal{H}_2\}$, so $\operatorname{ran} \sigma_1 \subseteq \mathcal{G} \oplus (\mathcal{G}_1 \ominus \mathcal{H}_1)$ and $\operatorname{ran} \rho_2 \subseteq \mathcal{G} \oplus (\mathcal{F}_2 \ominus \mathcal{H}_2)$.

Lemma 1.7 *The operators σ_1 and ρ_2 have the following properties:*

(a) σ_1 and ρ_2 are continuous isometries,
(b) $\mathcal{H} = CLS\,\{\sigma_1 \mathcal{G}_1, \rho_2 \mathcal{F}_2\}$,
(c) for all $g_1 \in \mathcal{G}_1$, $f_2 \in \mathcal{F}_2$, $\langle \sigma_1 g_1, \rho_2 f_2 \rangle_{\mathcal{H}} = \left\langle A P_1 g_1, P_{\mathcal{H}_2}^{\mathcal{F}_2} f_2 \right\rangle_{\mathcal{H}_2}$,
(d) ρ_2 is bicontractive if and only if A is contractive.

Proof. From their definitions it follows immediately that σ_1 and ρ_2 are continuous isometries and $\mathcal{H} = CLS\,\{\sigma_1 \mathcal{G}_1, \rho_2 \mathcal{F}_2\}$. Let $g_1 \in \mathcal{G}_1$ and $f_2 \in \mathcal{F}_2$. Then we have

$$\sigma_1 g_1 = \{\tau_1 P_1 g_1, (1 - P_1) g_1, 0\} \qquad \text{and} \qquad \rho_2 f_2 = \left\{\tau_2 P_{\mathcal{H}_2}^{\mathcal{F}_2} f_2, 0, (1 - P_{\mathcal{H}_2}^{\mathcal{F}_2}) f_2\right\}.$$

Therefore, by Lemma 1.2,

$$\langle \sigma_1 g_1, \rho_2 f_2 \rangle_{\mathcal{H}} = \left\langle \tau_1 P_1 g_1, \tau_2 P_{\mathcal{H}_2}^{\mathcal{F}_2} f_2 \right\rangle_{\mathcal{G}} = \left\langle A P_1 g_1, P_{\mathcal{H}_2}^{\mathcal{F}_2} f_2 \right\rangle_{\mathcal{H}_2}.$$

It remains to show the last assertion stated in the lemma. Now

$$\begin{aligned}
\mathcal{H} \ominus \rho_2 \mathcal{F}_2 \ &= \mathcal{H} \ominus \rho_2 \left(\mathcal{H}_2 \oplus (\mathcal{F}_2 \ominus \mathcal{H}_2) \right) \\
&= \mathcal{H} \ominus \left(\tau_2 \mathcal{H}_2 \oplus (\mathcal{F}_2 \ominus \mathcal{H}_2) \right) = (\mathcal{G} \ominus \tau_2 \mathcal{H}_2) \oplus (\mathcal{G}_1 \ominus \mathcal{H}_1).
\end{aligned}$$

But the space $\mathcal{G}_1 \ominus \mathcal{H}_1$ is a Hilbert space, and so ρ_2 is a bicontraction if and only if $\mathcal{G} \ominus \tau_2 \mathcal{H}_2$ is a Hilbert space, which, by Corollary 1.6, is equivalent to A being a contraction. ∎

By Theorem A (a), W_1 and V_2 are also liftings of T_1 and T_2^*, respectively. So, according to Lemma 1.7, if $g_1 \in \mathcal{G}_1$, $f_2 \in \mathcal{F}_2$, then

$$\begin{aligned}
\langle \sigma_1 W_1 g_1, \rho_2 f_2 \rangle_{\mathcal{H}} &= \left\langle A T_1 P_1 g_1, P_{\mathcal{H}_2}^{\mathcal{F}_2} f_2 \right\rangle_{\mathcal{H}_2} = \left\langle T_2 A P_1 g_1, P_{\mathcal{H}_2}^{\mathcal{F}_2} f_2 \right\rangle_{\mathcal{H}_2} \\
&= \left\langle A P_1 g_1, T_2^* P_{\mathcal{H}_2}^{\mathcal{F}_2} f_2 \right\rangle_{\mathcal{H}_2} = \left\langle A P_1 g_1, P_{\mathcal{H}_2}^{\mathcal{F}_2} V_2 f_2 \right\rangle_{\mathcal{H}_2} = \langle \sigma_1 g_1, \rho_2 V_2 f_2 \rangle_{\mathcal{H}}.
\end{aligned}$$

It follows that

$$\langle \sigma_1 g_1 + \rho_2 V_2 f_2, \sigma_1 g_1 + \rho_2 V_2 f_2 \rangle_{\mathcal{H}} = \langle \sigma_1 W_1 g_1 + \rho_2 f_2, \sigma_1 W_1 g_1 + \rho_2 f_2 \rangle_{\mathcal{H}}.$$

If $CLS \{\sigma_1 W_1 \mathcal{G}_1, \rho_2 \mathcal{F}_2\}$ is a nondegenerate subspace of $\mathcal{H}$, then this equality implies that the linear mapping V_0 with $\mathrm{dom}\, V_0 = LS \{\sigma_1 \mathcal{G}_1, \rho_2 V_2 \mathcal{F}_2\}$ and $\mathrm{ran}\, V_0 = LS \{\sigma_1 W_1 \mathcal{G}_1, \rho_2 \mathcal{F}_2\}$ determined by the relation

$$V_0(\sigma_1 g_1 + \rho_2 V_2 f_2) = \sigma_1 W_1 g_1 + \rho_2 f_2, \qquad g_1 \in \mathcal{G}_1, \ f_2 \in \mathcal{F}_2,$$

is well defined and isometric. We put

$$\mathcal{D} = CLS \{\sigma_1 \mathcal{G}_1, \rho_2 V_2 \mathcal{F}_2\} \quad \text{and} \quad \mathcal{R} = CLS \{\sigma_1 W_1 \mathcal{G}_1, \rho_2 \mathcal{F}_2\}.$$

Proposition 1.8 *Assume that A is a contraction. Then $\mathcal{D}$ and $\mathcal{R}$ are regular subspaces of $\mathcal{H}$ and V_0 can be extended to a continuous isometry V with $\mathrm{dom}\, V = \mathcal{D}$ and $\mathrm{ran}\, V = \mathcal{R}$. V has the following properties:*

(a) the defect subspaces $\mathcal{N} = \mathcal{H} \ominus \mathcal{D}$ and $\mathcal{M} = \mathcal{H} \ominus \mathcal{R}$ of V are Hilbert spaces,
(b) $V\sigma_1 = \sigma_1 W_1$,
(c) $V^{-1}\rho_2 = \rho_2 V_2$.

Proof. As A is a contraction, by Lemma 1.7 (d) ρ_2 is a bicontraction, which means that $\mathrm{ran}\, \rho_2$ contains a maximal uniformly negative subspace of $\mathcal{H}$. But since

$$\mathcal{R} = CLS \{\sigma_1 W_1 \mathcal{G}_1, \rho_2 \mathcal{F}_2\},$$

it too contains a maximal uniformly negative subspace. Hence $\mathcal{M} = \mathcal{H} \ominus \mathcal{R}$ is a Hilbert space and $\mathcal{R}$ is regular.

Now we show that $\mathcal{D}$ is regular. According to Theorem B (b), we have

$$V_2^* = P_{\mathcal{F}_2}^{\tilde{\mathcal{H}}_2} U_2 | \mathcal{F}_2,$$

where $(U_2, \tilde{\mathcal{H}}_2)$ is the canonical minimal unitary dilation of T_2. Since T_2 is a contraction, $\tilde{\mathcal{H}}_2 \ominus \mathcal{F}_2 = \tilde{\mathcal{D}}_2 \oplus \tilde{\mathcal{D}}_2 \oplus \tilde{\mathcal{D}}_2 \oplus \ldots$ is a Hilbert space. Hence for any $f_2 \in \mathcal{F}_2$,

$$\langle V_2^* f_2, V_2^* f_2 \rangle_{\mathcal{F}_2} = \langle f_2, f_2 \rangle_{\mathcal{F}_2} - \left\langle (1 - P_{\mathcal{F}_2}^{\tilde{\mathcal{H}}_2}) U_2 f_2, (1 - P_{\mathcal{F}_2}^{\tilde{\mathcal{H}}_2}) U_2 f_2 \right\rangle_{\tilde{\mathcal{H}}_2} \leq \langle f_2, f_2 \rangle_{\mathcal{F}_2}.$$

It follows that V_2 is a bicontractive isometry. So $\rho_2 V_2 : \mathcal{F}_2 \to \mathcal{H}$ is a bicontraction and the same argument used to prove the properties of $\mathcal{R}$ implies that $\mathcal{D}$ is regular and $\mathcal{N}$ in (a) is a Hilbert space.

Since $\mathcal{R}$ is regular, V_0 is a well defined isometric mapping. It is now easy to see that V_0 can be extended to a continuous isometry V with domain $\mathcal{D}$ and range $\mathcal{R}$. The properties (b) and (c) follow from the fact that V is an extension of V_0. $\blacksquare$

2. The commutant lifting theorem

In this section we assume that $\mathcal{H}_1$ and $\mathcal{H}_2$ are Kreĭn spaces, $T_1 \in \mathbf{L}(\mathcal{H}_1)$ and $T_2 \in \mathbf{L}(\mathcal{H}_2)$ are contractions with (canonical) minimal isometric dilations $(W_1, \mathcal{G}_1)$ and $(W_2, \mathcal{G}_2)$, respectively, and that $A \in \mathbf{L}(\mathcal{H}_1, \mathcal{H}_2)$ is a contraction satisfying $A T_1 = T_2 A$. Using the minimal isometric dilations $(W_1, \mathcal{G}_1)$ of T_1 and $(V_2, \mathcal{F}_2)$ of T_2^* we construct the Kreĭn space $\mathcal{H} = \mathcal{G} \oplus (\mathcal{G}_1 \ominus \mathcal{H}_1) \oplus (\mathcal{F}_2 \ominus \mathcal{H}_2)$. In $\mathcal{H}$ we consider the continuous isometry V with domain $\mathcal{D}$ and range $\mathcal{R}$ as given in the previous section.

Next we look at unitary Hilbert space extensions $(U, \tilde{\mathcal{H}})$ of V. By this we mean that $\mathcal{H}$ is a subspace of the Kreĭn space $\tilde{\mathcal{H}}$ such that $\tilde{\mathcal{H}} \ominus \mathcal{H}$ is a Hilbert space and that $U \in \mathbf{L}(\tilde{\mathcal{H}})$ is a unitary operator with $V \subset U$. To $(U, \tilde{\mathcal{H}})$ we associate an invariant subspace $\tilde{\mathcal{G}}_2$ with the property that U restricted to $\tilde{\mathcal{G}}_2$ is isomorphic to the canonical minimal isometric dilation W_2 of T_2. The orthogonal projection from $\tilde{\mathcal{H}}$ onto $\tilde{\mathcal{G}}_2$ will then be shown to be isomorphic to a contractive intertwining dilation of A, proving the commutant lifting theorem.

Lemma 2.1 *Let $(U, \tilde{\mathcal{H}})$ be a unitary Hilbert space extension of V. Put*

$$\tilde{\mathcal{G}}_2 := \mathop{CLS}_{n \in \mathbf{N}} \left\{ U^n \tau_2 \mathcal{H}_2 \right\}.$$

Then

(a) $\tilde{\mathcal{G}}_2$ is a regular U-invariant subspace of $\tilde{\mathcal{H}}$,

(b) $(U|\tilde{\mathcal{G}}_2, \tilde{\mathcal{G}}_2)$ is a minimal isometric dilation of T_2.

Consequently there exists a unitary operator $\sigma_2 : \mathcal{G}_2 \to \tilde{\mathcal{G}}_2$ such that $\sigma_2|\mathcal{H}_2 = \tau_2$ and $\sigma_2 W_2 = U\sigma_2$.

Proof. Since $\tau_2 \mathcal{H}_2$ is a regular subspace of $\mathcal{G}$, we have

$$\tilde{\mathcal{H}} \ominus \tau_2\mathcal{H}_2 = (\tilde{\mathcal{H}} \ominus \mathcal{H}) \oplus (\mathcal{G} \ominus \tau_2\mathcal{H}_2) \oplus (\mathcal{G}_1 \ominus \mathcal{H}_1) \oplus (\mathcal{F}_2 \ominus \mathcal{H}_2).$$

The first three summands on the right hand side are Hilbert spaces: $\tilde{\mathcal{H}} \ominus \mathcal{H}$ is a Hilbert space since U is a Hilbert space extension, $\mathcal{G} \ominus \tau_2\mathcal{H}_2$ since A is a contraction, and $\mathcal{G}_1 \ominus \mathcal{H}_1$ because T_1 is a contraction. It follows that any maximal uniformly negative subspace of $\mathcal{F}_2 \ominus \mathcal{H}_2$ is also maximal uniformly negative in $\tilde{\mathcal{H}} \ominus \tau_2\mathcal{H}_2$.

We claim that $\mathcal{F}_2 \ominus \mathcal{H}_2 \subseteq \tilde{\mathcal{H}} \ominus \tilde{\mathcal{G}}_2$. Indeed, by Proposition 1.8 (c), the canonical form of V_2 and the definition of ρ_2, we have for $f_2 \in \mathcal{F}_2 \ominus \mathcal{H}_2$, $h_2 \in \mathcal{H}_2$ and $n \in \mathbf{N}$,

$$\langle f_2, U^n\tau_2 h_2\rangle_{\tilde{\mathcal{H}}} = \langle \rho_2 f_2, U^n \rho_2 h_2\rangle_{\tilde{\mathcal{H}}} = \langle U^{-n}\rho_2 f_2, \rho_2 h_2\rangle_{\tilde{\mathcal{H}}}$$
$$= \langle V^{-n}\rho_2 f_2, \rho_2 h_2\rangle_{\tilde{\mathcal{H}}} = \langle \rho_2 V_2^n f_2, \rho_2 h_2\rangle_{\tilde{\mathcal{H}}} = \langle V_2^n f_2, h_2\rangle_{\mathcal{F}_2} = 0.$$

Since $\tilde{\mathcal{H}} \ominus \tilde{\mathcal{G}}_2 \subseteq \tilde{\mathcal{H}} \ominus \tau_2\mathcal{H}_2$, the claim implies that $\tilde{\mathcal{H}} \ominus \tilde{\mathcal{G}}_2$ contains a maximal uniformly negative subspace of $\tilde{\mathcal{H}} \ominus \tau_2\mathcal{H}_2$. Consequently, $\tilde{\mathcal{H}} \ominus \tilde{\mathcal{G}}_2$ is a regular subspace of $\tilde{\mathcal{H}} \ominus \tau_2\mathcal{H}_2$, or, equivalently, $\tilde{\mathcal{G}}_2$ is a regular subspace of $\tilde{\mathcal{H}}$ that contains $\tau_2\mathcal{H}_2$ as regular subspace.

For any $h_2, k_2 \in \mathcal{H}_2$ and $n \in \mathbf{N}$,

$$\langle T_2^n h_2, k_2\rangle_{\mathcal{H}_2} = \langle h_2, (T_2^*)^n k_2\rangle_{\mathcal{H}_2} = \langle h_2, V_2^n k_2\rangle_{\mathcal{F}_2} = \langle \tau_2 h_2, \rho_2 V_2^n k_2\rangle_{\mathcal{H}}$$
$$= \langle \tau_2 h_2, V^{-n}\tau_2 k_2\rangle_{\mathcal{H}} = \langle U^n\tau_2 h_2, U^n V^{-n}\tau_2 k_2\rangle_{\tilde{\mathcal{H}}} = \langle U^n\tau_2 h_2, \tau_2 k_2\rangle_{\tilde{\mathcal{H}}}.$$

Hence

$$T_2^n = \tau_2^{-1} P_{\tau_2\mathcal{H}_2}^{\tilde{\mathcal{H}}} U^n \tau_2 | \mathcal{H}_2$$

for all $n \in \mathbf{N}$. $\blacksquare$

Theorem 2.2 *Let $(U, \tilde{\mathcal{H}})$ be a unitary Hilbert space extension of V and let $\tilde{\mathcal{G}}_2$ and σ_2 be as in Lemma 2.1. Define the operator $\tilde{A} \in \mathbf{L}(\mathcal{G}_1, \mathcal{G}_2)$ by*

$$\tilde{A} = \sigma_2^{-1} P_{\sigma_2\mathcal{G}_2}^{\tilde{\mathcal{H}}} \sigma_1.$$

Then $\tilde{A} \in \mathbf{LIF}(A)$ (so, in particular, $\mathbf{LIF}(A)$ is nonempty), and

$$\mathrm{ind}_-(\mathcal{D}_{\tilde{A}}) = \mathrm{ind}_-(\mathcal{D}_A).$$

Proof. On account of Lemma 1.7, for $g_1 \in \mathcal{G}_1$ and $h_2 \in \mathcal{H}_2$,

$$\langle P_2\tilde{A}g_1, h_2\rangle_{\mathcal{H}_2} = \langle \tilde{A}g_1, h_2\rangle_{\mathcal{G}_2} = \langle \sigma_1 g_1, \tau_2 h_2\rangle_{\tilde{\mathcal{H}}} = \langle \sigma_1 g_1, \tau_2 h_2\rangle_{\mathcal{H}} = \langle AP_1 g_1, h_2\rangle_{\mathcal{H}_2}.$$

Hence $P_2\tilde{A} = AP_1$, that is, $\tilde{A}$ is a lifting of A.

Let $g_1 \in \mathcal{G}_1$, $h_2 \in \mathcal{H}_2$. Then for $n = 1, 2, \ldots$,

$$
\begin{aligned}
\left\langle W_2\tilde{A}g_1, W_2^n h_2 \right\rangle_{\mathcal{G}_2} &= \left\langle \tilde{A}g_1, W_2^{n-1}h_2 \right\rangle_{\mathcal{G}_2} = \left\langle \sigma_2\tilde{A}g_1, \sigma_2 W_2^{n-1}h_2 \right\rangle_{\mathcal{G}_2} \\
&= \langle \sigma_1 g_1, U^{n-1}\tau_2 h_2 \rangle_{\tilde{\mathcal{H}}} = \langle U\sigma_1 g_1, U^n\tau_2 h_2 \rangle_{\tilde{\mathcal{H}}} \\
&= \langle \sigma_1 W_1 g_1, \sigma_2 W_2^n h_2 \rangle_{\tilde{\mathcal{H}}} = \left\langle \tilde{A}W_1 g_1, W_2^n h_2 \right\rangle_{\mathcal{G}_2}.
\end{aligned}
$$

On the other hand, by Theorem A and Lemma 1.7,

$$
\begin{aligned}
\left\langle W_2\tilde{A}g_1, h_2 \right\rangle_{\mathcal{G}_2} &= \left\langle \tilde{A}g_1, W_2^* h_2 \right\rangle_{\mathcal{G}_2} = \langle \sigma_1 g_1, \tau_2 W_2^* h_2 \rangle_{\tilde{\mathcal{H}}} = \langle \sigma_1 g_1, \tau_2 W_2^* h_2 \rangle_{\mathcal{H}} \\
&= \langle AP_1 g_1, W_2^* h_2 \rangle_{\mathcal{H}_2} = \langle W_2 AP_1 g_1, h_2 \rangle_{\mathcal{G}_2} = \langle T_2 AP_1 g_1, h_2 \rangle_{\mathcal{H}_2} \\
&= \langle AT_1 P_1 g_1, h_2 \rangle_{\mathcal{H}_2} = \langle AP_1 W_1 g_1, h_2 \rangle_{\mathcal{H}_2} = \left\langle P_2\tilde{A}W_1 g_1, h_2 \right\rangle_{\mathcal{H}_2} \\
&= \left\langle \tilde{A}W_1 g_1, h_2 \right\rangle_{\mathcal{G}_2}.
\end{aligned}
$$

The minimality of $(W_2, \mathcal{G}_2)$ implies that $\tilde{A}W_1 = W_2\tilde{A}$, that is, $\tilde{A}$ intertwines W_1 and W_2.

We now show that $\tilde{A}$ is contractive. First we note that $\sigma_1\mathcal{G}_1$ and $\sigma_2\mathcal{G}_2 = \tilde{\mathcal{G}}_2$ are regular subspaces of $\mathcal{K} = (\tilde{\mathcal{H}} \ominus \mathcal{H}) \oplus \mathcal{G} \oplus (\mathcal{G}_1 \ominus \mathcal{H}_1)$. The latter follows, since, as we saw in the proof of the last lemma, $\mathcal{F}_2 \ominus \mathcal{H}_2 \subseteq \tilde{\mathcal{H}} \ominus \tilde{\mathcal{G}}_2$. It is also the case that $\sigma_2\mathcal{G}_2$ contains a maximal uniformly negative subspace of $\mathcal{K}$, for instance, $\tau_2\mathcal{M}_2$ where $\mathcal{M}_2$ is a maximal uniformly negative subspace of $\mathcal{H}_2$. Therefore $\mathcal{K} \ominus \sigma_2\mathcal{G}_2$ is a Hilbert space. Hence for any $g_1 \in \mathcal{G}_1$,

$$
\begin{aligned}
\left\langle \tilde{A}g_1, \tilde{A}g_1 \right\rangle_{\mathcal{G}_2} &= \left\langle P_{\sigma_2\mathcal{G}_2}^{\tilde{\mathcal{H}}}\sigma_1 g_1, P_{\sigma_2\mathcal{G}_2}^{\tilde{\mathcal{H}}}\sigma_1 g_1 \right\rangle_{\tilde{\mathcal{H}}} = \left\langle P_{\sigma_2\mathcal{G}_2}^{\tilde{\mathcal{H}}}\sigma_1 g_1, P_{\sigma_2\mathcal{G}_2}^{\tilde{\mathcal{H}}}\sigma_1 g_1 \right\rangle_{\mathcal{K}} \\
&= \langle g_1, g_1 \rangle_{\mathcal{G}_1} - \left\langle (1 - P_{\sigma_2\mathcal{G}_2}^{\tilde{\mathcal{H}}})\sigma_1 g_1, (1 - P_{\sigma_2\mathcal{G}_2}^{\tilde{\mathcal{H}}})\sigma_1 g_1 \right\rangle_{\mathcal{K}} \\
&\leq \langle g_1, g_1 \rangle_{\mathcal{G}_1}.
\end{aligned}
$$

It remains to prove the equality between the two negative indices. Since $\tilde{A}^* = \sigma_1^{-1} P_{\sigma_1\mathcal{G}_1}^{\tilde{\mathcal{H}}}\sigma_2$, we have that for all $g_2 \in \mathcal{G}_2$,

$$
\begin{aligned}
\left\langle (1 - \tilde{A}\tilde{A}^*)g_2, g_2 \right\rangle_{\mathcal{G}_2} &= \langle g_2, g_2 \rangle_{\mathcal{G}_2} - \left\langle \tilde{A}^* g_2, \tilde{A}^* g_2 \right\rangle_{\mathcal{G}_1} \\
&= \left\langle (1 - P_{\sigma_1\mathcal{G}_1}^{\tilde{\mathcal{H}}})\sigma_2 g_2, (1 - P_{\sigma_1\mathcal{G}_1}^{\tilde{\mathcal{H}}})\sigma_2 g_2 \right\rangle_{\tilde{\mathcal{H}}}.
\end{aligned}
$$

This equality, the decomposition

$$
\tilde{\mathcal{H}} \ominus \sigma_1\mathcal{G}_1 = (\tilde{\mathcal{H}} \ominus \mathcal{H}) \oplus (\mathcal{G} \ominus \tau_1\mathcal{H}_1) \oplus (\mathcal{F}_2 \ominus \mathcal{H}_2)
$$

and Corollary 1.4 (b) imply the inequality

$$
\mathrm{ind}_-(\mathcal{D}_{\tilde{A}}) \leq \mathrm{ind}_-(\mathcal{D}_A).
$$

Since $\tilde{A}$ is a lifting of A, we have $A^* = \tilde{A}^*|\mathcal{H}_2$, and hence the converse inequality also holds true. This completes the proof. $\blacksquare$

3. Characterization of the solutions

In the sequel we denote by $\mathcal{U}(V)$ the family of all minimal unitary Hilbert space extensions of V. Here a unitary extension $(U,\tilde{\mathcal{H}})$ of V is called *minimal* if

$$\tilde{\mathcal{H}} = \underset{n \in \mathbf{Z}}{CLS} \{U^n \mathcal{H}\} .$$

We claim that if $(U,\tilde{\mathcal{H}})$ belongs to $\mathcal{U}(V)$, then

$$\tilde{\mathcal{H}} = \underset{n \in \mathbf{Z},\, i=1,2}{CLS} \{U^n \tau_i \mathcal{H}_i\} . \tag{3.1}$$

Indeed, by Lemma 1.7, we have $\mathcal{H} = CLS\{\sigma_1 \mathcal{G}_1, \rho_2 \mathcal{F}_2\}$, where, by the minimality of $(W_1, \mathcal{G}_1)$ and $(V_2, \mathcal{F}_2)$,

$$\mathcal{G}_1 = \underset{n \in \mathbf{N}}{CLS} \{W_1^n \mathcal{H}_1\} \quad \text{and} \quad \mathcal{F}_2 = \underset{n \in \mathbf{N}}{CLS} \{V_2^n \mathcal{H}_2\} .$$

Now the claim follows from Proposition 1.8 (b), (c) which imply that for $n \in \mathbf{N}$,

$$U^n|\tau_1 \mathcal{H}_1 = \sigma_1 W_1^n|\mathcal{H}_1 \quad \text{and} \quad U^{-n}|\tau_2 \mathcal{H}_2 = \rho_2 V_2^n|\mathcal{H}_2, \tag{3.2}$$

and hence

$$\sigma_1 \mathcal{G}_1 = \underset{n \in \mathbf{N}}{CLS} \{U^n \tau_1 \mathcal{H}_1\} \quad \text{and} \quad \rho_2 \mathcal{F}_2 = \underset{n \in \mathbf{N}}{CLS} \left\{U^{-n} \tau_2 \mathcal{H}_2\right\} .$$

Note that, by Lemma 2.1, we also have

$$U^n|\tau_2 \mathcal{H}_2 = \sigma_2 W_2^n|\mathcal{H}_2, \qquad n \in \mathbf{N}. \tag{3.3}$$

We say that $(U,\tilde{\mathcal{H}})$ and $(U',\tilde{\mathcal{H}}')$ in $\mathcal{U}(V)$ are *isomorphic* if there exists a unitary operator ϕ from $\tilde{\mathcal{H}}$ onto $\tilde{\mathcal{H}}'$ such that $\phi|\mathcal{H} = 1$ and $\phi U = U'\phi$.

Theorem 3.1 *The correspondence*

$$(U,\tilde{\mathcal{H}}) \in \mathcal{U}(V) \;\mapsto\; \tilde{A} := \sigma_2^{-1} P_{\sigma_2 \mathcal{G}_2}^{\tilde{\mathcal{H}}} \sigma_1 \in \mathbf{LIF}(A) \tag{3.4}$$

is, up to isomorphism, a bijection between $\mathcal{U}(V)$ *and* $\mathbf{LIF}(A)$.

According to Theorem 2.2 the correspondence (3.4) is well defined. Note that if $(U,\tilde{\mathcal{H}})$ is a unitary Hilbert space extension of V then $CLS_{n \in \mathbf{Z}}\{U^n \mathcal{H}\}$ is a regular subspace of $\tilde{\mathcal{H}}$, as it contains $\mathcal{H}$ and hence a maximal uniformly negative subspace of $\tilde{\mathcal{H}}$. Furthermore, the restriction of U to $CLS_{n \in \mathbf{Z}}\{U^n \mathcal{H}\}$ belongs to $\mathcal{U}(V)$ and the operator $\tilde{A}$ associated with $(U,\tilde{\mathcal{H}})$ coincides with the one associated with this restriction of U.

Proof of Theorem 3.1. We first show that the correspondence (3.4) is, up to isomorphism, one-to-one. Let $(U, \tilde{\mathcal{H}})$ and $(U', \tilde{\mathcal{H}}')$ be two elements in $\mathcal{U}(V)$ such that

$$\tilde{A} := \sigma_2^{-1} P_{\sigma_2 \mathcal{G}_2}^{\tilde{\mathcal{H}}} \sigma_1 = (\sigma_2')^{-1} P_{\sigma_2' \mathcal{G}_2}^{\tilde{\mathcal{H}}'} \sigma_1 =: \tilde{A}'.$$

From (3.2), (3.3) and their versions for $(U', \tilde{\mathcal{H}}')$ we obtain that for $n \in \mathbf{N}$,

$$U^n|_{\tau_1 \mathcal{H}_1} = U'^n|_{\tau_1 \mathcal{H}_1} \qquad \text{and} \qquad U^n|_{\tau_2 \mathcal{H}_2} = U'^n|_{\tau_2 \mathcal{H}_2}.$$

These equalities imply that for $h_i, k_i \in \mathcal{H}_i$ and $n, m \in \mathbf{Z}$,

$$\left\langle U'^m \tau_i h_i, U'^m \tau_i k_i \right\rangle_{\tilde{\mathcal{H}}'} = \left\langle U'^{n+|n|+|m|} \tau_i h_i, U'^{m+|n|+|m|} \tau_i k_i \right\rangle_{\tilde{\mathcal{H}}'} = \left\langle U^n \tau_i h_i, U^m \tau_i k_i \right\rangle_{\tilde{\mathcal{H}}}$$

for $i = 1, 2$. Moreover, since $\tilde{A} = \tilde{A}'$, we find that for $h_1 \in \mathcal{H}_1$, $h_2 \in \mathcal{H}_2$ and $n, m \in \mathbf{Z}$,

$$\begin{aligned}
\langle U'^n \tau_1 h_1, U'^m \tau_2 h_2 \rangle_{\tilde{\mathcal{H}}'} &= \left\langle \sigma_1 W_1^{n+|n|+|m|} h_1, \sigma_2' W_2^{m+|n|+|m|} h_2 \right\rangle_{\tilde{\mathcal{H}}'} \\
&= \left\langle \tilde{A}' W_1^{n+|n|+|m|} h_1, W_2^{m+|n|+|m|} h_2 \right\rangle_{\mathcal{G}_2} \\
&= \left\langle \tilde{A} W_1^{n+|n|+|m|} h_1, W_2^{m+|n|+|m|} h_2 \right\rangle_{\mathcal{G}_2} = \langle U^n \tau_1 h_1, U^m \tau_2 h_2 \rangle_{\tilde{\mathcal{H}}}.
\end{aligned}$$

These formulas and (3.1) and its version for $(U', \tilde{\mathcal{H}}')$ imply that the linear mapping

$$\psi_0 : \underset{n \in \mathbf{Z}; \ i=1,2}{LS} \{U^n \tau_i \mathcal{H}_i\} \ \to \ \underset{n \in \mathbf{Z}; \ i=1,2}{LS} \{U'^n \tau_i \mathcal{H}_i\}$$

determined by $\psi_0 U^n \tau_i h_i = U'^n \tau_i h_i$, $n \in \mathbf{Z}$, $h_i \in \mathcal{H}_i$, $i = 1, 2$, defines a weak isomorphism between $\tilde{\mathcal{H}}$ and $\tilde{\mathcal{H}}'$. From the proof of (3.1) we see that

$$\mathcal{M} := \underset{n \in \mathbf{N}}{LS} \{\sigma_1 W_1^n \mathcal{H}_1, \rho_2 V_2^n \mathcal{H}_2\}$$

is a dense subspace of $\mathcal{H}$ and that $U^m \mathcal{M} \subseteq \operatorname{dom} \psi_0$ for any $m \in \mathbf{Z}$. Thus for $h, k \in \mathcal{H}$ there exist sequences $\{x_p\}, \{y_q\} \subseteq \mathcal{M}$ such that $\lim_{p \to \infty} x_p = h$, $\lim_{q \to \infty} y_q = k$ and hence for $n, m \in \mathbf{Z}$,

$$\begin{aligned}
\langle U'^n h, U'^m k \rangle_{\tilde{\mathcal{H}}'} &= \lim_{p,q \to \infty} \langle U'^n x_p, U'^m y_q \rangle_{\tilde{\mathcal{H}}'} \\
&= \lim_{p,q \to \infty} \langle \psi_0 U^n x_p, \psi_0 U^m y_q \rangle_{\tilde{\mathcal{H}}'} \\
&= \lim_{p,q \to \infty} \langle U^n x_p, U^m y_q \rangle_{\tilde{\mathcal{H}}} = \langle U^n h, U^m k \rangle_{\tilde{\mathcal{H}}}.
\end{aligned}$$

It follows that the linear mapping

$$\phi_0 : \underset{n \in \mathbf{Z}}{LS} \{U^n \mathcal{H}\} \to \underset{n \in \mathbf{Z}}{LS} \{U'^n \mathcal{H}\}$$

with $\phi_0 U^n h = U'^n h$, $n \in \mathbf{Z}$, $h \in \mathcal{H}$, gives a weak isomorphism between $\tilde{\mathcal{H}}$ and $\tilde{\mathcal{H}}'$. Since U is a Hilbert space extension of V, $\tilde{\mathcal{H}} \ominus \mathcal{H}$ is a Hilbert space and hence any maximal uniformly negative subspace of $\mathcal{H}$ is also maximal uniformly negative in

$\tilde{\mathcal{H}}$. From $\mathcal{H} \subseteq \operatorname{dom} \phi_0$ it follows that $\operatorname{dom} \phi_0$ contains a maximal uniformly negative subspace. Hence ϕ_0 is continuous and its closure is a unitary operator ϕ from $\tilde{\mathcal{H}}$ onto $\tilde{\mathcal{H}}'$. Clearly, $\phi U = U' \phi$ and $\phi | \mathcal{H} = 1$.

Next we show that the correspondence (3.4) is onto. Let $\tilde{A} \in \mathbf{LIF}(A)$ be any fixed lifting of A. We repeat the construction in Section 1 but now for $\{W_1, W_2, \tilde{A}\}$ instead of $\{T_1, T_2, A\}$. We obtain a Kreĭn space $\left(\tilde{\mathcal{G}}, \langle \cdot, \cdot \rangle_{\tilde{\mathcal{G}}} \right)$ associated with the Kreĭn space $\mathcal{G}_1 \oplus \mathcal{G}_2$ and the selfadjoint operator

$$\tilde{B} = \left(\begin{array}{cc} 1 & \tilde{A}^* \\ \tilde{A} & 1 \end{array} \right) \in \mathbf{L}(\mathcal{G}_1 \oplus \mathcal{G}_2).$$

This construction also yields two continuous isometries $\tilde{\tau}_1 : \mathcal{G}_1 \to \tilde{\mathcal{G}}$ and $\tilde{\tau}_2 : \mathcal{G}_2 \to \tilde{\mathcal{G}}$ such that $\tilde{\mathcal{G}} = CLS\{\tilde{\tau}_1 \mathcal{G}_1, \tilde{\tau}_2 \mathcal{G}_2\}$ and

$$\langle \tilde{\tau}_1 g_1, \tilde{\tau}_2 g_2 \rangle_{\tilde{\mathcal{G}}} = \left\langle \tilde{A} g_1, g_2 \right\rangle_{\mathcal{G}_2}, \qquad g_1 \in \mathcal{G}_1, g_2 \in \mathcal{G}_2,$$

cf. Lemma 1.1. Note that $(W_1, \mathcal{G}_1)$ is the minimal isometric dilation of W_1 itself. Let $(U_2, \tilde{\mathcal{H}}_2)$ be the minimal unitary dilation of T_2. Then, according to Theorem B (a), $(U_2^{-1}, \tilde{\mathcal{H}}_2)$ is the minimal isometric dilation of W_2^*. Furthermore, by Theorem B, we have

$$\tilde{\mathcal{H}}_2 = (\mathcal{F}_2 \ominus \mathcal{H}_2) \oplus \mathcal{H}_2 \oplus (\mathcal{G}_2 \ominus \mathcal{H}_2). \tag{3.5}$$

In this case the analog of the space $\mathcal{H}$ considered in Section 1 is the Kreĭn space

$$\tilde{\mathcal{H}} := \tilde{\mathcal{G}} \oplus (\tilde{\mathcal{H}}_2 \ominus \mathcal{G}_2).$$

Also, analogous to the definition in Section 1, we define the operator $\tilde{\rho}_2 : \tilde{\mathcal{H}}_2 \to \tilde{\mathcal{H}}$ by setting $\tilde{\rho}_2 | \mathcal{G}_2 := \tilde{\tau}_2$ and $\tilde{\rho}_2 | \tilde{\mathcal{H}}_2 \ominus \mathcal{G}_2 := 1$. Then $\tilde{\rho}_2$ is a bicontractive isometric operator, $\tilde{\mathcal{H}} = CLS\left\{ \tilde{\tau}_1 \mathcal{G}_1, \tilde{\rho}_2 \tilde{\mathcal{H}}_2 \right\}$ and for all $g_1 \in \mathcal{G}_1, \tilde{h}_2 \in \tilde{\mathcal{H}}_2$,

$$\left\langle \tilde{\tau}_1 g_1, \tilde{\rho}_2 \tilde{h}_2 \right\rangle_{\tilde{\mathcal{H}}} = \left\langle \tilde{A} g_1, P_{\mathcal{G}_2}^{\tilde{\mathcal{H}}_2} \tilde{h}_2 \right\rangle_{\mathcal{G}_2}, \tag{3.6}$$

cf. Lemma 1.6. Finally, the operator $\tilde{V}$ in $\tilde{\mathcal{H}}$ defined by

$$\tilde{V}(\tilde{\tau}_1 g_1 + \tilde{\rho}_2 U_2^{-1} \tilde{h}_2) = \tilde{\tau}_1 W_1 g_1 + \tilde{\rho}_2 \tilde{h}_2, \qquad g_1 \in \mathcal{G}_1, \tilde{h}_2 \in \tilde{\mathcal{H}}_2,$$

turns out to be a continuous isometry defined on the whole space $\tilde{\mathcal{H}}$ (since U_2 is unitary) and, moreover, its remaining defect subspace $\tilde{\mathcal{H}} \ominus \operatorname{ran} \tilde{V}$ is a Hilbert space and hence $\tilde{V}$ is even a bicontraction.

We claim that $(\tilde{V}, \tilde{\mathcal{H}})$ is a regular isometric extension of $(V, \mathcal{H})$, that is, there exists a continuous isomorphism $\sigma : \mathcal{H} \to \tilde{\mathcal{H}}$ such that $\tilde{\mathcal{H}} \ominus \sigma \mathcal{H}$ is a Hilbert space and $\sigma V = \tilde{V} \sigma$. To prove this, we first show that

$$\sigma_0(\sigma_1 g_1 + \rho_2 f_2) = \tilde{\tau}_1 g_1 + \tilde{\rho}_2 f_2, \qquad g_1 \in \mathcal{G}_1, f_2 \in \mathcal{F}_2,$$

defines a continuous isometric mapping σ_0 from $\mathcal{H}$ to $\tilde{\mathcal{H}}$ with dom $\sigma_0 = LS\{\sigma_1\mathcal{G}_1, \rho_2\mathcal{F}_2\}$ and ran $\sigma_0 = LS\{\tilde{\tau}_1\mathcal{G}_1, \tilde{\rho}_2\mathcal{F}_2\}$. For $g_1, y_1 \in \mathcal{G}_1$ and $f_2, x_2 \in \mathcal{F}_2$,

$$\langle \sigma_1 g_1, \sigma_1 y_1 \rangle_{\mathcal{H}} = \langle g_1, y_1 \rangle_{\mathcal{G}_1} = \langle \tilde{\tau}_1 g_1, \tilde{\tau}_1 y_1 \rangle_{\tilde{\mathcal{H}}}$$

and

$$\langle \rho_2 f_2, \rho_2 x_2 \rangle_{\mathcal{H}} = \langle f_2, x_2 \rangle_{\mathcal{F}_2} = \langle \tilde{\rho}_2 f_2, \tilde{\rho}_2 x_2 \rangle_{\tilde{\mathcal{H}}},$$

while, by Lemma 1.6 and formulas (3.5) and (3.6),

$$\begin{aligned}
\langle \sigma_1 g_1, \rho_2 f_2 \rangle_{\mathcal{H}} &= \langle AP_1 g_1, f_2 \rangle_{\mathcal{F}_2} = \langle P_2 \tilde{A} g_1, f_2 \rangle_{\mathcal{F}_2} = \left\langle P_{\mathcal{H}_2}^{\tilde{\mathcal{H}}_2} \tilde{A} g_1, f_2 \right\rangle_{\tilde{\mathcal{H}}_2} \\
&= \left\langle \tilde{A} g_1, P_{\mathcal{H}_2}^{\tilde{\mathcal{H}}_2} f_2 \right\rangle_{\tilde{\mathcal{H}}_2} = \left\langle \tilde{A} g_1, P_{\mathcal{G}_2}^{\tilde{\mathcal{H}}_2} f_2 \right\rangle_{\mathcal{G}_2} = \langle \tilde{\tau}_1 g_1, \tilde{\rho}_2 f_2 \rangle_{\tilde{\mathcal{H}}}.
\end{aligned}$$

These formulas show that σ_0 is well defined, continuous and isometric provided it happens that $CLS\{\tilde{\tau}_1\mathcal{G}_1, \tilde{\rho}_2\mathcal{F}_2\}$ contains a maximal uniformly negative subspace of $\tilde{\mathcal{H}}$. The latter is true by (3.5) (note that $\mathcal{G}_2 \ominus \mathcal{H}_2$ is a Hilbert space) and the fact that $\tilde{\rho}_2 : \tilde{\mathcal{H}}_2 \to \tilde{\mathcal{H}}$ is bicontractive. These results and Lemma 1.7 (b) imply that σ_0 can be extended to a continuous isomorphism σ from $\mathcal{H}$ into $\tilde{\mathcal{H}}$ such that $\tilde{\mathcal{H}} \ominus \sigma\mathcal{H}$ is a Hilbert space. Furthermore, for $g_1 \in \mathcal{G}_1$ and $f_2 \in \mathcal{F}_2$ we have

$$\begin{aligned}
\sigma V(\sigma_1 g_1 + \rho_2 V_2 f_2) &= \sigma(\sigma_1 W_1 g_1 + \rho_2 f_2) = \tilde{\tau}_1 W_1 g_1 + \tilde{\rho}_2 f_2 \\
&= \tilde{V}(\tilde{\tau}_1 g_1 + \tilde{\rho}_2 V_2 f_2) = \tilde{V}\sigma(\sigma_1 g_1 + \rho_2 V_2 f_2),
\end{aligned}$$

that is, $\sigma V = \tilde{V}\sigma$. This completes the proof of the claim. We identify $(V, \mathcal{H})$ with its isomorphic copy $(\sigma V, \sigma\mathcal{H})$.

Let $(U_{\tilde{V}}, \mathcal{H}_{\tilde{V}})$ be the minimal unitary Hilbert space extension of $\tilde{V}$ (which is unique up to isomorphism, and, since $\tilde{V}$ is bicontractive, coincides with the canonical minimal unitary dilation of $\tilde{V}$). Put

$$\mathcal{K} = \underset{n \in \mathbf{Z}}{CLS}\left\{U_{\tilde{V}}^n \sigma\mathcal{H}\right\}.$$

Then $\mathcal{K}$ is a regular subspace of $\mathcal{H}_{\tilde{V}}$ and $(U_{\tilde{V}}|\mathcal{K}, \mathcal{K})$ belongs to $\mathcal{U}(V)$.

Let $A_{\mathcal{K}} = \sigma_2^{-1} P_{\sigma_2 \mathcal{G}_2}^{\mathcal{K}} \sigma\sigma_1$, where σ_2 is a unitary operator from $\mathcal{G}_2$ onto

$$\tilde{\mathcal{G}}_{\mathcal{K}} := \underset{n \in \mathbf{N}}{CLS}\left\{U_{\tilde{V}}^n \tilde{\tau}_2 \mathcal{H}_2\right\}$$

such that $\sigma_2|\mathcal{H}_2 = \tilde{\tau}_2$ and $\sigma_2 W_2 = U_{\tilde{V}}\sigma_2$, cf. Lemma 2.1. Note that $A_{\mathcal{K}} \in \mathbf{LIF}(A)$ and that $A_{\mathcal{K}}$ corresponds to $(U_{\tilde{V}}|\mathcal{K}, \mathcal{K})$ by means of (3.4). Finally, for $h_1 \in \mathcal{H}_1$, $h_2 \in \mathcal{H}_2$ and $n, m \in \mathbf{N}$ we find that

$$\begin{aligned}
\langle A_{\mathcal{K}} W_1^n h_1, W_2^m h_2 \rangle_{\mathcal{G}_2} &= \langle \tilde{\tau}_1 W_1^n h_1, \sigma_2 W_2^m h_2 \rangle_{\mathcal{K}} \\
&= \left\langle \tilde{\tau}_1 W_1^n h_1, U_{\tilde{V}}^m \tilde{\tau}_2 h_2 \right\rangle_{\mathcal{K}} = \left\langle \tilde{\tau}_1 W_1^n h_1, \tilde{V}^m \tilde{\tau}_2 h_2 \right\rangle_{\mathcal{K}} \\
&= \left\langle \tilde{\tau}_1 W_1^n h_1, \tilde{V}^m \tilde{\tau}_2 h_2 \right\rangle_{\tilde{\mathcal{H}}} = \langle \tilde{\tau}_1 W_1^n h_1, \tilde{\tau}_2 U_2^m h_2 \rangle_{\tilde{\mathcal{H}}} \\
&= \langle \tilde{\tau}_1 W_1^n h_1, \tilde{\tau}_2 W_2^m h_2 \rangle_{\tilde{\mathcal{H}}} = \left\langle \tilde{A} W_1^n h_1, W_2^m h_2 \right\rangle_{\mathcal{G}_2}.
\end{aligned}$$

For the last three equalities we used the definition of $\tilde{V}$, Theorem B (a) and formula (3.6). From the minimality of $(W_i, \mathcal{G}_i)$, i=1,2, it now follows that

$$\langle A_K g_1, g_2 \rangle_{\mathcal{G}_2} = \left\langle \tilde{A} g_1, g_2 \right\rangle_{\mathcal{G}_2}, \qquad g_i \in \mathcal{G}_i, \ i = 1, 2,$$

that is, $\tilde{A} = A_K$. ∎

We remark that to every $(U, \tilde{\mathcal{H}}) \in \mathcal{U}(V)$ there corresponds a unitary colligation $\{\mathcal{H}_0, \mathcal{N}, \mathcal{M}; X\}$, where $\mathcal{H}_0 = \tilde{\mathcal{H}} \ominus \mathcal{H}$, $\mathcal{N}$ and $\mathcal{M}$ are the defect subspaces of V, and $X := U|(\mathcal{H}_0 \oplus \mathcal{N})$ is a unitary operator from $\mathcal{H}_0 \oplus \mathcal{N}$ onto $\mathcal{H}_0 \oplus \mathcal{M}$. Every such colligation is uniquely determined by its characteristic function which is a holomorphic function on the open unit disc in the complex plane whose values are contractions in $\mathbf{L}(\mathcal{N}, \mathcal{M})$. Theorem 3.1 implies there exists a bijection between the class of all these functions and the set $\mathbf{LIF}(A)$. This bijection will be treated elsewhere.

References

[AN]   T. Andô, *Linear Operators on Kreĭn Spaces,* Hokkaido University, Research Institute of Applied Electricity, Division of Applied Mathematics, Sapporo, 1979. MR 81e:47032.

[AI]   T. Azizov and I.S. Iokhvidov, *Foundations of the Theory of Linear Operators in Spaces with Indefinite Metric,* "Nauka", Moscow, 1986; English transl. *Linear Operators in Spaces with Indefinite Metric,* Wiley, New York, 1989. MR 88g:47070.

[AR]   R. Arocena, *Unitary extensions of isometries and contractive intertwining dilations,* Operator Theory: Adv. Appl., Vol. 41 (1989), Birkhäuser, Basel, 13–23. MR 91c:47009.

[BO]   J. Bognár, *Indefinite Inner Product Spaces,* Springer, Berlin, 1974. MR 57#7125.

[C]   T. Constantinescu, *On a general extrapolation problem,* Rev. Roumaine Math. Pures Appl. **32** (1987), 509–521.

[CG]   T. Constantinescu and A. Gheondea, *Elementary rotations of linear operators in Kreĭn spaces,* INCREST, Bucharest, preprint, 1991.

[D]   M.A. Dritschel, *The essential uniqueness property for operators on Kreĭn spaces,* preprint, 1990.

[DR]   M.A. Dritschel and J. Rovnyak, *Extension theorems for contraction operators on Kreĭn spaces*, Operator Theory: Adv. Appl., Vol. 47 (1990), Birkhäuser, Basel, 221–305.

[FF]   C. Foiaş and A. Frazho, *The Commutant Lifting Approach to Interpolation Problems*, Operator Theory: Adv. Appl., Vol. 44, Birkhäuser, Basel, 1990.

[MO]   M.D. Morán, *On Intertwining Dilations*, J. Math. Anal. Appl. **141** (1989), 219–234. MR 90e:47003.

[NF]   B. Sz.-Nagy and C. Foiaş, *Harmonic Analysis of Operators on Hilbert Space*, North Holland, New York, 1970. MR 43#947.

[SA1] D. Sarason, *Generalized interpolation in H^∞*, Trans. Amer. Math. Soc. **127** (1967), 179–203. MR 34#8193.

[SA2] D. Sarason, *New Hilbert Spaces from Old*, in *"Paul Halmos. Celebrating 50 years of Mathematics"* (J. Ewing and F.W. Gehring, eds.), Springer, Berlin, 1991, 195–204.

**Aad Dijksma, Michael Dritschel,
Stefania Marcantognini, Henk de Snoo**

Department of Mathematics
University of Groningen
P.B. 800
9700 AV Groningen
The Netherlands.

Operator Theory:
Advances and Applications, Vol. 61
© 1993 Birkhäuser Verlag Basel

A Method for Constructing Invariant Subspaces for Some Operators on Kreĭn Spaces

Michael A. Dritschel[*]

Abstract. A generalization of the notion of a Julia operator is used to produce models to give a straightforward construction of invariant, and in fact reducing, subspaces for a class of operators on Kreĭn spaces known as weakly definitizable operators. This includes, among others, the definitizable selfadjoint and definitizable unitary operators. A simple proof of the existence of an orthogonal pair of maximal definite invariant subspaces for positive operators on a Kreĭn space is found, and applied to show the existence of such invariant subspaces for definitizable selfadjoint operators. A new proof of the Spectral Theorem for continuous definitizable selfadjoint operators is also given, along with a functional calculus which includes functions that are Borel measurable and bounded after division by the definitizing polynomial on a suitably large subset of the real line.

1. Introduction and preliminaries

Determining the existence of nontrivial invariant subspaces for a variety of operators on Kreĭn spaces has been the subject of intense research for more than 40 years. Even now it is unknown whether every selfadjoint or unitary operator on a Kreĭn space has nontrivial invariant subspaces. Indications are that this problem is as difficult as the invariant subspace problem for operators on a Hilbert space, since the classes of selfadjoint and unitary operators are quite large. For example, if $\mathcal{H}$ is a Hilbert space, we may make $\mathcal{H} \oplus \mathcal{H}$ into a Kreĭn space using the fundamental symmetry $\begin{pmatrix} 0 & 1 \\ 1 & 0 \end{pmatrix}$. Then, as was observed in [18], for any bounded operator $A \in \mathbf{B}(\mathcal{H})$, the operator $\begin{pmatrix} A & 0 \\ 0 & A^* \end{pmatrix}$, is Kreĭn space selfadjoint (here "A^*" is used for the adjoint of A as a Hilbert space operator). Unitary operators may be obtained from selfadjoint operators by means of the Caley transform (see, for example, [5] or [4]), which means that the situation for unitary operators is no better.

In any case, a certain amount of progress has been made for some special subclasses, and the reader is referred to the such sources as [17], [18], [5], and [4] for a

---

[*]The author was supported by the National Science Foundation.

partial list of references. In this work we approach the problem of invariant subspaces of various classes of operators on Kreĭn spaces and their spectral theory through a generalization of the notion of a Julia operator, an object which, prior to this, has been most useful in the study of the properties of contractions (see [10], [12]). The idea is to associate with the given operator a Hilbert space operator by means of this generalized Julia operator, and then to transcribe reducing subspaces for the Hilbert space operator to reducing subspaces for the Kreĭn space operator. The method of transcription insures the existence of reducing subspaces, since it yields a Hilbert space selfadjoint operator that commutes with the Hilbert space operator. The technique is connected with the theory of representations of Banach $*$-algebras. In fact, the transcription mentioned above is carried out via a special kind of representation of a Banach $*$-algebra containing the given Kreĭn space operator.

A corollary of the above will be the existence of reducing subspaces for Kreĭn space selfadjoint and unitary operators when such operators are definitizable. A certain amount of the invariant subspace lattice structure for the operators for which this process works is also ascertained. In the situation that the operator under consideration is positive, a simple proof of the existence of a pair of orthogonal maximal positive and maximal negative invariant subspaces is given.

Many of the results stated above were originally proved by Langer [16] (see also [17] and [18]) using a spectral theory for definitizable selfadjoint operators. The method used here has, in the author's opinion, the advantage of achieving these ends more directly, and provides a new way of viewing the problem of the structure of selfadjoint operators on Kreĭn spaces. In the last section it is shown how a spectral theory for selfadjoint operators may be derived by means of the tools that have been developed. A large functional calculus is found as well. The technique bears a resemblance to that used by Andô [1] and Bognár ([6], [7]) for certain special cases.

For basic information on Kreĭn spaces and operators acting on them, as well as the notation used, the reader is referred to [12]. Throughout the paper, all Kreĭn and Hilbert spaces are assumed to be over the complex field.

The author would like to thank Peter Jonas and Erik Thomas for enlightening discussions.

2.   Factorization and Julia operators

We begin with a key lemma due in large part to Bognár and Krámli [5]. It is basically an application of the polar decomposition in Hilbert spaces. A proof of the lemma as stated here may be found in [12], Theorems 1.2.2 and B.1.

Lemma 1. *Let $\mathcal{H}$ be a Kreĭn space with a fixed fundamental decomposition $\mathcal{H}_+ \oplus \mathcal{H}_-$,* *and $H \in \mathbf{B}(\mathcal{H})$ selfadjoint. Then there is a Kreĭn space $\mathcal{A}$, a fundamental decomposition $\mathcal{A}_+ \oplus \mathcal{A}_-$ of $\mathcal{A}$, and an operator $A \in \mathbf{B}(\mathcal{A}, \mathcal{H})$ with the following properties:*

(i) $\ker A = \{0\}$ *and* $H = AA^*$;

(ii) $\operatorname{ran} A^* = \mathcal{A}'_+ + \mathcal{A}'_-$ *where* $\mathcal{A}'_\pm \subseteq \mathcal{A}_\pm$ *densely;*

(iii) Let $\mathcal{H}_J$ and $\mathcal{A}_J$ be the Hilbert spaces associated to $\mathcal{H}$ and $\mathcal{A}$ through the fixed fundamental decompositions, and let $A = RW$ the polar decomposition of A as an operator on $\mathcal{A}_J$ to $\mathcal{H}_J$, then

$$AA^* = R^2 WW^* \qquad \text{and} \qquad R^2 = AA^*(WW^*)^\times.$$

Suppose $T \in \mathbf{B}(\mathcal{H}, \mathcal{K})$. Then the operators $1 - T^*T$ and $1 - TT^*$ are selfadjoint. Using Lemma 1 we factor them as $\tilde{D}\tilde{D}^*$ and DD^*, respectively, where $\tilde{D} \in \mathbf{B}(\tilde{\mathcal{D}}, \mathcal{H})$ and $D \in \mathbf{B}(\mathcal{D}, \mathcal{K})$ both have zero kernel. The operator $\tilde{D}$ is called the defect operator for T and D the defect operator for T^*. The name is suggested by the fact that $\left(\begin{smallmatrix} T \\ \tilde{D}^* \end{smallmatrix}\right)$ and $\left(\begin{smallmatrix} T^* \\ D^* \end{smallmatrix}\right)$ are isometries. A Julia operator $U(T)$ for the operator T is a unitary operator of the form $\left(\begin{smallmatrix} T & D \\ \tilde{D}^* & L \end{smallmatrix}\right) \in \mathbf{B}(\mathcal{H} \oplus \mathcal{D}, \mathcal{K} \oplus \tilde{\mathcal{D}})$, where $\tilde{D}$ and D are defect operators. The elements of the Julia operator satisfy the following equalities:

$$
\begin{aligned}
\tilde{D}\tilde{D}^* &= 1 - T^*T & DD^* &= 1 - TT^* \\
\tilde{D}^*\tilde{D} &= 1 - LL^* & D^*D &= 1 - L^*L \\
T\tilde{D} &= -DL^* & T^*D &= -\tilde{D}L.
\end{aligned}
\tag{1}
$$

Perhaps the easiest way to construct a Julia operator for a given operator $T \in \mathbf{B}(\mathcal{H}, \mathcal{K})$ is as follows. First choose a defect operator $\tilde{D}$ for T by factoring $1 - T^*T$. Set $V = \left(\begin{smallmatrix} T \\ \tilde{D}^* \end{smallmatrix}\right)$. Let D_V be a defect operator for V^*, obtained by factoring $1 - VV^*$. Then $U(T) = (V \quad D_V)$ is a Julia operator for T. Details may be found in [12], Theorem 1.2.4.

In the above construction of a Julia operator, the defect operator for T^*, which equals $P_{\mathcal{K}} D_V$, is determined in some sense by the choice of $\tilde{D}$. One might wonder if the defect operators $\tilde{D}$ and D of T and T^* were fixed, whether it would always possible to find an operator L so that $\left(\begin{smallmatrix} T & D \\ \tilde{D}^* & L \end{smallmatrix}\right) \in \mathbf{B}(\mathcal{H} \oplus \mathcal{D}, \mathcal{K} \oplus \tilde{\mathcal{D}})$ is a Julia operator. This happens to be of particular interest in the case where $T \in \mathbf{B}(\mathcal{H})$ is normal (that is, $TT^* = T^*T$), since in this situation we might choose $\tilde{D} = D$. Using the method outlined above for constructing a Julia operator, there is a priori no guarantee that L can be found once we fix both $\tilde{D}$ and D. However, it turns out that by an alternate construction, a Julia operator for a normal operator T may be found with $\tilde{D} = D$.

We defer the proof until the next section, where this will be seen to be a corollary of a more general result.

In the case that $T \in \mathbf{B}(\mathcal{H})$ is selfadjoint and D and $\tilde{D}$ are chosen to be equal, and under the assumption that the fourth element L of the Julia operator exists, the third pair of equations in (1) show that L and L^* are equal on a dense subset of $\mathcal{D}$ and so by continuity are identical; that is, L is selfadjoint. Hence for a selfadjoint operator T, the Julia operator constructed with $\tilde{D} = D$, in addition to being unitary, is selfadjoint. The existence of such a Julia operator for a particular choice of D is a consequence of B.3 from [12], or Theorem 5 below.

Selfadjoint unitary operators have an especially simple structure, as the next lemma indicates. Such operators are power-bounded, and the proof that power-bounded unitary operators have an invariant orthogonal pair of subspaces that are, respectively, maximal uniformly positive and maximal uniformly negative, may be found, for example, in Chapter 2, section 5, of [4] or Chapter 3 of [1]. A simpler proof, due to James Rovnyak, of an equivalent result in the case that the unitary operator is selfadjoint is given below. See also [14].

Lemma 2. *Let $U \in \mathbf{B}(\mathcal{H})$ be unitary and selfadjoint. Then there is a regular subspace $\mathcal{A}$ of $\mathcal{H}$ such that with respect to the decomposition $\mathcal{H} = \mathcal{A} \oplus \mathcal{A}^\perp$, U has the form $U = \begin{pmatrix} 1 & 0 \\ 0 & -1 \end{pmatrix}$.*

Proof. Let $\mathcal{A} = \ker(1 - U)$ and $\mathcal{B} = \ker(1 + U)$. Then if $f \in \mathcal{H}$, we may write

$$f = (1 - U)\frac{1}{2}f + (1 + U)\frac{1}{2}f.$$

Since U is unitary and selfadjoint, $(1 - U)(1 + U) = (1 + U)(1 - U) = 0$, which means that $(1 - U)\frac{1}{2}f \in \ker(1 + U)$ and $(1 + U)\frac{1}{2}f \in \ker(1 - U)$. Hence $\mathcal{H} = \mathcal{A} + \mathcal{B}$.

Suppose $f \in \mathcal{A}$, $g \in \mathcal{B}$, and write $f = (1 - U)\frac{1}{2}f + (1 + U)\frac{1}{2}f = (1 + U)\frac{1}{2}f$. Then

$$\langle f, g \rangle = \left\langle (1 + U)\frac{1}{2}f, g \right\rangle = \left\langle \frac{1}{2}f, (1 + U)g \right\rangle = 0.$$

Thus $\mathcal{B} \subseteq \mathcal{A}^\perp$, and so $\mathcal{A} + \mathcal{A}^\perp = \mathcal{H}$. But then $\mathcal{A}$ is regular and $\mathcal{B} = \mathcal{A}^\perp$. ∎

The next result depends on being able to choose a selfadjoint Julia operator for a selfadjoint operator T, which as was noted, is a consequence of Lemma B.3 of [12] or Theorem 5 below.

Corollary 3. *Let $T \in \mathbf{B}(\mathcal{H})$ be selfadjoint. Then there is a Kreĭn space $\mathcal{K}$ containing $\mathcal{H}$ as a regular subspace, a regular subspace $\mathcal{A}$ of $\mathcal{K}$, and a unitary operator $U \in \mathbf{B}(\mathcal{K})$ of the form $U = \begin{pmatrix} 1 & 0 \\ 0 & -1 \end{pmatrix}$ on $\mathcal{A} \oplus \mathcal{A}^\perp$ such that A is the compression of U to $\mathcal{H}$.*

As intriguing as the last result is, it does not appear to be the most useful tool for attacking the invariant subspace problem for selfadjoint operators on Kreĭn spaces. Instead we take another approach by generalizing the notion of a Julia operator.

3. The generalized Julia operator

Let $S, T \in \mathbf{B}(\mathcal{H}, \mathcal{K})$, and suppose $A \in \mathbf{B}(\mathcal{H})$, $B \in \mathbf{B}(\mathcal{K})$ are selfadjoint with $TA = BS$. Factoring A and B as in Lemma 1, we write $A = \tilde{D}\tilde{D}^*$ and $B = DD^*$, where $\tilde{D} \in \mathbf{B}(\tilde{\mathcal{D}}, \mathcal{H})$ and $D \in \mathbf{B}(\mathcal{D}, \mathcal{K})$ have zero kernel. Fix fundamental decompositions on $\mathcal{H}$, $\mathcal{K}$, $\tilde{\mathcal{D}}$, $\mathcal{D}$ and let $\mathcal{H}_J$, $\mathcal{K}_J$, $\tilde{\mathcal{D}}_J$, $\mathcal{D}_J$ be the associated Hilbert spaces. Write the polar decompositions of $\tilde{D}$ and D as $\tilde{D} = \tilde{R}\tilde{W}$ and $D = RW$ as operators on $\tilde{\mathcal{D}}_J$ to $\mathcal{H}_J$ and $\mathcal{D}_J$ to $\mathcal{K}_J$, respectively. We will next construct an operator $L \in \mathbf{B}(\tilde{\mathcal{D}}, \mathcal{D})$ satisfying $T\tilde{D} = DL$, following a method essentially due to Arsene, Constantinescu, and Gheondea [2]. The proof of a theorem related to the more general result stated here may be found in Agnes Yang's dissertation [23]. An alternate, though equivalent approach is taken by Michael Gebel in [13] where the equation $TA = BS$ is replaced by the condition that T maps the range of A into the range of B.

One possible proof of the existence of L with the required properties uses a theorem of M.G. Kreĭn (see [12], Lemma B.2 and the proof of Theorem B.3). This turns out to be equivalent to a tool more suited to our needs, which is a version of a standard C^*-algebra result, an elementary proof of which may be found for example in [19] (Theorem 2.2.6). It may also be viewed as part of a more general result on monotone operator functions due to C. Loewner (see Chapter 2 of [22]). The equivalence is shown in Appendix A.

Lemma 4. *(Version I, M.G. Kreĭn). Let $\mathcal{H}$ be a Hilbert space, $C, X \in \mathbf{B}(\mathcal{H})$. If C is nonnegative and CX is selfadjoint, then for all $f \in \mathcal{H}$,*

$$|\langle CXf, f\rangle| \le \|X\|\langle Cf, f\rangle.$$

Lemma 4. *(Version II, C. Loewner). Let $\mathcal{H}$ be a Hilbert space, $A, B \in \mathbf{B}(\mathcal{H})$ positive selfadjoint operators. Then $A^2 \le B^2$ implies $A \le B$.*

The construction of L depends crucially on the the relation $TA = BS$. Using Lemma 1 to write A and B in terms of the polar decompositions of D and $\tilde{D}$, this relation becomes $T\tilde{R}^2\tilde{W}\tilde{W}^* = R^2WW^*S$. The operator $\tilde{W}\tilde{W}^*$ is a Hilbert space partial isometry with range equal to the Hilbert space orthogonal complement of the kernel of $\tilde{R}$, so

$$T\tilde{R}^2T^{\times} = R^2WW^*S(\tilde{W}\tilde{W}^*)^{\times}T^{\times},$$

which is a nonnegative Hilbert space operator. Let $X = WW^*T(\tilde{W}\tilde{W}^*)^{\times}T^{\times}$ and $C = R^2$. Clearly $C \geq 0$ and $\|X\| \leq \|S\|\|T\|$. By the Douglas-Shmul'yan lemma, $(CX)^2 = CXX^{\times}C \leq \|X\|^2C^2 = \|S\|^2\|T\|^2C^2$, and so by Version II of Lemma 4, $CX \leq \|S\|\|T\|C$, or equivalently, $T\tilde{R}^2T^{\times} \leq \|S\|\|T\|R^2$. Therefore there exists F with $\|F\| \leq \|S\|^{1/2}\|T\|^{1/2}$ and $T\tilde{R} = RF$. We can choose F so that $\operatorname{ran} F \subseteq \overline{\operatorname{ran} R}$, and then $T\tilde{R}\tilde{W} = RWW^{\times}F\tilde{W}$. Set $L = W^{\times}F\tilde{W}$. Then we have $T\tilde{D} = DL$. Furthermore,

$$\tilde{D}L^*D^* = \tilde{D}\tilde{D}^*T^* = S^*B = S^*DD^*.$$

Since $\operatorname{ran} D^*$ is dense in $\mathcal{D}$, it follows that $S^*D = \tilde{D}L^*$.

Observe that the choice of D and $\tilde{D}$ in the above construction is independent of the intertwining operators S and T. We summarize with the following theorem.

Theorem 5. *Let $A \in \mathbf{B}(\mathcal{H})$, $B \in \mathbf{B}(\mathcal{K})$ be selfadjoint, $\{A, B\}'$ the set of pairs $\{S, T\}$, $S, T \in \mathbf{B}(\mathcal{H}, \mathcal{K})$ with $TA = BS$. Then there are factorizations $A = \tilde{D}\tilde{D}^*$, $\tilde{D}$ a Kreĭn space, $\tilde{D} \in \mathbf{B}(\tilde{\mathcal{D}}, \mathcal{H})$ with zero kernel, $B = DD^*$, $\mathcal{D}$ a Kreĭn space, $D \in \mathbf{B}(\mathcal{D}, \mathcal{K})$ with zero kernel, such that for any $\{S, T\} \in \{A, B\}'$, there exists an operator $L \in \mathbf{B}(\tilde{\mathcal{D}}, \mathcal{D})$ satisfying*

$$T\tilde{D} = DL \qquad\qquad S^*D = \tilde{D}L^*, \tag{2}$$

and

$$DLL^* = TS^*D \qquad\qquad \tilde{D}L^*L = S^*T\tilde{D}. \tag{3}$$

Furthermore, if we let $\hat{A} = \tilde{D}^*\tilde{D}$ and $\hat{B} = D^*D$, then in the case that $S = T$,

$$\hat{A}L^* = L^*\hat{B}. \tag{4}$$

Finally, when $A = B$ it is possible to set $\tilde{D} = D$.

Equations (3) and (4) are proved via repeated applications of (2).

When $S = T$ call the operator

$$U_{A,B}(T) = \begin{pmatrix} T & D \\ \tilde{D}^* & -L^* \end{pmatrix} \in \mathbf{B}(\mathcal{H} \oplus \mathcal{D}, \mathcal{K} \oplus \tilde{\mathcal{D}})$$

a *generalized Julia operator* for T, $\tilde{D}$ and D the *generalized defect operators* for T and T^*, and L^* the *generalized link operator*. We write $U_A(T)$ for $U_{A,A}(T)$. Appropriate subscripts will be appended to D, $\tilde{D}$, and L^* if this is necessary to avoid confusion. In the case that $A = 1 - T^*T$ and $B = 1 - TT^*$, the generalized Julia operator is seen to reduce to the usual Julia operator with L replaced by $-L^*$.

For certain generalized Julia operators, the elements T and L^* share a number of traits, as the next theorem demonstrates.

Theorem 6. *Let $T \in \mathbf{B}(\mathcal{H})$, $A, B \in \mathbf{B}(\mathcal{H})$ selfadjoint with $TA = BT$. If T is normal, we assume $B = A$. Let $U_{A,B}(T) = \begin{pmatrix} T & D \\ \tilde{D}^* & -L^* \end{pmatrix}$ be the associated generalized Julia operator with $\tilde{D} = D$ if T is normal.*

(i) If T is an isometry, then L is an isometry.

(ii) If T is normal, then L is normal.

(iii) If T is unitary, then L is unitary.

(iv) If T is selfadjoint, then L is selfadjoint.

(v) If T is a projection, then L is a projection.

Proof. First consider the case where T is an isometry, A, B selfadjoint with $TA = BT$. Choose $\tilde{D}$, D, and L as in Theorem 5. Then by (3) we see that L is an isometry. If instead T is unitary (that is both T and T^* are isometries), then it clearly follows that L is unitary.

Next consider the case where $T \in \mathbf{B}(\mathcal{H})$ is normal, A selfadjoint with $TA = AT$. We choose $\tilde{D} = D$ and L as in Theorem 5. Since the range of D^* is dense in $\mathcal{D}$, it follows easily that L is a normal operator (just use (3)). If T happens to be selfadjoint, that is, $T = T^*$, by (2) it is clear that L must also be selfadjoint. Finally if T is a projection, by (iv) L is selfadjoint. So it is only necessary to check that L is idempotent. This immediately follows from (2). ∎

The above theorem may also be interpreted in part from the view point of operator algebras. Let $T \in \mathbf{B}(\mathcal{H})$, $\mathcal{H}$ a Kreĭn space, and fix a fundamental symmetry J on $\mathcal{H}$ and associated Hilbert space $\mathcal{H}_J$ for computing norms. The adjoint of T as a Kreĭn space operator, T^*, is related to the adjoint of T as a Hilbert space operator, $T^\times$ by means of the relation $T^* = JT^\times J$. Consequently $\|T^*\| = \|T\|$ since J is Hilbert space unitary. Thus if $\mathcal{A}$ is a $*$-closed algebra of operators contained in $\mathbf{B}(\mathcal{H})$, it is a Banach $*$-algebra. If $A \in \mathbf{B}(\mathcal{H})$ is selfadjoint and $\mathcal{A}$ is a $*$-subalgebra of $\{A\}'$, the commutant of A, then the map π defined by

$$\pi(T) = L$$

where $-L^*$ is the $\{2, 2\}$ component of $U_A(T)$, is a $*$-homomorphism with range in the commutant of D^*D. Linearity is easily seen from the relation $TD = DL$. Suppose $T_1, T_2 \in \{A\}'$, $T = T_1 T_2$, $T_i D = DL_i$ for $i = 1, 2$, and $TD = DL$. Then $0 = (T - T_1 T_2)D = D(L - L_1 L_2)$; and since $\ker D = \{0\}$, $L = L_1 L_2$, or $\pi(T_1 T_2) = \pi(T_1)\pi(T_2)$. The situation where A is a positive operator, so that the closure of the range of π

is a Hilbert space C^*-algebra (that is, a representation of $\{A\}'$) will be of particular interest further on.

The next proposition relates the spectrum of T and its link operator.

Proposition 7. *Let* $T, A \in \mathbf{B}(\mathcal{H})$, A *selfadjoint, and* $TA = AT$. *Let* $U_A(T) = \begin{pmatrix} T & D \\ D^* & -L^* \end{pmatrix} \in \mathbf{B}(\mathcal{H} \oplus \mathcal{D}, \mathcal{K} \oplus \tilde{\mathcal{D}})$ *be a generalized Julia operator for* T. *Then*

$$\sigma(T) \supseteq \sigma(L).$$

Proof. Suppose $\lambda \in \rho(T)$, the resolvent of T. Then $(T - \lambda)^{-1} \in \mathbf{B}(\mathcal{H})$. Furthermore $(T - \lambda)^{-1}$ commutes with A. Consequently, by Theorem 5, there exists $X \in \mathbf{B}(\mathcal{D})$ such that $(T - \lambda)^{-1} D = DX$. Hence

$$D = (T - \lambda)(T - \lambda)^{-1} D = D(L - \lambda)X.$$

Since the kernel of D is zero, this implies X is a right inverse of $L - \lambda$. Similarly it is a left inverse, and so $\lambda \in \rho(L)$, or equivalently, $\sigma(T) \supseteq \sigma(L)$. $\blacksquare$

In constructing the generalized Julia operator of an operator T, we recover only the last pair of equations in (1) for the ordinary Julia operator since we have made no assumptions about the form of the selfadjoint operators that intertwine T. As the next two lemmas show, a form of the other equations in (1) may be obtained under certain conditions.

Lemma 8. *Let* $T \in \mathbf{B}(\mathcal{H}, \mathcal{K})$, p *a polynomial in a single variable. Assume that* $A = p(T^*T)$ *and* $B = p(TT^*)$ *are self adjoint. Then* $TA = BT$. *Let* $U_{A,B}(T) = \begin{pmatrix} T & D \\ \tilde{D}^* & -L^* \end{pmatrix} \in \mathbf{B}(\mathcal{H} \oplus \mathcal{D}, \mathcal{K} \oplus \tilde{\mathcal{D}})$ *be a generalized Julia operator. Then*

$$\tilde{D}^* \tilde{D} = p(L^* L) \qquad \text{and} \qquad D^* D = p(LL^*).$$

Proof. By Theorem 5, and in particular, equation (3), $\tilde{D} L^* L = T^* T \tilde{D}$. Thus

$$\tilde{D}^* \tilde{D} \tilde{D}^* = \tilde{D}^* p(T^* T) = p(L^* L) \tilde{D}^*.$$

The range of $\tilde{D}^*$ is dense, so by continuity, $\tilde{D}^* \tilde{D} = p(L^* L)$. $\blacksquare$

Lemma 9. *Let* $T \in \mathbf{B}(\mathcal{H})$, $p(x, y)$ *a polynomial in two noncommuting variables such that* $A = p(T^*, T)$ *is selfadjoint with* $TA = AT$. *If* $U_A(T) = \begin{pmatrix} T & D \\ D^* & -L^* \end{pmatrix} \in \mathbf{B}(\mathcal{H} \oplus \mathcal{D})$ *is a generalized Julia operator, then*

$$D^* D = p(L^*, L).$$

Proof. This is an immediate consequence of Theorem 5, and in particular, equation (2). ∎

Corollary 10. *Let* $T \in \mathbf{B}(\mathcal{H})$ *be normal,* $p(x,y)$ *any polynomial such that* $A = p(T^*,T)$ *is selfadjoint. Then* $AT = TA$. *If* $U_A(T) = \begin{pmatrix} T & D \\ D^* & -L^* \end{pmatrix} \in \mathbf{B}(\mathcal{H} \oplus \mathcal{D})$ *is a generalized Julia operator, then*

$$D^*D = p(L^*, L).$$

Corollary 11. *If* $T \in \mathbf{B}(\mathcal{H})$ *is selfadjoint,* p *is a polynomial in a single variable with real coefficients, and* $A = p(T)$, *then* $AT = TA$. *If* $U_A(T) = \begin{pmatrix} T & D \\ D^* & -L \end{pmatrix} \in \mathbf{B}(\mathcal{H} \oplus \mathcal{D})$ *is a generalized Julia operator, then we have* $D^*D = p(L)$.

Definition. *Let* $\mathcal{H}$ *be a Kreĭn space. We say that an operator* $T \in \mathbf{B}(\mathcal{H})$ *weakly definitizable if there exist nonzero selfadjoint operators* $B, C \in \mathbf{B}(\mathcal{H})$ *which are not both constant multiples of the identity such that* $A = BC \geq 0$ *and* T *commutes with both* B *and* C.

We call A the *definitizing operator.* Observe that in the case that A is not zero, the definition is equivalent to $T \in \{A\}'$. Note that $T^* \in \{A\}'$ as well.

Definition. *Let* $\mathcal{H}$ *be a Kreĭn space. We say that an operator* $T \in \mathbf{B}(\mathcal{H})$ *is definitizable if there are polynomials* $r(x,y)$ *and* $s(x,y)$ *such that* $r(T^*,T)$ *and* $s(T^*,T)$ *are nonzero selfadjoint operators commuting with* T *which are not both constant multiples of the identity, and* $q(T^*,T) = r(T^*,T)s(T^*,T) \geq 0$.

Clearly a definitizable operator is weakly definitizable. Here q is the so-called *definitizing polynomial.* Traditionally, a definitizable operator T is assumed to be either selfadjoint or unitary and q is a polynomial of a single variable with $\langle q(T)f, f \rangle \geq 0$ for all $f \in \mathcal{H}$. Since $\mathcal{H}$ is over the complex field, this is equivalent to $q(T) \geq 0$. Except in the trivial case that T is a constant multiple of the identity, operators T that satisfy the original condition for being definitizable are definitizable in the new sense; and even in the trivial case, they are weakly definitizable since they then commute with a fundamental symmetry, which is always positive.

The property of being definitizable is in some sense transferable as well, at least in one direction.

Lemma 12. *Let* $T, A \in \mathbf{B}(\mathcal{H})$ $A = p(T^*,T)$ *selfadjoint with* $TA = AT$. *Let* $U_A(T) = \begin{pmatrix} T & D \\ D^* & -L^* \end{pmatrix} \in \mathbf{B}(\mathcal{H} \oplus \mathcal{D})$ *be a generalized Julia operator. If* T *is definitizable then* L *is definitizable.*

Proof. Suppose T is definitizable with definitizing operator $B = q(T^*, T)$. By Theorems 5 and 6, there is a selfadjoint operator $\hat{B} = q(L^*, L) \in \mathbf{B}(\mathcal{D})$ for which $BD = D\hat{B}$. Since $TB = BT$, it is easy to see $L\hat{B} = \hat{B}L$, and by Theorem 5, $LD^*D = D^*DL$ and $\hat{B}D^*D = \hat{B}D^*D$, where $D^*D = p(L^*, L)$. Thus $D^*D\hat{B}$ is selfadjoint and commutes with L. But $0 \leq D^*BD = D^*D\hat{B}$, so pq is a definitizing polynomial for L. ∎

A converse to the above theorem could be proved if it is assumed that $\ker q(T^*, T) = \{0\}$, since a similar argument to that given in the proof could then be used to show that if L is definitizable, there is a polynomial in T and T^* which is positive on the closure of the range of $q(T^*, T)$.

4. A correspondence principle for invariant subspaces

Throughout this section, the dimension of the space $\mathcal{H}$ is assumed to be greater than one.

A subspace $\mathcal{N} \subset \mathcal{H}$ is said to be reducing for T if $T\mathcal{N} \subset \mathcal{N}$ and $T\mathcal{N}^\perp \subset \mathcal{N}^\perp$. It is not hard to see that this is equivalent to $\mathcal{N}$ being a reducing subspace for T^*, as well as the statement that $\mathcal{N}$ is invariant for both T and T^*. On a Kreĭn space, the fact that a subspace reduces an operator may not be as strong a condition as on a Hilbert space, since, for example, a subspace of a Kreĭn space may be contained in its orthogonal complement.

Theorem 13. *Let $T \in \mathbf{B}(\mathcal{H})$ and $A \in \mathbf{B}(\mathcal{H})$ selfadjoint with $AT = TA$. Let $U_A = \begin{pmatrix} T & D \\ D^* & -L^* \end{pmatrix} \in \mathbf{B}(\mathcal{H} \oplus \mathcal{D})$ be a generalized Julia operator.*

(i) Suppose that L has a nontrivial invariant subspace $\mathcal{L}$ or that $\dim \mathcal{D} = 1$, in which case set $\mathcal{L} = \mathcal{D}$. Assume that $\mathcal{L}$ is also invariant for D^*D. Then $\mathcal{N} = \overline{D\mathcal{L}}$ is a nontrivial invariant subspace for T.

(ii) If $\mathcal{L}$ is also a reducing subspace for L, then $\mathcal{N}$ is a reducing subspace for T.

(iii) If $\tilde{\mathcal{L}}$ is another invariant subspace for L with $\tilde{\mathcal{L}} \perp \mathcal{L}$, then $\tilde{\mathcal{N}} = \overline{D\tilde{\mathcal{L}}}$ is invariant for T with $\tilde{\mathcal{N}} \perp \mathcal{N}$.

Proof. Suppose $\mathcal{L}$ is a nontrivial invariant subspace for L that is also invariant for D^*D. Let $\mathcal{N} = \overline{D\mathcal{L}}$. Then $\mathcal{N}$ is an invariant subspace for T. To see this, first note that if $f \in \mathcal{L}$,

$$TDf = DLf \in D\mathcal{L} \subseteq \mathcal{N}.$$

By continuity of T, $T\mathcal{N} \subseteq \mathcal{N}$. A similar argument shows that if $\mathcal{L}$ is invariant for L^*, and so reducing for L, then $\mathcal{N}$ is invariant for T^* as well.

To see that $\mathcal{N}$ is nontrivial, first observe that it is nonzero since $\ker D = \{0\}$. Suppose now that $\mathcal{N} = \mathcal{H}$. Then $D^*D\mathcal{L} \subset \mathcal{L}$ is dense in $\mathcal{D}$, a contradiction. So $\mathcal{N}$ is a proper subset of $\mathcal{H}$.

If $\dim \mathcal{D} = 1$, the same arguments as above show that $\overline{D\mathcal{D}}$ is reducing for T and is nonzero. Since $\dim \mathcal{N} = 1$ and $\dim \mathcal{H} > 1$, $\mathcal{N} \neq \mathcal{H}$.

Now suppose that $\tilde{\mathcal{L}}$ is another invariant subspace for L and that $\tilde{\mathcal{L}} \perp \mathcal{L}$. Let $\tilde{\mathcal{N}} = \overline{D\tilde{\mathcal{L}}}$. If $f \in \mathcal{L}$ and $g \in \tilde{\mathcal{L}}$,

$$\langle Df, Dg \rangle_{\mathcal{H}} = \langle D^*Df, g \rangle_{\mathcal{H}} = 0,$$

since $\mathcal{L}$ is invariant for D^*D. By continuity, $\tilde{\mathcal{N}} \perp \mathcal{N}$. ∎

5. The weakly definitizable case

We again assume $\dim \mathcal{H} > 1$.

Theorem 14. *All weakly definitizable operators have nontrivial reducing subspaces. In the case that $A \in \mathbf{B}(\mathcal{H})$ is a nonzero positive selfadjoint operator, $T \in \{A\}'$, and $U_A = \begin{pmatrix} T & D \\ D^* & -L^* \end{pmatrix} \in \mathbf{B}(\mathcal{H} \oplus \mathcal{D})$ a generalized Julia operator, then there is a nonnegative subspace $\mathcal{N}_+$ and a nonpositive subspace $\mathcal{N}_-$ with $\mathcal{N}_- \perp \mathcal{N}_+$, $\mathcal{N}_+ + \mathcal{N}_- \supseteq \operatorname{ran} D \supseteq \operatorname{ran} A$, and $\mathcal{N}_\pm^\perp \cap \operatorname{ran} D \subseteq \mathcal{N}_\mp$, which are reducing subspaces for T.*

Proof. The case where $\mathcal{H}$ is a Hilbert space or the antispace of a Hilbert space is handled by means of the standard theory for selfadjoint operators on a Hilbert space, so we assume $\mathcal{H}$ is not of this form.

We initially consider the case where there exists nonzero and selfadjoint $B, C \in \mathbf{B}(\mathcal{H})$ with $BC = 0$ and T commutes with both B and C. Then either the kernel of B or the kernel of C is nonzero, and these are not all of $\mathcal{H}$ since it is assumed that neither B nor C are zero. But these spaces are invariant for T. The fact that B and C are selfadjoint insures that they invariant for T^* as well.

Next consider the situation where $A \in \mathbf{B}(\mathcal{H})$ is a nonzero positive operator and T commutes with A. Let $U_A = \begin{pmatrix} T & D \\ D^* & -L^* \end{pmatrix} \in \mathbf{B}(\mathcal{H} \oplus \mathcal{D})$ be a generalized Julia operator for T. First consider the case when D^*D is a constant c times the identity. If $c = 0$, then $\overline{\operatorname{ran} D} = \overline{\operatorname{ran} A}$ is a neutral subspace since $\langle Df, Df \rangle = 0$ for all $f \in \mathcal{D}$. Likewise, if $c > 0$, then the closure of the range of D is a nonnegative subspace, and if $c < 0$, then this is a nonpositive subspace. Since $\overline{\operatorname{ran} A}$ is reducing for T, $A \neq 0$, and $\mathcal{H}$ is not a Hilbert space or the antispace of a Hilbert space, the theorem is true in this situation.

So now assume that D^*D is not a constant multiple of the identity. Let $E(\cdot)$ be a spectral measure for D^*D. Then there is an interval $\Delta \subset \mathbf{R}$ so that $0 < E(\Delta) < 1$

and ran $E(\Delta)$ is a nontrivial invariant subspace for D^*D. Also $E(\Delta)$ is in the double commutant of D^*D, and so in particular, commutes with L^* and L. But then $\mathcal{N} = \overline{\operatorname{ran} DE(\Delta)}$ is a nontrivial reducing subspace for T by Theorem 13.

If we choose $\Delta = (-\infty, 0]$, then

$$\langle DE(\Delta)f, DE(\Delta)f \rangle = \langle D^*DE(\Delta)f, E(\Delta)f \rangle \le 0.$$

So $\mathcal{N}_- = \overline{\operatorname{ran} DE(\Delta)}$ is a negative reducing subspace for T. Likewise we obtain that $\mathcal{N}_+ = \overline{\operatorname{ran} DE([0,\infty))}$ is a positive reducing subspace for T. Since

$$\langle DE((-\infty,0])f, DE([0,\infty))g \rangle = \langle D^*DE((-\infty,0])f, E([0,\infty))g \rangle = 0,$$

$\mathcal{N}_+$ is orthogonal to $\mathcal{N}_-$. It is clear that $\mathcal{N}_+ + \mathcal{N}_- \supseteq \operatorname{ran} D$. If $f = Dg$ is orthogonal to $\mathcal{N}_+$, then for all $h \in \mathcal{D}$,

$$\langle Dg, DE([0,\infty))h \rangle = \langle g, AE([0,\infty))h \rangle,$$

meaning that $g \in \operatorname{ran} E((-\infty,0])$, or $f \in \mathcal{N}_-$. Likewise, $\mathcal{N}_-^{\perp} \cap \operatorname{ran} D \subseteq \mathcal{N}_+$. ∎

It is not difficult to construct examples of weakly definitizable operators that are not definitizable. For example, the Kreĭn subnormal operators considered by Cowen and Li [9] are weakly definitizable where the operator A in the statement of Theorem 14 is a fundamental symmetry. Examples where T is selfadjoint or unitary can also be constructed. One might hope that all selfadjoint operators on a Kreĭn space were weakly definitizable. As it happens, they are not. However, it is not clear how broad the class of weakly definitizable operators really is.

A corollary of Theorem 14 is that definitizable operators have nontrivial reducing subspaces. We explore some further consequences of this theorem in the subsequent sections.

6. Maximal dual pairs of invariant subspaces

Let $\mathcal{N}_+$, $\mathcal{N}_-$ be closed orthogonal subspaces of a Kreĭn space $\mathcal{H}$ that are positive and negative, respectively. Such a pair is referred to as a *dual pair*. If $\mathcal{N}_+$ is maximal positive and $\mathcal{N}_-$ is maximal negative, then we have a *maximal dual pair*. An important result regarding dual pairs is called Phillips' Theorem [20].

Phillips' Theorem. *Let $\tilde{\mathcal{N}}_+$, $\tilde{\mathcal{N}}_-$ be a dual pair in a Kreĭn space $\mathcal{H}$. Then there is a maximal dual pair $\mathcal{N}_+$, $\mathcal{N}_-$ with $\mathcal{N}_\pm \supseteq \tilde{\mathcal{N}}_\pm$.*

A nice proof of this theorem using the two-by-two extension theorem for contractions on Hilbert spaces may be found in [3]. There is a simple proof of the extension theorem in [21].

A fundamental result due to Langer in the theory of definitizable selfadjoint operators is the existence of invariant maximal dual pairs for such operators. A proof using the spectral theory occurs in [17], for example. For such a proof in the special case of positive operators, see also [18].

We next give a proof of this result for positive operators avoiding the spectral theory for definitizable operators.

Theorem 15. *Let $T \in \mathbf{B}(\mathcal{H})$ be a positive operator. Then there exists a maximal dual pair of reducing subspaces for T.*

By Theorem 14 there is a dual pair $\tilde{\mathcal{N}}_+, \tilde{\mathcal{N}}_-$ of reducing subspaces for T. Moreover, these are orthogonal, their sum contains the range of T, and $\tilde{\mathcal{N}}_\pm \cap \operatorname{ran} T \subseteq \tilde{\mathcal{N}}_\mp$. These may be extended by Phillip's Theorem to a maximal dual pair $\mathcal{N}_+, \mathcal{N}_-$. Now

$$T\mathcal{N}_+ \subseteq T(\tilde{\mathcal{N}}_-^\perp) \subseteq \tilde{\mathcal{N}}_-^\perp \cap \operatorname{ran} T \subseteq \tilde{\mathcal{N}}_+ \subseteq \mathcal{N}_+.$$

Invariant subspaces for selfadjoint operators are reducing, and $\mathcal{N}_- = \mathcal{N}_+^\perp$, so this reduces T as well. ∎

We shall also consider the case where some power of a selfadjoint operator is positive, but before doing so, we prove the existence of neutral reducing subspaces for normal operators that are maximal in the sense that they are not properly contained in any other neutral invariant subspace. It is closely related to results of Langer's for families of commuting selfadjoint operators [17] (see sections 2 and 3, especially Lemma 2.4), and enables us to perform a reduction to a simpler case for the selfadjoint operators we are about to study, in a fashion similar to that of Langer's paper.

Theorem 16. *Let $T \in \mathbf{B}(\mathcal{H})$ be a normal operator. Fix a fundamental symmetry J on $\mathcal{H}$. Then there is a (possibly zero) neutral subspace $\mathcal{N}$ of $\mathcal{H}$ which reduces T and is maximal in the sense that it is not contained in any larger neutral invariant subspace of T. Let $\tilde{\mathcal{H}} = \mathcal{H} \ominus (\mathcal{N} + J\mathcal{N})$, which is a regular subspace, and set $\tilde{T} = P_{\tilde{\mathcal{H}}}^{\mathcal{H}} T|\tilde{\mathcal{H}}$. Then $\tilde{T}$ is normal, and is also unitary or selfadjoint whenever T has those properties. If $p(T^*, T) \geq 0$ for some polynomial p, then $p(\tilde{T}^*, \tilde{T}) \geq 0$. Furthermore, neither $\tilde{T}$ nor $\tilde{T}^*$ has neutral invariant subspaces. Finally, if $\mathcal{L} \subseteq \tilde{\mathcal{H}}$ is an invariant subspace for $\tilde{T}$, then $\mathcal{L} + \mathcal{N}$ is an invariant subspace for T, and $\mathcal{L} + \mathcal{N}$ is a (maximal) positive or negative subspace of $\mathcal{H}$ in the case that $\mathcal{L}$ is (maximal) positive or negative subspace of $\tilde{\mathcal{H}}$.*

Proof. Let

$$\mathcal{A} = \{\mathcal{N} \subset \mathcal{H} : \mathcal{N} \text{ is neutral and } T\mathcal{N} \subseteq \mathcal{N}\}.$$

The collection $\mathcal{A}$ may be partially ordered by inclusion and is nonempty since it contains the zero space. Any linearly ordered chain has a least upper bound given by the closed span of the elements of the chain. So by Zorn's lemma, there exists a maximal $\mathcal{N}$ in $\mathcal{A}$.

Set $\mathcal{M} = \bigvee_0^\infty T^{*n}\mathcal{N}$. Clearly $\mathcal{M}$ is invariant under T^*. Since T is normal, for $n = 0, 1, 2, \ldots$, $TT^{*n}\mathcal{N} = T^{*n}T\mathcal{N} \subseteq T^{*n}\mathcal{N}$, and so $\mathcal{M}$ is invariant under T as well; that is, $\mathcal{M}$ reduces T. Suppose $f, g \in \mathcal{N}$. Then for $n = 0, 1, 2, \ldots$, $\langle T^{*n}f, g \rangle = \langle f, T^n g \rangle = 0$. Thus $\mathcal{N} \subseteq \mathcal{M} \subseteq \mathcal{N}^\perp$. Taking orthogonal complements, we see that the same is true with $\mathcal{M}$ replaced by $\mathcal{M}^\perp$. Therefore $\mathcal{N}$ is contained in $\mathcal{M} \cap \mathcal{M}^\perp$, which is invariant under T. By maximality of $\mathcal{N}$ these two spaces are identical, and so $\mathcal{N}$ reduces T.

Let $\tilde{\mathcal{H}} = \mathcal{H} \ominus (\mathcal{N} + J\mathcal{N})$. It is not difficult to show that this is a regular subspace (see, for example, [17], or [11], Lemma 3.2). Furthermore, $\mathcal{N}^\perp = \mathcal{N} + \tilde{\mathcal{H}}$. Set $\tilde{T} = P_{\tilde{\mathcal{H}}}^{\mathcal{H}} T | \tilde{\mathcal{H}}$. For $f \in \mathcal{N}^\perp$, $f = g + h$, where $g \in \mathcal{N}$, $h \in \tilde{\mathcal{H}}$. Then

$$P_{\tilde{\mathcal{H}}}^{\mathcal{H}} T f = P_{\tilde{\mathcal{H}}}^{\mathcal{H}} T h = P_{\tilde{\mathcal{H}}}^{\mathcal{H}} T P_{\tilde{\mathcal{H}}}^{\mathcal{H}} f,$$

since $Tg \in \mathcal{N}$. Likewise $P_{\tilde{\mathcal{H}}}^{\mathcal{H}} T^* f = P_{\tilde{\mathcal{H}}}^{\mathcal{H}} T P_{\tilde{\mathcal{H}}}^{\mathcal{H}} f$. But since $\tilde{\mathcal{H}} \subseteq \mathcal{N}^\perp$,

$$\tilde{T}\tilde{T}^* = P_{\tilde{\mathcal{H}}}^{\mathcal{H}} T P_{\tilde{\mathcal{H}}}^{\mathcal{H}} T^* | \tilde{\mathcal{H}} = P_{\tilde{\mathcal{H}}}^{\mathcal{H}} T T^* | \tilde{\mathcal{H}} = P_{\tilde{\mathcal{H}}}^{\mathcal{H}} T^* T | \tilde{\mathcal{H}} = P_{\tilde{\mathcal{H}}}^{\mathcal{H}} T^* P_{\tilde{\mathcal{H}}}^{\mathcal{H}} H T | \tilde{\mathcal{H}} = \tilde{T}^* \tilde{T};$$

that is, $\tilde{T}$ is normal. If T is selfadjoint, then for any $f, g \in \tilde{\mathcal{H}}$,

$$\langle \tilde{T}f, g \rangle = \langle Tf, g \rangle = \langle f, Tg \rangle = \langle f, \tilde{T}g \rangle,$$

or $\tilde{T}$ is selfadjoint. It is readily noted from this that if T is positive, then so is $\tilde{T}$. In the case that T is unitary, for $f, g \in \tilde{\mathcal{H}}$,

$$\langle \tilde{T}f, \tilde{T}g \rangle = \langle Tf, Tg \rangle - \langle P_{\tilde{\mathcal{H}}^\perp}^{\mathcal{H}} f, P_{\tilde{\mathcal{H}}^\perp}^{\mathcal{H}} g \rangle = \langle f, g \rangle,$$

since $P_{\tilde{\mathcal{H}}^\perp}^{\mathcal{H}} T\tilde{\mathcal{H}} \subset \mathcal{N}$, which is neutral. Likewise $\tilde{T}^*$ is isometric, and so $\tilde{T}$ is unitary. By the same sorts of arguments, it is not difficult to see that for any polynomial p, $\mathcal{N}$ reduces $p(T^*, T)$, and so therefore $p(\tilde{T}^*, \tilde{T}) \geq 0$ if $p(T^*, T) \geq 0$.

Finally, suppose $\mathcal{L}$ is an invariant subspace for $\tilde{T}$. Then since $\mathcal{N}$ and $\tilde{\mathcal{H}}$ are orthogonal in the Hilbert space associated to the fundamental symmetry J, $\mathcal{L} + \mathcal{N}$ is closed. Furthermore, for $f \in \mathcal{L}$, $Tf = \tilde{T}f + P_{\tilde{\mathcal{H}}^\perp}^{\mathcal{H}} Tf \in \mathcal{L} + \mathcal{N}$, since $\mathcal{L} \subseteq \mathcal{N}^\perp$, which is invariant under T, and $P_{\tilde{\mathcal{H}}^\perp}^{\mathcal{H}} \mathcal{N}^\perp = \mathcal{N}$. Thus $\mathcal{L} + \mathcal{N}$ is invariant under T. If $\mathcal{L}$ were neutral, then $\mathcal{L} + \mathcal{N}$ would be neutral, violating the maximality of $\mathcal{N}$ (a similar argument shows that $\tilde{T}^*$ has no neutral invariant subspaces). Clearly, $\mathcal{L} + \mathcal{N}$ is positive or negative whenever $\mathcal{L}$ is. The orthogonal complement of $\mathcal{L} + \mathcal{N}$ in $\mathcal{H}$ is $\mathcal{L}^\perp \cap \mathcal{N}^\perp$. But $\mathcal{L}^\perp$ is orthogonal sum of the orthogonal complement of $\mathcal{L}$ in $\tilde{\mathcal{H}}$ and

$\tilde{\mathcal{H}}^\perp$, while $\mathcal{N}^\perp$ is the orthogonal sum of $\tilde{\mathcal{H}}$ and $\mathcal{N} \subset \tilde{\mathcal{H}}^\perp$. Consequently $(\mathcal{L} + \mathcal{N})^\perp$ is the orthogonal sum of $\mathcal{N}$ and the orthogonal complement of $\mathcal{L}$ in $\tilde{\mathcal{H}}$. If $\mathcal{L}$ is maximal positive in $\tilde{\mathcal{H}}$, then $(\mathcal{L} + \mathcal{N})^\perp$ is a negative subspace. But then $\mathcal{L} + \mathcal{N}$ is maximal positive in $\mathcal{H}$ (see, for example, [5], V.4.4 and V.4.5). The case where $\mathcal{L}$ is maximal negative in $\tilde{\mathcal{H}}$ is handled identically. $\blacksquare$

It is interesting to note in the above theorem that for all complex λ, the kernel of $\tilde{T} - \lambda$ must be strictly definite, since otherwise it it would contain a neutral subspace invariant for $\tilde{T}$, which would lead to a contradiction of the maximality of $\mathcal{N}$.

The proof of the next result is essentially due to Langer (see [17], Satz 3.1), with some minor modifications.

Theorem 17. *Let $T \in \mathbf{B}(\mathcal{H})$ be a selfadjoint operator such that for some complex number λ and positive integer n, $\pm(T - \lambda)^n \geq 0$. Then there exists a maximal dual pair of reducing subspaces for T.*

Proof. Since invariant subspaces for $T - \lambda$ are also invariant for T, we may assume $\lambda = 0$. If $-T^n \geq 0$, then replacing the scalar product on $\mathcal{H}$ by its negative, we get a new Kreĭn space on which $T^n \geq 0$. Since positive and negative subspaces of the new Kreĭn space are negative and positive, respectively, in the old Kreĭn space, the problem reduces to considering the case where $T^n \geq 0$ for some positive integer n.

So suppose this is the case. We first apply Theorem 16 to find a maximal neutral subspace $\mathcal{N}$ and construct $\tilde{T} \in \mathbf{B}(\tilde{\mathcal{H}})$ where $\tilde{T}$ has the property that it is also selfadjoint and $\tilde{T}^n \geq 0$. Furthermore, $\tilde{T}$ has no neutral invariant subspaces, and thus since both $\ker \tilde{T}$ and $(\ker \tilde{T})^\perp$ are invariant, $\ker \tilde{T} \cap (\ker \tilde{T})^\perp = \{0\}$; that is, $\ker \tilde{T}$ and $\operatorname{ran} \tilde{T} = (\ker \tilde{T})^\perp$ are nondegenerate.

Claim that $\overline{\operatorname{ran} \tilde{T}^k} = \overline{\operatorname{ran} \tilde{T}}$ for $k = 1, 2, 3, \ldots$. This is trivial for $k = 1$. Suppose it is true for $k - 1$, where $k > 1$ is fixed. Then since

$$\ker \tilde{T}^k = \ker \tilde{T} \cup \left\{ f \in \tilde{\mathcal{H}} : \tilde{T} f \in \ker \tilde{T}^{k-1} \right\},$$

and $\ker \tilde{T}^{k-1} = (\operatorname{ran} \tilde{T}^{k-1})^\perp = (\operatorname{ran} \tilde{T})^\perp$, it follows that $\ker \tilde{T}^k = \ker \tilde{T}$, and thus $\operatorname{ran} \tilde{T}^k = \operatorname{ran} \tilde{T}$. The claim then follows by induction.

Now consider the case where $\tilde{T}^n \geq 0$ and n is even. Then $\left\langle \tilde{T}^{n/2} f, \tilde{T}^{n/2} f \right\rangle \geq 0$ for all $f \in \tilde{\mathcal{H}}$, and so $\overline{\operatorname{ran} \tilde{T}}$ is a positive subspace. As was noted following the proof of Theorem 16, $\ker \tilde{T}$ is strictly definite. If it is positive, then since $\ker \tilde{T} + \operatorname{ran} \tilde{T}$ is dense in $\tilde{\mathcal{H}}$, $\tilde{\mathcal{H}}$ must be a Hilbert space, which is naturally maximal positive and invariant for $\tilde{T}$. If $\ker \tilde{T}$ is negative, then it is maximal negative since its orthogonal complement, $\overline{\operatorname{ran} \tilde{T}}$, is positive. So likewise we have $\overline{\operatorname{ran} \tilde{T}}$ is maximal positive. In either case, Theorem 16 then gives us a maximal dual pair for T.

If n is odd, then for all $f \in \tilde{\mathcal{H}}$, $g \in \ker \tilde{T}$,

$$\left\langle \tilde{T}(\tilde{T}^{(n-1)/2} f + g), \tilde{T}^{(n-1)/2} f + g \right\rangle \geq 0,$$

and since the vectors $\tilde{T}^{(n-1)/2} f + g$ span a dense subset of $\tilde{\mathcal{H}}$, it follows that $\tilde{T} \geq 0$. By Theorem 15, $\tilde{T}$ has a maximal dual pair $\mathcal{L}_+$, $\mathcal{L}_-$ of invariant subspaces, and by Theorem 16, $\mathcal{L}_+ + \mathcal{N}$ and $\mathcal{L}_- + \mathcal{N}$ are a maximal dual pair of invariant subspaces for T. $\blacksquare$

7. Spectral theory regained

In this section we show how Langer's spectral theory for bounded selfadjoint definitizable operators on Kreĭn spaces can be obtained by means of the methods developed above. There is a close connection between the approach we take and that of Andô [1] for positive operators and Bognár [7] for selfadjoint operators with definitizing polynomials of the form $p(x) = (x - \lambda)^n$, for λ real and n a positive integer. Though it is not done here, the spectral theory for definitizable unitary operators could be derived in a similar manner.

We begin by stating the spectral theorem.

Theorem 18. *Let $T \in \mathbf{B}(\mathcal{H})$ be a selfadjoint operator with definitizing polynomial p, and assume that the set $\mathcal{Z}$ of the zeros of p is contained in the real line $\mathbf{R}$. Let Ω the semi-ring of all Borel subsets of $\mathbf{R}$ the boundaries of which are bounded away from $\mathcal{Z}$. Then there is a unique spectral function $E : \Omega \to \mathbf{B}(\mathcal{H})$ with the following properties:*

(i) For $\Delta \in \Omega$, $E(\Delta) \in \mathbf{B}(\mathcal{H})$ is an orthogonal projection.

(ii) $E(\emptyset) = 0$ and $E(\mathbf{R}) = 1$.

(iii) For $\Delta, \Delta' \in \Omega$, $E(\Delta \cap \Delta') = E(\Delta)E(\Delta')$.

(iv) If $\{\Delta_n\}_{n=1}^{\infty} \subset \Omega$ are pairwise disjoint and $\bigcup_{n=1}^{\infty} \Delta_n \in \Omega$, then

$$E\left(\bigcup_{n=1}^{\infty} \Delta_n\right) = \sum_{n=1}^{\infty} E(\Delta_n).$$

(v) Let $\Delta \in \Omega$ and $\mathcal{E}(\Delta) = \operatorname{ran} E(\Delta)$. If for all $t \in \Delta$ we have $p(t) > 0$, then $\mathcal{E}(\Delta)$ is uniformly positive, and if $p(t) < 0$, then $\mathcal{E}(\Delta)$ is uniformly negative.

(vi) For $\Delta \in \Omega$, $E(\Delta) \in \{T\}''$, the double commutant of T.

(vii) If $\Delta \in \Omega$ and $\mathcal{E}(\Delta) = \operatorname{ran} E(\Delta)$, then $\sigma(T|\mathcal{E}(\Delta)) \subset \overline{\Delta}$.

Moreover, if ϕ is a Borel measurable function which is bounded on $\sigma(T)$, then the integral

$$\int_{\mathcal{X}} \phi(\lambda)p(\lambda)\,E(d\lambda)$$

is a strongly convergent improper integral, where $\mathcal{X} = \mathbf{R} \setminus \mathcal{Z}$. Finally, the operator $\phi(T)p(T)$ may be expressed as

$$\phi(T)p(T) = \int_{\mathcal{X}} \phi(\lambda)p(\lambda)\,E(d\lambda) + \sum_{\nu \in \mathcal{Z}} \phi(\nu)N_{\nu}, \tag{6}$$

where $N_{\nu} \in \mathbf{B}(\mathcal{H})$ is a positive selfadjoint operator, $N_{\nu}^2 = 0$, and $E(\Delta)N_{\nu} = N_{\nu}E(\Delta) = 0$ for every $\Delta \in \Omega$ and $\nu \in \mathcal{Z}$, $\nu \notin \Delta$.

The spectrum of a selfadjoint operator is always symmetric with respect to the real line. It may be assumed (see, for example, [15], Lemma 1 or [18], Lemma 2.1) that the nonreal spectrum of a definitizable selfadjoint operator corresponds to the nonreal zeros of the definitizing polynomial. Also, the spectral subspace associated to a nonreal point in the spectrum is neutral and finite dimensional, and when summed with the corresponding subspace for the conjugate point, yields a regular subspace. Consequently, it is possible to reduce the study of general definitizable selfadjoint operators to the consideration of definitizable selfadjoint operators with real spectrum and definitizing polynomial with real zeros.

Proof of Theorem 18. Let $\mathcal{X}$ be the complement of $\mathcal{Z}$ in $\mathbf{R}$. To begin with, we construct the spectral function E for those subsets $\Delta \in \Omega$ such that Δ either contains the set $\mathcal{Z}$ or is disjoint from it (that is, $\Delta \subset \mathcal{X}$), and prove that properties (i)–(iii), (vi), and (vii) hold for this E. We refer to the collection of these sets as Ω_0. We will then extend the definition of E to all of Ω and prove these properties, as well (iv) and (v), hold in this case.

Since $p(T)$ is a positive operator, by Theorem 5 we may factor it as DD^*, $D \in \mathbf{B}(\mathcal{D}, \mathcal{H})$, $\ker D = \{0\}$, $\mathcal{D}$ a Hilbert space, and furthermore, there exists $L \in \mathbf{B}(\mathcal{D})$ with $TD = DL$. By Theorem 6, L is (Hilbert space) selfadjoint. By the spectral theorem for such operators, it is possible to write L in terms of a spectral function F mapping the σ-algebra of Borel subsets of the real line $\mathbf{R}$ into the bounded orthogonal projections on $\mathcal{D}$ as $L = \int t\,F(dt)$.

Fix $\Delta \in \Omega_0$ and in $\mathcal{X}$. We will show that there exists a orthogonal projection $E(\Delta) \in \mathbf{B}(\mathcal{H})$ with the property that $E(\Delta)D = DF(\Delta)$. Since Δ does not contain a zero of p, we may write $\Delta = \Delta_+ \cup \Delta_-$, where $\Delta_+ \cap \Delta_- = \emptyset$, where for $t \in \Delta_+$ we have $p(t) > 0$, and for $t \in \Delta_-$, $p(t) < 0$.

Let $\mathcal{F}_{\pm} = \operatorname{ran} F(\Delta_{\pm})$ and $\mathcal{F}(\Delta) = \operatorname{ran} F(\Delta)$. The subspaces $\mathcal{F}_+$, $\mathcal{F}_-$ are orthogonal and invariant subspaces for L, so $\mathcal{E}_{\pm} = \overline{D\mathcal{F}_{\pm}}$ are orthogonal invariant subspaces for T by Theorem 12.

We show that $\mathcal{E}_+$, $\mathcal{E}_-$ are uniformly positive and uniformly negative, respectively. The set Δ is bounded away from the zeros of p, and so there is a positive constant γ such that for all $t \in \Delta_+$, $p(t) \geq \gamma$. Fix a fundamental decomposition for $\mathcal{H}$ in order to calculate norms. Then for $f \in \mathcal{F}_+$,

$$\langle Df, Df \rangle = \langle p(L)f, f \rangle \geq \gamma \|f\|^2 \geq \frac{\gamma}{\|D\|^2} \|Df\|^2. \tag{7}$$

It follows that $\mathcal{E}_+$ is uniformly positive. An identical argument shows that $\mathcal{E}_-$ is uniformly negative. Consequently, $\mathcal{E}(\Delta) = \overline{D\mathcal{F}(\Delta)}$ is a regular subspace of $\mathcal{H}$.

Actually, equation (7) tells us a little more. Since $\|Df\|^2 \geq |\langle Df, Df \rangle| \geq \gamma \|f\|^2$, we have in fact that $\mathcal{E}_+ = D\mathcal{F}_+$. Likewise $\mathcal{E}_- = D\mathcal{F}_-$, and thus

$$\mathcal{E}(\Delta) = D\mathcal{F}(\Delta), \qquad \Delta \subset \mathcal{X}. \tag{8}$$

Define an operator $E(\Delta)$ on $\mathcal{H}$ by $E(\Delta)(g + h) = g$, where $g \in \mathcal{E}(\Delta)$ and $h \in \mathcal{E}(\Delta)^\perp$. It is readily verified that $E(\Delta)$ is an everywhere defined and bounded orthogonal projection. Furthermore, by Theorem 12 (iii), $D(\mathcal{D} \ominus \mathcal{F}(\Delta)) \subset \mathcal{E}(\Delta)^\perp$, so for $f = g + h \in \mathcal{D}$, $g \in \mathcal{F}(\Delta)$, $h \in \mathcal{D} \ominus \mathcal{F}(\Delta)$, we have

$$E(\Delta)Df = E(\Delta)Dg = Dg = DF(\Delta)f. \tag{9}$$

Now consider a subset $\Delta' \in \Omega_0$ which is the complement of a subset $\Delta \subset \mathcal{X}$ and define $E(\Delta') = 1 - E(\Delta)$. Then the trivial observation that $1D = D1$ implies that (9) holds for all $\Delta \in \Omega_0$. Furthermore, (ii) is readily seen to hold.

Now let us consider (vi). Suppose $A \in \{T\}'$, the commutant of T. It is sufficient to prove that $E(\Delta)$ commutes with A for $\Delta \in \Omega_0$ with $\Delta \subset \mathcal{X}$. Since A commutes with $p(T)$, by Theorem 5, there exists an operator $B \in \mathbf{B}(\mathcal{D})$ such that $AD = DB$. But $DBL = ATD = TAD = DLB$, and so since $\ker D = \{0\}$, $B \in \{L\}'$. But $F(\Delta) \in \{L\}''$, hence

$$E(\Delta)AE(\Delta)D = DF(\Delta)BF(\Delta) = DBF(\Delta) = AE(\Delta)D.$$

By (8),

$$E(\Delta)AE(\Delta) = AE(\Delta). \tag{10}$$

Since T is selfadjoint, $A^* \in \{T\}'$ as well. Thus $E(\Delta)A^*E(\Delta) = A^*E(\Delta)$. Taking adjoints in this equation and comparing with (10) yields the result. In particular then, for $\Delta, \Delta' \in \Omega_0$, $E(\Delta)E(\Delta') = E(\Delta')E(\Delta)$.

We show next that if $\Delta, \Delta' \in \Omega_0$ and $\Delta \cap \Delta' = \emptyset$, then

$$E(\Delta \cup \Delta') = E(\Delta) + E(\Delta'). \tag{11}$$

For $\Delta, \Delta' \subset \mathcal{X}$,

$$E(\Delta \cup \Delta')D = DF(\Delta \cup \Delta') = D(F(\Delta) + F(\Delta')) = (E(\Delta) + E(\Delta'))D.$$

But (8) implies ran $E(\Delta)$, ran $E(\Delta')$, and ran $E(\Delta \cup \Delta')$ are contained in ran D, and so (11) follows by continuity in this case.

The only other possibility is for $\Delta, \Delta'^c \subset \mathcal{X}$, where again $\Delta \cap \Delta' = \emptyset$. Then

$$\begin{aligned}
E(\Delta') &= 1 - E(\Delta'^c) = 1 - E(\Delta \cup (\Delta \cup \Delta')^c) \\
&= 1 - E(\Delta) - E((\Delta \cup \Delta')^c) = 1 - E(\Delta) - (1 - E(\Delta \cup \Delta')) \\
&= -E(\Delta) + E(\Delta \cup \Delta'),
\end{aligned}$$

and (11) is seen to hold in this case as well.

We now prove (iii) for $\Delta, \Delta' \in \Omega_0$. First note that if $\Delta_1, \Delta_2 \in \Omega_0$ and $\Delta_1 \cap \Delta_2 = \emptyset$, then using (11), we have

$$\begin{aligned}
E(\Delta_1 \cup \Delta_2) &= E(\Delta_1 \cup \Delta_2)^2 = E(\Delta_1) + E(\Delta_2) + 2E(\Delta_1)E(\Delta_2) \\
&= E(\Delta_1 \cup \Delta_2) + 2E(\Delta_1)E(\Delta_2),
\end{aligned}$$

and so $E(\Delta_1)E(\Delta_2) = 0$. If on the other hand, $\Delta_3, \Delta_4 \in \Omega_0$ and $\Delta_3 \subseteq \Delta_4$ then

$$E(\Delta_3)E(\Delta_4) = E(\Delta_3)(1 - E(\Delta_4)) = E(\Delta_3).$$

And so it follows that

$$E(\Delta)E(\Delta') = E(\Delta \cap \Delta')E(\Delta') + E(\Delta \cap \Delta'^c)E(\Delta') = E(\Delta \cap \Delta').$$

For (vii), we show that if $\Delta \in \Omega_0$ and $\Delta \subset \mathcal{X}$, then

$$\sigma(T|\mathrm{ran}\, E(\Delta)) = \sigma(L|\mathrm{ran}\, F(\Delta)) \subset \overline{\Delta} \cap \sigma(T). \tag{12}$$

In the proof of this part, we make use of the following simple observation: If A is a continuous selfadjoint operator on a Kreĭn space $\mathcal{K}$ and if $\mathcal{N}$ is a regular subspace that reduces A, then

$$\sigma(A) = \sigma(A|\mathcal{N}) \cup \sigma(A|\mathcal{K} \ominus \mathcal{N}) \supset \sigma(A|\mathcal{N}) \tag{13}$$

(just note that there is a fundamental symmetry J on $\mathcal{K}$ that leaves $\mathcal{N}$ invariant, and so $\mathcal{N}$ and $\mathcal{K} \ominus \mathcal{N}$ are Hilbert space orthogonal in $\mathcal{K}_J$).

Write $\mathcal{E}(\Delta)$ for ran $E(\Delta)$ and $\mathcal{F}(\Delta)$ for ran $F(\Delta)$. Define $G \in \mathbf{B}(\mathcal{F}(\Delta), \mathcal{E}(\Delta))$ by $G = D|\mathcal{F}(\Delta)$. By (8), G has a bounded inverse. Hence the equation $T|\mathcal{E}(\Delta)G = GL|\mathcal{F}(\Delta)$ is a similarity relation, and consequently $\sigma(T|\mathcal{E}(\Delta)) = \sigma(L|\mathcal{F}(\Delta))$, which is contained in $\overline{\Delta}$ as a consequence of the spectral theory for selfadjoint operators on a Hilbert space. Applying (13) yields (12).

To finish off (vii), we must now consider the case where $\Delta \in \Omega_0$ and the complement of Δ is in $\mathcal{X}$. In this situation, Δ contains the zeros of p. Let $\mathcal{E}(\Delta) = \operatorname{ran} E(\Delta)$. Using the fact that if $A, B \in \mathbf{B}(\mathcal{H})$, then $\sigma(AB) \subseteq \sigma(BA) \cup \{0\}$, we find

$$
\begin{aligned}
\sigma(p(T)|\mathcal{E}(\Delta)) &= \sigma(p(T)E(\Delta)|\mathcal{E}(\Delta)) \\
&\subseteq \sigma(p(T)E(\Delta)) = \sigma(DF(\Delta)D^*) \\
&\subseteq \{0\} \cup \sigma(F(\Delta)D^*D) = \{0\} \cup \sigma(F(\Delta)p(L)) \\
&\subseteq p(\overline{\Delta}).
\end{aligned}
\tag{14}
$$

Here again we have made use of (13).

Let $z \notin \overline{\Delta}$. We will show that $z \in \rho(T|\mathcal{E}(\Delta))$. Since $p(z) \neq 0$ and p is continuous, we can find $\Delta', \Delta'' \in \Omega_0$, for which $\Delta' \subset \mathcal{X}$, $\Delta' \cap \Delta'' = \emptyset$, $\Delta' \cup \Delta'' = \overline{\Delta}$, and $|p(t)| < \frac{1}{2}|p(z)|$ for $t \in \Delta''$. It follows by the same set of containments as in (14) with Δ'' rather than Δ, that $p(z) \notin \sigma(p(T)|\mathcal{E}(\Delta''))$. By the Spectral Mapping Theorem, $\sigma(p(T)|\mathcal{E}(\Delta'')) = p(\sigma(T|\mathcal{E}(\Delta'')))$, and so $z \notin \sigma(T|\mathcal{E}(\Delta''))$. But by the the previous arguments, $\sigma(T|\mathcal{E}(\Delta')) \subseteq \overline{\Delta'} \subset \overline{\Delta}$, and so $z \notin \sigma(T|\mathcal{E}(\Delta'))$. The spaces $\mathcal{E}(\Delta')$ and $\mathcal{E}(\Delta'')$ are regular and orthogonal, and so are Hilbert space orthogonal in one of the Hilbert spaces associated to $\mathcal{H}$. Thus

$$
\sigma(T|\mathcal{E}(\Delta)) = \sigma(T|\mathcal{E}(\Delta')) \cup \sigma(T|\mathcal{E}(\Delta'')) \subseteq \overline{\Delta},
$$

finishing the proof of (vii) when $\Delta \in \Omega_0$.

We now extend the function E to all $\Delta \in \Omega$. If Δ is an arbitrary member of Ω, there is a closed set Δ' consisting of a finite number of intervals in Ω such that $\Delta' \cap \overline{\Delta} = \emptyset$ and $\tilde{\Delta} = \Delta \cup \Delta' \supset \mathcal{Z}$ since the boundary of Δ is bounded away from $\mathcal{Z}$. Let $\tilde{T} = T|\mathcal{E}(\tilde{\Delta})$. By what we have just shown, $\sigma(\tilde{T}) \subseteq \tilde{\Delta} = \overline{\Delta} \cup \Delta'$. Hence $\sigma(\tilde{T})$ is contained in two disjoint pieces, one in $\overline{\Delta}$ and the other in Δ'. Now consider a finite collection of positively oriented circle $\{\gamma_k\}$ with center on the real line chosen so that $\overline{\Delta}$ is in the interior of the set of disks bounded by these circles and Δ' is outside of all of the circles. Using the Riesz functional calculus (see, eg., [8], VII.4) we define an idempotent $E(\Delta) = (2\pi i)^{-1} \sum_k \int_{\gamma_k} (z - \tilde{T})^{-1} dz$. Since $\tilde{T}$ is selfadjoint, it is not difficult to see that $E(\Delta)$ is also selfadjoint. Hence it is a bounded orthogonal projection and its range, $\mathcal{E}(\Delta)$ is a regular subspace of $\mathcal{E}(\tilde{\Delta})$. Furthermore, $\sigma(T|\mathcal{E}(\Delta)) \subseteq \overline{\Delta}$, so (vii) holds in general. It is useful to note that $E(\tilde{\Delta}) = E(\Delta) + E(\Delta')$, and indeed, if Δ' is the disjoint union of closed sets Δ_1 and Δ_2, then $E(\tilde{\Delta}) = E(\Delta) + E(\Delta_1) + E(\Delta_2)$, where the sum is orthogonal (that is, the ranges of the projections involved are orthogonal, or equivalently, the product of any two of the projections is zero). Observe that for $z \in \gamma_k$, $(z - \tilde{T})^{-1}$ commutes with $p(T)$, and so $E(\Delta)D = D(2\pi i)^{-1} \sum_k \int_\gamma (z - L)^{-1} dz\, F(\tilde{\Delta}) = DF(\Delta)$, since L is Hilbert space selfadjoint.

We need to show that the definition of $E(\Delta)$ is independent of the choice of $\tilde{\Delta}$. So let Δ'' be another closed subset of the real line with the same properties as Δ', and set $\tilde{\Delta}' = \Delta \cup \Delta''$. We may assume that $\tilde{\Delta}' \supset \tilde{\Delta}$ (and consequently $\Delta'' \supset \Delta'$), since the general case then follows by showing that $\tilde{\Delta}$ and $\tilde{\Delta}'$ define the same projection as $\tilde{\Delta} \cap \tilde{\Delta}'$, and hence as each other. By (11), $E(\tilde{\Delta}') = E(\tilde{\Delta}) + E(\tilde{\Delta}^c \cap \tilde{\Delta}')$, where this sum is orthogonal. Since the ranges of these operators reduce T, we have $T|\mathcal{E}(\tilde{\Delta}') = T|\mathcal{E}(\tilde{\Delta}) \oplus T|\mathcal{E}(\tilde{\Delta}' \cap \tilde{\Delta}^c)$. If z is in the resolvent of $T|\mathcal{E}(\tilde{\Delta}')$, it is in the resolvent of its components, and

$$(T|\mathcal{E}(\tilde{\Delta}') - z)^{-1} = \begin{pmatrix} (T|\mathcal{E}(\tilde{\Delta}) - z)^{-1} & 0 \\ 0 & (T|\mathcal{E}(\tilde{\Delta}' \cap \tilde{\Delta}^c) - z)^{-1} \end{pmatrix}.$$

But since $\sigma(T|\mathcal{E}(\tilde{\Delta}' \cap \tilde{\Delta}^c)) \subseteq \overline{\tilde{\Delta}' \cap \tilde{\Delta}^c} \subset \Delta''$, it follows that we get the same projection for $E(\Delta)$ using either $\tilde{\Delta}$ or $\tilde{\Delta}'$.

Viewing $\tilde{T}$ as an operator on all of $\mathcal{H}$, we have for $A \in \{T\}'$, $\tilde{T}A = TE(\tilde{\Delta})A = ATE(\tilde{\Delta}) = A\tilde{T}$, since $E(\tilde{\Delta}) \in \{T\}''$. But $E(\Delta) \in \{\tilde{T}\}''$, and so $E(\Delta) \in \{T\}''$; that is, (vi) holds for all $\Delta \in \Omega$.

We will prove (iii) is valid for sets in Ω by first proving (11) holds for nonintersecting sets in Ω. To begin with, assume that $\Delta = \Delta_1 \cup \Delta_2 \in \Omega$ where $\Delta_1 \cap \Delta_2$ and $\Delta_2 \cap \mathcal{Z}$ are both empty. Fix $\Delta' \in \Omega$ a closed set consisting of a finite number of intervals such that $\overline{(\Delta_1 \cup \Delta_2)} \cap \Delta' = \emptyset$ and $\Delta_1 \cup \Delta_2 \cup \Delta' \supset \mathcal{Z}$. Then since $\mathcal{Z} \subset \Delta_1 \cup \Delta' \in \Omega_0$ and $\Delta_2 \in \Omega_0$, it follows that

$$E(\Delta_1 \cup \Delta_2 \cup \Delta') = E(\Delta_1 \cup \Delta') + E(\Delta_2) = E(\Delta_1 \cup \Delta_2) + E(\Delta')$$

and

$$E(\Delta_1 \cup \Delta') = E(\Delta_1) + E(\Delta').$$

Combining these equations gives $E(\Delta_1 \cup \Delta_2) = E(\Delta_1) + E(\Delta_2)$, where the sum is orthogonal.

Now consider the same situation as in the last paragraph, but this time do not assume that $\Delta_2 \cap \mathcal{Z} = \emptyset$. Choose Δ' as before, and set $\mathcal{Z}_i = \mathcal{Z} \cap \Delta_i$, $i = 1, 2$. Since the boundary of Δ_i is bounded away from $\mathcal{Z}$, there exists a closed set $K_i \subseteq \Delta_i$ with $\mathcal{Z}_i \subset K_i$, $i = 1, 2$. Let $\Sigma_i = \Delta_i \cap K_i^c$. Note that $\Sigma_i \cap \mathcal{Z} = \emptyset$, and so $\Sigma_i \in \Omega_0$, $i = 1, 2$. Hence by (11), $E(\Sigma_1 \cup \Sigma_2) = E(\Sigma_1) + E(\Sigma_2)$. By what was shown in the last paragraph, $E(\Delta_i) = E(K_i) + E(\Sigma_i)$, $i = 1, 2$. Consequently,

$$E(\tilde{\Delta}) \;=\; E(\Delta_1 \cup \Delta_2) + E(\Delta')$$

$$
\begin{aligned}
&= E(K_1 \cup K_2) + E(\Sigma_1 \cup \Sigma_2) + E(\Delta') \\
&= E(K_1) + E(K_2) + E(\Sigma_1) + E(\Sigma_2) + E(\Delta') \\
&= E(\Delta_1) + E(\Delta_2) + E(\Delta').
\end{aligned}
$$

Thus $E(\Delta_1 \cup \Delta_2) = E(\Delta_1) + E(\Delta_2)$, or (11) holds.

Using (11) we find that $E(\Delta^c) = 1 - E(\Delta)$ for $\Delta \in \Omega$. Property (iii) then follows for sets in Ω by the same arguments that were used to prove (iii) for sets in Ω_0.

We next show that (v) holds. Suppose that $p(t) > 0$ for all $t \in \Delta$, $\Delta \in \Omega$. Then $\Delta \subset \mathcal{X}$ and consequently Δ (and so $\overline{\Delta}$) is bounded away from $\mathcal{Z}$. By continuity, $p(t)$ achieves a minimum value γ on $\overline{\Delta}$ which is greater than zero. Choose a fundamental decomposition on $\mathcal{H}$ for computing norms. Then (7) holds, proving that ran $E(\Delta)$ is uniformly positive. The other case is handled in an identical manner.

Next consider (iv). Suppose $\{\Delta_n\}_{n=1}^\infty \subset \Omega$ are pairwise disjoint and $\bigcup_{n=1}^\infty \Delta_n \in \Omega$. Since $\mathcal{Z}$ is a finite set, at most finitely many Δ_n have $\Delta_n \cap \mathcal{Z} \neq \emptyset$. Applying (11) finitely many times, we have

$$
E\left(\bigcup_{n=1}^\infty \Delta_n\right) = \sum_{n=1}^N E(\Delta_n) + E\left(\bigcup_{n=N+1}^\infty \Delta_n\right),
$$

where $\Delta_n \cap \mathcal{Z} = \emptyset$ for all $n \geq N+1$, and where naturally the term involving the summation is zero if $N = 0$.

Consequently, it now suffices to show that (iv) holds in the case that all the sets Δ_n, and hence $\Delta = \bigcup_n \Delta_n$, are contained in $\mathcal{X}$. For each n we write $\Delta_n = \Delta_{n+} + \Delta_{n-}$, where $\Delta_{n+} \cap \Delta_{n-} = \emptyset$ and $p(t) > 0$ for $t \in \Delta_{n+}$, $p(t) < 0$ for Δ_{n-}. Set $\Delta_+ = \bigcup_n \Delta_{n+}$, and $\Delta_- = \bigcup_n \Delta_{n-}$. Observe that $\Delta = \Delta_+ \cup \Delta_-$.

Since Δ is bounded away from $\mathcal{Z}$, the same is obviously true for $\Delta_\pm$. But then by (v), ran $E(\Delta_+)$ is uniformly positive and ran $E(\Delta_-)$ is uniformly negative. By (iii) they are orthogonal. Let $\mathcal{H} = \mathcal{H}_+ \oplus \mathcal{H}_-$ be a fundamental decomposition with ran $E(\Delta_\pm) \subseteq \mathcal{H}_\pm$. Write $\mathcal{H}_J$ for the associated Hilbert space. For any n, ran $E(\Delta_{n+}) \subseteq$ ran $E(\Delta_+)$. Consequently $E(\Delta_{n+})$ is Hilbert space selfadjoint in $\mathcal{H}_J$. Likewise for $E(\Delta_{n-})$, and since $\Delta_{n+} \cap \Delta_{n-} = \emptyset$, $E(\Delta_n) = E(\Delta_{n+}) + E(\Delta_{n-})$ is also a Hilbert space projection. Hence $\|E(\Delta_n)\|$ is uniformly bounded in $\mathcal{H}_J$. In addition, ran $E(\Delta_n)$ is Hilbert space orthogonal to ran $E(\Delta_m)$ for $m \neq n$.

Since F is a Hilbert space spectral measure, we know that $\sum_{n=1}^N F(\Delta_n)$ converges to $F(\sum_{n=1}^\infty \Delta_n)$ in the strong operator topology. By (8) we may write any $f \in \mathcal{H}$ as $f = Dg + h$ where $g \in$ ran $F(\Delta)$ and $h \in \ker E(\Delta)\}$. Then we have

$$
\left\|\left(\sum_{n=1}^N E(\Delta_n) - E(\Delta)\right)(Dg + h)\right\| = \left\|\left(\sum_{n=1}^N E(\Delta_n) - E(\Delta)\right) Dg\right\|
$$

$$= \left\| D \left(\sum_{n=1}^{N} F(\Delta_n) - F(\Delta) \right) g \right\|$$

$$\leq \|D\| \left\| \left(\sum_{n=1}^{N} F(\Delta_n) - F(\Delta) \right) g \right\|.$$

Consequently, $\sum_{n=1}^{N} E(\Delta_n)$ converges to $E(\Delta)$ in the strong operator topology, proving (iv).

We now show that the spectral family is uniquely determined by properties (i)–(vii). Let $\tilde{E}(\Delta)$ be another family of projections satisfying (i)–(vii). Consider closed sets $\Delta, \Delta' \in \Omega$. By (vi) $\tilde{E}(\Delta')$ commutes with T and $E(\Delta)$. Thus $\tilde{E}(\Delta')E(\Delta)$ is an orthogonal projection, and in particular, the range is regular. Likewise for $\tilde{E}(\Delta')E(\Delta^c)$, $\tilde{E}(\Delta'^c)E(\Delta)$, and $\tilde{E}(\Delta'^c)E(\Delta^c)$. Since $1 = E(\Delta) + E(\Delta^c) = \tilde{E}(\Delta') + \tilde{E}(\Delta'^c)$, we may write

$$E(\Delta)\mathcal{H} = E(\Delta)\tilde{E}(\Delta')\mathcal{H} \oplus E(\Delta)\tilde{E}(\Delta'^c)\mathcal{H}$$

and

$$\tilde{E}(\Delta)\mathcal{H} = E(\Delta)\tilde{E}(\Delta')\mathcal{H} \oplus E(\Delta^c)\tilde{E}(\Delta')\mathcal{H}.$$

Then by (13),

$$\sigma(T|\tilde{E}(\Delta')\mathcal{H}) = \sigma(T|\tilde{E}(\Delta')E(\Delta^c)\mathcal{H}) \cup \sigma(T|\tilde{E}(\Delta')E(\Delta)\mathcal{H})$$

and

$$\sigma(T|E(\Delta)\mathcal{H}) = \sigma(T|E(\Delta)\tilde{E}(\Delta'^c)\mathcal{H}) \cup \sigma(T|E(\Delta)\tilde{E}(\Delta')\mathcal{H}).$$

Consequently by (vii),

$$\sigma(T|\tilde{E}(\Delta')E(\Delta)\mathcal{H}) \subset \sigma(T|\tilde{E}(\Delta')\mathcal{H}) \cap \sigma(T|E(\Delta)\mathcal{H}) \subseteq \Delta' \cap \Delta = \emptyset.$$

Since $\mathcal{H}$ was assumed to be over the field of complex numbers, this implies that $E(\Delta)\tilde{E}(\Delta') = 0$.

Next write Δ^c as the union of closed subsets Δ_n contained in Ω. Let $\Delta'_1 = \Delta_1$ and

$$\Delta'_n = \Delta_n \cap \left(\bigcup_{j=1}^{n-1} \Delta_j \right)^c = \Delta_n \cap \left(\bigcap_{j=1}^{n-1} \Delta_j^c \right), \qquad n = 2, 3, \ldots. \tag{15}$$

By property (iii), $E(\Delta)\tilde{E}(\Delta'_n) = E(\Delta)\tilde{E}(\Delta_n \cap \Delta'_n) = E(\Delta)\tilde{E}(\Delta_n)\tilde{E}(\Delta'_n) = 0$. By construction, $\Delta'_n \cap \Delta'_m = \emptyset$ when $m \neq n$ and $\Delta^c = \bigcup_n \Delta'_n$. Using property (iv), we find that $E(\Delta)\tilde{E}(\Delta^c) = 0$. Hence

$$\tilde{E}(\Delta)E(\Delta) = (1 - \tilde{E}(\Delta^c))E(\Delta) = E(\Delta).$$

Interchanging the roles of E and $\tilde{E}$ in the above arguments also gives $\tilde{E}(\Delta)E(\Delta) = \tilde{E}(\Delta)$, and thus $\tilde{E}(\Delta) = E(\Delta)$ for closed $\Delta \in \Omega$.

Let Γ be the collection of subsets Δ in Ω for which $\tilde{E}(\Delta) = E(\Delta)$ is true. Then for $\Delta \in \Gamma$, $\tilde{E}(\Delta^c) = 1 - \tilde{E}(\Delta) = 1 - E(\Delta) = E(\Delta^c)$, meaning that Δ^c is also in Γ. Next suppose that $\{\Delta_n\}$ is a countable collection of subsets of Γ such that $\bigcup_n \Delta_n \in \Omega$. Let $\Delta_1' = \Delta_1$ and define Δ_n' as in (15). Then $\bigcup_n \Delta_n' = \bigcup_n \Delta_n$. By (iii), $\tilde{E}(\Delta_n') = E(\Delta_n')$, and thus by (iv), $\tilde{E}\left(\bigcup_n \Delta_n\right) = E\left(\bigcup_n \Delta_n\right)$. Since Γ contains all of the closed subsets of Ω, we have $\Gamma = \Omega$, proving that $E(\cdot)$ is uniquely defined.

Let ϕ be a Borel measurable function which is bounded on $\sigma(T)$. By assumption, $p(T) = DD^*$. We define

$$\phi(T)p(T) = D\phi(L)D^*.$$

By Proposition 7, $\sigma(L) \subseteq \sigma(T)$. Then by the standard spectral theory (see, for example, Theorem IX.2.3 of [8]), $\phi(L)$ is uniquely defined in terms of the spectral measure of L, and so $\phi(T)p(T)$ is uniquely defined since $\ker D = \{0\}$.

Next we show that equation (6) in the statement of the proof is valid. Let $\Delta = \Delta_+ \cup \Delta_-$ in Ω bounded and closed and contained in $\mathcal{X}$, Δ_+ and Δ_- disjoint, and $p(t)$ strictly positive or negative, depending on whether t is in Δ_+ or Δ_-. Note that the sets $\Delta_\pm$ are also bounded and closed. Define $E_\pm(\Gamma_\pm) = E(\Gamma_\pm)$ for $\Gamma_\pm$ in the σ-algebra of subsets contained in $\Delta_\pm$. Since each such $\Gamma_\pm$ is necessarily in Ω, this is well-defined. By proper choice of fundamental symmetry and (v), $E_\pm(\cdot)$ is a Hilbert space spectral measure. By (12), $\sigma(T|E(\Delta_\pm)) \subset \sigma(T)$, and consequently the integrals $\int_{\Delta_\pm} \phi(\lambda)p(\lambda)\,E_\pm(d\lambda)$ are well defined with

$$\int_{\Delta_+} \phi(\lambda)p(\lambda)\,E_+(d\lambda) = \phi(T)p(T)E(\Delta_+)$$

and

$$\int_{\Delta_-} \phi(\lambda)p(\lambda)\,E_-(d\lambda) = \phi(T)p(T)E(\Delta_-).$$

Thus

$$\begin{aligned}
\int_\Delta \phi(\lambda)p(\lambda)\,E(d\lambda) &= \phi(T)p(T)E(\Delta) = \phi(T)DD^*E(\Delta) = D\phi(L)F(\Delta)D^* \\
&= D\int_\Delta \phi(\lambda)\,F(d\lambda)\,D^*.
\end{aligned}$$

Letting Δ represent a member of a nested sequence of sets in Ω contained in $\mathcal{X}$ and converging to $\mathcal{X}$, it is clear that the right hand entry, converges in the strong operator topology, and that this implies the same for the left hand side.

Consider the operators $N_\nu = DF(\nu)D^*$ for $\nu \in \mathcal{Z}$. These are obviously bounded and positive, $N_\nu^2 = DF(\nu)p(L)F(\nu)D^* = 0$ by the Spectral Mapping Theorem, and

$$E(\Delta)N_\nu = DF(\Delta)F(\nu)D^* = 0, \qquad \Delta \in \Omega, \ \nu \notin \Delta.$$

Likewise $N_\nu E(\Delta) = 0$ for such sets. Furthermore,

$$
\begin{aligned}
D \sum_{\nu \in \mathcal{Z}} \phi(\nu) F(\nu) D^* &= D \int_{\mathbf{R} \backslash \mathcal{X}} \phi(\lambda)\, F(d\lambda) D^* \\
&= D \int_{\mathbf{R}} \phi(\lambda) F(d\lambda) D^* - \int_{\mathcal{X}} \phi(\lambda) p(\lambda)\, E(d\lambda) \\
&= D \phi(L) D^* - \int_{\mathcal{X}} \phi(\lambda) p(\lambda)\, E(d\lambda) \\
&= \phi(T) p(T) - \int_{\mathcal{X}} \phi(\lambda) p(\lambda)\, E(\lambda),
\end{aligned}
$$

proving (6). ∎

It is interesting to note that the integrals (and so the functional calculus) in the statement of Theorem 18 only depend on the spectral function E being defined on the Borel subsets Ω_0 of $\mathbf{R}$, which were the sets in Ω that either contain $\mathcal{Z}$ (the set of zeros of the definitizing polynomial p) or have empty intersection with that set. However the functional calculus given in this way is not the best possible. For example, any polynomial of the operator could also be added (and not just those which are divisible by the definitizing polynomial).

Another factor that should be considered in trying to extend the functional calculus is that the definitizing polynomial is not unique, and so we might be able to do better if a polynomial of lower degree can be found. To see this, let K be the smallest closed interval of the real line containing $\sigma(T)$ (which as in Theorem 18 we assume is real). If $\lambda \notin \sigma(T)$ is a root of the definitizing polynomial, and $p(z) = (z - \lambda)^k q(z)$ where λ is not a root of q, then it is not difficult to see that we may replace k by 1 if k is odd and 0 if k is even.

We can go further in the case where $k = 1$, and additionally, $\lambda \notin K$. Write $q(T) = (T - \lambda)^{-1} p(T)$. By the spectral mapping theorem, every point in the spectrum of $(T - \lambda)^{-1}$ is bounded away to the right (or left) of 0. Assume the former, since the latter could likewise be handled by replacing $q(T)$ and $(T - \lambda)^{-1}$ by their negatives. Then $(T - \lambda)^{-1}$ has a square root commuting with $p(T)$ by the Riesz functional calculus. It is not hard to deduce from this then that there is a definitizing polynomial with no roots outside of K. See [18] for details.

Finally, a more refined analysis shows that any Borel measurable function which is bounded on $\sigma(T)$ and satisfies a kind of differentiability condition at those zeros of the definitizing polynomial that are in the spectrum of the operator T may also be included in the functional calculus [15].

Even though it is not needed for our functional calculus, it is useful to have E defined on all of Ω in order to prove the existence of a maximal dual pair of invariant subspaces. We sketch here a proof due to Langer ([17], Satz 3.2). First, split the real line into disjoint intervals, each containing one of the zeros of p in its interior.

If there is only one zero, we are done, since Theorem 17 applies directly. Otherwise the sets we have are members of Ω (and clearly not in Ω_0). The spectral projections associated with each of these intervals give rise to orthogonal reducing subspaces for T, the sum of which is all of $\mathcal{H}$. It suffices to show that T restricted to any one of these subspaces has a maximal dual pair (in that subspace). Let $\mathcal{H}_i$ be one of these subspaces, $T_i = T|\mathcal{H}_i$, and λ the zero of p in the associated interval. Write $p(z) = (z-\lambda)^n q(z)$, where λ is not a zero of q. Obviously p is a definitizing polynomial for T_i. Furthermore, by (vii) of Theorem 18, the roots of q are outside of the smallest closed interval containing the spectrum of T_i. Applying the argument given in the last paragraph, we may take either $(z - \lambda)^n$ or $-(z - \lambda)^n$ as the definitizing polynomial for T_i. Theorem 17 then gives us the desired result, which is stated below.

Theorem 19. *Let $T \in \mathbf{B}(\mathcal{H})$ be a definitizable selfadjoint operator. Then T has a maximal dual pair of invariant subspaces.*

Clearly the full force of Theorem 18 is not used in the proof of Theorem 19. In fact we really only need the definition of $E(\Delta)$ for Δ in Ω but not in Ω_0. This in turn uses property (vii) for sets in Ω_0, the proof of which depends on everything up through equation (11).

8. Appendix

In this appendix we show that the following two lemmas are equivalent.

Lemma 4. *(Version I). Let $\mathcal{H}$ be a Hilbert space, $C, X \in \mathbf{B}(\mathcal{H})$. If C is nonnegative and CX is selfadjoint, then for all $f \in \mathcal{H}$,*

$$|\langle CXf, f\rangle| \leq \|X\|\langle Cf, f\rangle.$$

Lemma 4. *(Version II). Let $\mathcal{H}$ be a Hilbert space, $A, B \in \mathbf{B}(\mathcal{H})$ positive selfadjoint operators. Then $A^2 \leq B^2$ implies $A \leq B$.*

Assume first that Version I holds and that $B, C \in \mathbf{B}(\mathcal{H})$ are nonnegative operators such that $B^2 \leq C^2$. By the Douglas-Shmul'yan Lemma, there exists an operator $G \in \mathbf{B}(\mathcal{H})$ with $\|G\| \leq 1$ such that $B = CG$. Identify C with A and G with X in Version I to conclude that $\langle Bf, f\rangle \leq \langle Cf, f\rangle$ for all $f \in \mathcal{H}$; or $B \leq C$.

Conversely, assume Version II holds, $C, X \in \mathbf{B}(\mathcal{H})$, with $C \geq 0$ and CX self-adjoint. Since $\operatorname{ran} C \supseteq \operatorname{ran} CX$, again by the Douglas-Shmul'yan Lemma, $C^2 \geq$

$(1/\|X\|^2)CXX^*C = |(1/(\|X\|)CX|^2$. By Version II then, $C \geq |1/(\|X\|)CX|$. Equivalently, for all $f \in \mathcal{H}$,

$$\frac{1}{\|X\|}|\langle CXf, f\rangle| \leq \frac{1}{\|X\|}\langle|CX|f, f\rangle \leq \langle Cf, f\rangle.$$

Curiously, in the proof that Version II implies Version I, the fact that CX is selfadjoint is used merely to give $|\langle CXf, f\rangle| \leq \langle|CX|f, f\rangle$, which would also be true under milder restrictions; for example, CX normal.

References

[1] T. Andô, *Linear Operators on Kreĭn Spaces*, Hokkaido University, Research Institute of Applied Electricity, Division of Applied Mathematics, Sapporo, 1979. MR 81e:47032.

[2] Gr. Arsene, T. Constantinescu, and A. Gheondea, *Lifting of operators and prescribed numbers of negative squares*, Michigan J. Math. **34** (1987), 201–216. MR 88j:47008.

[3] Gr. Arsene and A. Gheondea, *Completing matrix contractions*, J. Operator Theory **7** (1982), 179–189. MR 83i:4710.

[4] T.Ya. Azizov and I.S. Iokhvidov, *Foundations of the Theory of Linear Operators in Spaces with Indefinite Metric*, "Nauka", Moscow, 1986; English transl. *Linear Operators in Spaces with Indefinite Metric*, Wiley, New York, 1989. MR 88g:47070.

[5] J. Bognár, *Indefinite Inner Product Spaces*, Springer, Berlin-New York, 1974. MR 57#7125.

[6] J. Bognár, *A proof of the spectral theorem for J-positive operators* Acta Sci. Math. (Szeged) **45** (1983), 75–80. MR 85a:47041

[7] J. Bognár, *An approach to the spectral decomposition of J-positizable operators*, J. Operator Theory **17** (1987), 309–326. MR 88g:47071

[8] J.B. Conway, *A course in functional analysis*, Graduate Texts in Mathematics, 96, Springer-Verlag, New York-Berlin, 1985. MR 86h:46001

[9] C.C. Cowen and S. Li, *Hilbert space operators that are subnormal in the Kreĭn space sense*, J. Operator Theory **20** (1988), 165–181. MR 90b:47063

[10] M.A. Dritschel, *Extension Theorems for Operators on Kreĭn Spaces*, Dissertation, University of Virginia, 1989.

[11] M.A. Dritschel, *The essential uniqueness property for operators on Kreĭn spaces*, J. Funct. Anal., to appear.

[12] M.A. Dritschel and J. Rovnyak, *Extension theorems for contraction operators on Kreĭn spaces*, in *Extension and Interpolation of Linear Operators and Matrix Functions*, Oper. Theory: Adv. Appl., Vol. 47, pp. 221–305, Birkhäuser, Basel-Boston, 1990. MR 92m:47068

[13] M. Gebel, *Spektraleigenschaften Adjungierbarer Operatoren*, Habilitationsschrift, Martin-Luther-Universität, Halle-Wittenberg, 1992.

[14] A. Gheondea, *Canonical forms of unbounded unitary operators in Kreĭn spaces*, Publ. Res. Inst. Math. Sci. **24** (1988), 205–224. MR 90a:47087.

[15] P. Jonas, *On the functional calculus and the spectral function for definitizable operators in Kreĭn space*, Beiträge Anal. **16** (1981), 121–135. MR 83i:47047

[16] H. Langer, *Eine Verallgemeinerung eines Satzes von L.S. Pontrjagin*, Math. Ann. **152** (1963), 434–436.

[17] H. Langer *Invariante Teilräume definisierbarer J-selbstadjungierter Operatoren*, Ann. Acad. Sci. Fennicae, Series A I **475** (1971), 1–23.

[18] H. Langer, *Spectral functions of definitizable operators in Krein spaces*, *Functional Analysis (Dubrovnik, 1981)*, pp. 1–46, Lecture Notes in Math., Vol. 948, Springer, Berlin-New York, 1982. MR 84g:47034

[19] G.J. Murphy, *C*-Algebras and Operator Theory*, Academic Press, San Diego, 1990.

[20] R.S. Phillips, *The extension of dual subspaces invariant under an algebra*, in *Proc. Internat. Sympos. Linear Spaces (Jerusalem, 1960)*, pp. 366–398, Jerusalem Academic Press, Jerusalem, Pergamon, Oxford, 1961. MR 24#A3512

[21] V. Pták and P. Vrbová, *Lifting intertwining relations*, Integral Equations and Operator Theory **11** (1988), 128–147.

[22] M. Rosenblum and J. Rovnyak, *Hardy Classes and Operator Theory*, Oxford Math. Monographs, Oxford University Press, New York, 1985. MR 87e:47001.

[23] Agnes Yang, *A Construction of Krein spaces of Analytic Functions*, Dissertation, Purdue University, 1990.

Michael A. Dritschel

Department of Mathematics
College of William and Mary
Williamsburg, VA 23187-8795
U.S.A.

Operator Theory:
Advances and Applications, Vol. 61
© 1993 Birkhäuser Verlag Basel

Applications of the Furuta Inequality to Operator Inequalities and Norm Inequalities Preserving Some Orders

Takayuki Furuta*

Dedicated to Professor Szökefalvi-Nagy Béla at his 80th birthday

Abstract. The Furuta inequality is as follows: $A \geq B \geq 0$ assures $(B^r A^p B^r)^{1/q} \geq B^{(p+2r)/q}$ for $r \geq 0, p \geq 0, q \geq 1$ with $(1+2r)q \geq p+2r$. We shall give two applications of this inequality. At first we give characterizations of operators satisfying $\log A \geq \log B$. Then we give norm inequalities preserving some order.

1. Introduction

A capital letter means a bounded linear operator on a complex Hilbert space H. An operator T is said to be positive if $(Tx, x) \geq 0$ for all $x \in H$. As an extension of Löwner-Heinz theorem [17], [15], we established the following inequality in [5].

If $A \geq B \geq 0$, then for each $r \geq 0$

(i) $(B^r A^p B^r)^{1/q} \geq B^{(p+2r)/q}$ and

(ii) $A^{(p+2r)/q} \geq (A^r B^p A^r)^{1/q}$ hold each p and q such that $p \geq 0$ and $q \geq 1$ with $(1 + 2r)q \geq p + 2r$.

Alternative proofs of this inequality are given in [2],[7],[16] and an elementary proof is shown in [8].

We remark that the Furuta inequality yields the Löwner-Heinz theorem [17],[15] when we put $r = 0$ in (i) or (ii) stated above, that is, if $A \geq B \geq 0$, then $A^\alpha \geq B^\alpha$ for each $\alpha \in [0, 1]$. There are a lot of proofs of this theorem, among them, an elegant proof of this theorem is given in [18].

---

*I would like to express my sincere appreciation to Professor A. Gheondea for inviting me to the 14th International Conference on Operator Theory at Timişoara and for his hospitality to me during this Conference which has been been excellently organized.

We cite applications of this inequality in [3],[4],[6], [9],[10],[11],[12],[13] and [14]. For example, some of them are results related to the relative operator entropy and an application to some extended result of Ando [3],[4],[10],[11] and some others are applications of the Heinz-Kato theorem [13],[14] or related results on the operator equations [6] and so on.

In this paper we first give an application of the Furuta inequality to operator inequalities preserving some order, that is, we shall give other characterizations of operators satisfying $\log A \geq \log B$ for invertible positive operators A and B. Then we shall give other application of the Furuta inequality to norm inequalities preserving some order.

It is well known that $A \geq B \geq 0$ does not always ensure $A^{1/2} X A^{1/2} \geq B^{1/2} X B^{1/2}$ in general for any operator X, but $A \geq B \geq 0$ surely ensures $\|A^{1/2} X A^{1/2}\| \geq \|B^{1/2} X B^{1/2}\|$ for any operator X. We shall generalize this norm inequality as nice application of the Furuta inequality as a continuation of [12].

2. Statement of results

Theorem 2.1 *Let A and B be invertible positive operators. Then the following assertions are mutually equivalent.*

(1) $\log A \geq \log B$.

(2) $A^p \geq (A^{p/2} B^p A^{p/2})^{p/2}$ for all $p \geq 0$.

(3) $A^p \geq (A^{p/2} B^s A^{p/2})^{p/(p+s)}$ for all $p \geq 0$ and $s \geq 0$.

(4) $A^{-p/2}(A^{p/2} B^p A^{p/2})^{1/2} A^{-p/2}$ is a decreasing function of $p \geq 0$.

(5) For any fixed $t_0 \geq 0$, $G_{(p,r)} = A^{-r}(A^r B^p A^r)^{(t_0+2r)/(p+2r)} A^{-r}$ is a decreasing function of both of $p \geq t_0 \geq 0$ and $r \geq 0$.

(6) For any $p \geq 0$, there exists the unique invertible positive contraction $T_{(p,n)}$ depending on both of p and the natural number n such that

$$T_{(p,n)}(A^{p/n} T_{(p,n)})^n = B^p$$

for any natural number n. Moreover $T_{(p,n)}$ is a decreasing function of p and also $T_{(p,n)}$ is an increasing function of n.

(7) For any $p \geq 0$, there exists some natural number n_0 such that $T_{(p,n_0)}$ is the unique invertible positive contraction satisfying $T_{(p,n_0)}(A^{p/n_0} T_{(p,n_0)})^{n_0} = B^p$.

(8) There exists the unique invertible positive contraction T_p depending on $p \geq 0$ such that $T_p A^p T_p = B^p$ for any $p \geq 0$.

Remark 2.2 Other characterizations of operators satisfying $\log A \geq \log B$ are cited in [4],[10],[11]. It is interesting to remark that $A^p \geq B^p$ does not always hold for any $p > 1$ even if $A \geq B \geq 0$, but (1) and (8) of Theorem 1 tell us that $A \geq B \geq 0$ surely ensures the existence of the unique positive invertible contraction T_p depending on $p \geq 0$ such that $T_p A^p T_p = B^p$ for any $p \geq 0$ when A and B are invertible because $\log t$ is an operator monotone function, so that $A \geq B \geq 0$ ensures $\log A \geq \log B$ which is the condition (1) of Theorem 2.1.

Theorem 2.3 *If $A \geq B \geq 0$ and $C \geq D \geq 0$, then the following norm inequality (2.1) holds for any X*

$$\|A^{\alpha/2} X C^{\beta/2}\| \geq \|B^{\alpha/2} X D^{\beta/2}\| \tag{2.1}$$

for any $\alpha, \beta \in [0,1]$.

The following Corollary 2.4 is a well known result.

Corollary 2.4 *If $A \geq B \geq 0$ and $C \geq D \geq 0$, then $\|A^{1/2} X C^{1/2}\| \geq \|B^{1/2} X D^{1/2}\|$ holds for any X.*

Theorem 2.5 *If $A \geq B \geq 0$ and $C \geq D \geq 0$, then for fixed $t_0, u_0 \in [0,1]$ and for each $r \geq 0, s \geq 0$, the following norm inequality (2.2) holds for any X:*

$$\|A^{\alpha(t_0+r)/2} X C^{\beta(u_0+s)/2}\|$$

$$\geq \|(A^{r/2} B^p A^{r/2})^{\alpha(t_0+r)/2(p+r)} X (C^{s/2} D^q C^{s/2})^{\beta(u_0+s)/2(q+s)}\| \tag{2.2}$$

for any $p \geq t_0 \geq 0, q \geq u_0 \geq 0$ and $\alpha, \beta \in [0,1]$.

Theorem 2.6 *Let A, B, C and D be invertible positive operators. If $\log A \geq \log B$ and $\log C \geq \log D$, then for each $r \geq 0, s \geq 0$, the following norm inequality (2.3) holds for any X:*

$$\|A^{\alpha r/2} X C^{\beta s/2}\|$$

$$\geq \|(A^{r/2} B^p A^{r/2})^{\alpha r/2(p+r)} X (C^{s/2} D^q C^{s/2})^{\beta s/2(q+s)}\| \tag{2.3}$$

for any $p \geq 0, q \geq 0$ and $\alpha, \beta \in [0,1]$.

Remark 2.7 When we put $r = s = 0, t_0 = u_0 = 1$ and $p = q = 1$ in Theorem 2.5, then we have Theorem 2.3, so Theorem 2.5 can be considered as an extension of Theorem 2.3.

Remark 2.8 The item (2.3) of Theorem 2.6 is just the corresponding assertion to (2.2) of Theorem 2.5 when we put $t_0 = u_0 = 0$ in (2.2) of Theorem 2.5 for invertible positive operators A, B, C and D. This fact is naturally understood because $\log t$ is an operator monotone function, so that the hypotheses $A \geq B \geq 0$ and $C \geq D \geq 0$ in Theorem 2.5 ensure $\log A \geq \log B$ and $\log C \geq \log D$ in the hypotheses of Theorem 2.6 for invertible operators A, B, C, and D.

3. Proofs of the results

In order to give proofs of the results, we need the following Theorem 3.1, Theorem 3.2 and Theorem 3.3 obtained by Furuta inequality [5].

Theorem 3.1 *If $A \geq B \geq 0$, then for a fixed $t_0 \in [0,1]$ and for each $r \geq 0$ the following two operator inequalities (i) and (ii) hold*

(i) $(B^{r/2} A^p B^{r/2})^{(t_0+r)/(p+r)} \geq B^{t_0+r}$ *and*

(ii) $A^{t_0+r} \geq (A^{r/2} B^p A^{r/2})^{(t_0+r)/(p+r)}$

hold for any p such that $p \geq t_0 \geq 0$.

Theorem 3.2 [4],[10],[11]. *If $\log A \geq \log B$, then for each $r \geq 0$ the following two operator inequalities (i) and (ii) hold*

(i) $(B^{r/2} A^p B^{r/2})^{r/(p+r)} \geq B^r$ *and*

(ii) $A^r \geq (A^{r/2} B^p A^{r/2})^{r/(p+r)}$

hold for any $p \geq 0$.

Theorem 3.3 [6]. *Let H and K be positive operators and assume that H is nonsingular. If there exists a positive operator T such that $T(H^{1/n}T)^n = K$ for some natural number n, then for any natural number m such that $m \leq n$ there exists a positive operator T_1 such that $T_1(H^{1/m}T_1)^m = K$. In each case, there is at most one positive solution T and T_1.*

Remark 3.4 It is interesting to remark that (i) and (ii) of Theorem 3.2 are just the corresponding results to (i) and (ii) of Theorem 3.1 respectively when we put $t_0 = 0$ in Theorem 3.1 because the logarithmic function $\log t$ is an operator monotone function, so that the hypothesis $A \geq B \geq 0$ in Theorem 3.1 yields $\log A \geq \log B$ the hypothesis in Theorem 3.2. Theorem 3.3 is proved by using the Furuta inequality.

For the sake of convenience, we cite a proof of Theorem 3.1.

Proof of Theorem 3.1. By Theorem 2 in [9], if $A \geq B \geq 0$, then for a fixed $t_0 \geq 0$ and for each $r \geq 0$,

(*) $F_{t_0}(p) = (B^{r/2} A^p B^{r/2})^{(t_0+r)/(p+r)}$ is an increasing functionof p for $p \geq t_0 \geq 0$.

(**) $G_{t_0}(p) = (A^{r/2} B^p A^{r/2})^{(t_0+r)/(p+r)}$ is a decreasing function of p for $p \geq t_0 \geq 0$.

Then by (*) we have $F_{t_0}(p) \geq F_{t_0}(t_0)$ for $p \geq t_0 \geq 0$, that is, for each $r \geq 0$

$$(B^{r/2}A^pB^{r/2})^{(t_0+r)/(p+r)} \geq B^{r/2}A^{t_0}B^{r/2} \geq B^{r/2}B^{t_0}B^{r/2} = B^{t_0+r}$$

holds for any p such that $p \geq t_0 \geq 0$ because the second inequality holds by the Löwner-Heinz theorem since $t_0 \in [0,1]$, so we have (i). By using (**) and the same way as one in the proof of (i), we have (ii) which is equivalent to (i). Whence the proof of Theorem 3.1 is complete. ∎

Proof of Theorem 2.1. The equivalence relation of (1),(2) and (4) is shown in [1] and also the equivalence of (2),(3) and (5) is shown in [4],[10] and [11].

(5) $\Longrightarrow$ (6). In (5), we put $t_0 = 0$ and $r = p/2n$, then let $T_{(p,n)}$ be defined by the following

$$T_{(p,n)} = G_{(p,r)} = A^{-p/2n}(A^{p/2n}B^pA^{p/2n})^{1/(n+1)}A^{-p/2n}.$$

By (5), $T_{(p,n)}$ is a decreasing function of p and also $T_{(p,n)}$ is an increasing function of n since $r = p/2n$. Then for any $p \geq 0$ and any natural number n, by (5) we have

$$T_{(p,n)} \leq T_{(0,n)} = I,$$

so $T_{(p,n)}$ is an invertible positive contraction for any $p \geq 0$ and any natural number n. Also we have

$$A^{p/2n}T_{(p,n)}A^{p/2n} = (A^{p/2n}B^pA^{p/2n})^{1/(n+1)},$$

that is,

$$(A^{p/2n}T_{(p,n)}A^{p/2n})^{n+1} = A^{p/2n}B^pA^{p/2n}. \tag{3.1}$$

In fact the left hand side of (3.1) turns out to be $A^{p/2n}T_{(p,n)}(A^{p/n}T_{(p,n)})^nA^{p/2n}$ by an easy calculation. Consequently by (3.1) we have

$$A^{p/2n}T_{(p,n)}(A^{p/n}T_{(p,n)})^nA^{p/2n} = A^{p/2n}B^pA^{p/2n},$$

so we get

$$T_{(p,n)}(A^{p/n}T_{(p,n)})^n = B^p$$

since $A^{p/2n}$ is invertible.

Assume that there exists another positive contraction Z such that

$$Z(A^{p/n}Z)^n = T_{(p,n)}(A^{p/n}T_{(p,n)})^n = B^p$$

so that we have

$$(A^{p/2n}ZA^{p/2n})^{n+1} = (A^{p/2n}T_{(p,n)}A^{p/2n})^{n+1},$$

and the invertibility of A yields $Z = T_{(p,n)}$.

The proof of (6) is complete.

$(6) \implies (7)$ is obvious.

$(7) \implies (8)$. Assume (7). Then for any $p \geq 0$ there exists some natural number n_0 such that $T_{(p,n_0)}$ is the unique invertible positive contraction satisfying

$$T_{(p,n_0)}(A^{p/n_0}T_{(p,n_0)})^{n_0} = B^p.$$

Then we put $m = 1$ and $n = n_0 \geq 1$ in Theorem 3.3, so that Theorem 3.3 yields the existence of the unique positive contraction T_p depending on $p \geq 0$ such that $T_p A^p T_p = B^p$ for any $p \geq 0$.

$(8) \implies (2)$. By the hypothesis of (8), we have

$$(A^{p/2}TA^{p/2})^2 = A^{p/2}T_p A^p T_p A^{p/2} = A^{p/2}B^p A^{p/2},$$

so that

$$A^p \geq A^{p/2}TA^{p/2} = (A^{p/2}B^p A^{p/2})^{1/2}$$

holds for all $p \geq 0$ since T is positive contraction, so we have (2). The proof is complete. $\blacksquare$

Proof of Theorem 2.3. The hypotheses $A \geq B \geq 0$ and $C \geq D \geq 0$ ensure the following inequalities for any $\alpha, \beta \in [0,1]$ by the Löwner-Heinz Theorem

$$A^\alpha \geq B^\alpha \tag{3.2}$$

$$C^\beta \geq D^\beta. \tag{3.3}$$

Then we have

$$\begin{aligned}
\|A^{\alpha/2}XC^{\beta/2}\|^2 &= \|C^{\beta/2}X^*A^\alpha XC^{\beta/2}\| \geq \|C^{\beta/2}X^*B^\alpha XC^{\beta/2}\| \quad \text{by (3.2)}\\
&= \|B^{\alpha/2}XC^{\beta/2}\|^2 = \|B^{\alpha/2}XC^\beta X^*B^{\alpha/2}\|\\
&\geq \|B^{\alpha/2}XD^\beta X^*B^{\alpha/2}\| \quad \text{by (3.3)}\\
&= \|B^{\alpha/2}XD^{\beta/2}\|^2
\end{aligned}$$

for any $\alpha, \beta \in [0,1]$, so the proof is complete. $\blacksquare$

Proof of Theorem 2.5. By (ii) of Theorem 3.1, the hypothesis $A \geq B \geq 0$ ensures the following inequality for fixed $t_0 \in [0,1]$ and for each $r \geq 0$

$$A^{(t_0+r)} \geq (A^{r/2}B^p A^{r/2})^{(t_0+r)/(p+r)} \tag{3.4}$$

for any $p \geq t_0 \geq 0$.

Also by the same way, the hypothesis $C \geq D \geq 0$ ensures the following inequality for fixed $u_0 \in [0,1]$ and for each $s \geq 0$

$$C^{(u_0+s)} \geq (C^{s/2}D^q C^{s/2})^{(u_0+s)/(q+s)} \tag{3.5}$$

for any $q \geq u_0 \geq 0$. Then by (3.4), (3.5) and Theorem 2.3, we have for fixed $t_0, u_0 \in [0,1]$ and for each $r \geq 0, s \geq 0$

$$\|A^{\alpha(t_0+r)/2} X C^{\beta(u_0+s)/2}\|$$
$$\geq \|(A^{r/2}B^p A^{r/2})^{\alpha(t_0+r)/2(p+r)} X (C^{s/2}D^q C^{s/2})^{\beta(u_0+s)/2(q+s)}\|$$

for any $p \geq t_0 \geq 0, q \geq u_0 \geq 0$ and $\alpha, \beta \in [0,1]$, so the proof is complete. ∎

Proof of Theorem 2.6. The hypothesis $\log A \geq \log B$ yields for each $r \geq 0$

$$A^r \geq (A^{r/2}B^p A^{r/2})^{r/(p+r)} \tag{3.6}$$

holds for any $p \geq 0$ by (ii) of Theorem 3.2. Also by the same way, the hypothesis $\log C \geq \log D$ yields for each $s \geq 0$

$$C^s \geq (C^{s/2}D^q C^{s/2})^{s/(q+s)} \tag{3.7}$$

holds for any $q \geq 0$. Then by (3.6), (3.7) and Theorem 2.3, we have the following norm inequality

$$\|A^{\alpha r/2} X C^{\beta s/2}\|$$

$$\geq \|(A^{r/2}B^p A^{r/2})^{\alpha r/2(p+r)} X (C^{s/2}D^q C^{s/2})^{\beta s/2(q+s)}\|,$$

for any $p \geq 0, q \geq 0$ and $\alpha, \beta \in [0,1]$, so the proof is complete. ∎

References

[1] T. Ando: Some operator inequalities, *Math. Ann.*, **279**(1987), 157-159.

[2] M. Fujii: Furuta's inequality and its mean theoretic approach, *J. Operator Theory*, **23**(1990), 67-72.

[3] M. Fujii, T. Furuta and E. Kamei: Operator functions associated with Furuta's inequality, *Linear Alg. and Its Appl.*, **149**(1991), 91-96.

[4] M. Fujii, T. Furuta and E. Kamei: Furuta's inequality and its application to Ando's theorem, in press in *Linear Alg. and Its Appl.*.

[5] T. Furuta: $A \geq B \geq 0$ assures $(B^r A^p B^r)^{1/q} \geq B^{(p+2r)/q}$ for $r \geq 0, p \geq 0, q \geq 1$ with $(1+2r)q \geq p+2r$. *Proc. Amer. Math. Soc.* **101** (1987), 85-88.

[6] T. Furuta: The Operator equation $T(H^{1/n}T)^n = K$, *Linear Alg. and Its Appl.*, **109** (1988), 149-152.

[7] T. Furuta: A proof via operator means of an order preserving inequality, *Linear Alg. and Appl.*, **113** (1989), 129-130.

[8] T. Furuta: Elementary proof of an order preserving inequality, *Proc. Japan Acad.*, **65** (1989), 126.

[9] T. Furuta: Two operator functions with monotone property, *Proc. Amer. Math. Soc.*, **111** (1991), 511-516.

[10] T. Furuta: Furuta's inequality and its application to the relative operator entropy, to appear in *J. Operator Theory*.

[11] T. Furuta: Applications of order preserving operator inequalities, to appear in *Proc. of WOTCA* in Birkhäuser.

[12] T. Furuta: Some norm inequalities and operator inequalities via Furuta inequality, to appear in *Acta Sci. Math. (Szeged)*.

[13] T. Furuta: Generalization of Heinz-Kato theorem via Furuta inequality, to appear in *Proc. Amer. Math. Soc.*.

[14] T. Furuta: Determinant type generalizations of Heinz- Kato theorem via Furuta inequality, to appear in *Proc. Amer. Math. Soc.*.

[15] E. Heinz: Beiträge zur Störungstheorie der Spektralzerlegung, *Math. Ann.*, **123** (1951), 415-438.

[16] E. Kamei: A satellite to Furuta's inequality, *Math. Japon*, **33** (1988), 883-886.

[17] K. Löwner: Über monotone Matrixfunktionen, *Math. Z.*, **38** (1934), 177-216.

[18] G. K. Pedersen: Some operator monotone function, *Proc. Amer. Math. Soc.*, **36** (1972), 309-310.

Takayuki Furuta

Department of Applied Mathematics
Faculty of Science, Science University of Tokyo
1-3 Kagurazaka, Shinjuku-ku
Tokyo 162
Japan

Operator Theory:
Advances and Applications, Vol. 61
© 1993 Birkhäuser Verlag Basel

Quasi–Contractions on Kreĭn Spaces

Aurelian Gheondea

Abstract. A quasi-contraction, acting between Kreĭn spaces, is a linear bounded operator which is contractive when restricted to some subspace of finite codimension. If, in addition, the same property is satisfied by the adjoint of the operator, then it is called a double quasi-contraction.
This paper is devoted to the investigation of basic properties of quasi–contractions and double quasi-contractions. Among others, we prove that the compression of a quasi-contraction (double quasi-contraction) to maximal uniformly negative subspaces is a semi-Fredholm operator (respectively, Fredholm operator) and we give a formula to compute its index. A spectral characterization of double quasi-contractions within the class of quasi-contractions is also obtained.

1. Introduction

An operator $T \in \mathcal{L}(\mathcal{K}_1, \mathcal{K}_2)$ is called a quasi–contraction if the quadratic form

$$[x, x] - [Tx, Tx] , \quad x \in \mathcal{K}_1, \tag{1.1}$$

has a finite number of negative squares that we denote by $\kappa^-(I - T^\sharp T)$. If the number $\kappa^-(I - T^\sharp T)$ vanishes then T is called contraction. Beginning with the celebrated paper of V. P. Potapov [18] contractive operators have been intensively studied by Yu. P. Ginzburg [9], M. G. Kreĭn and Yu. L. Shmulyan [14],[15] in connection with interpolation of functions but also for operator theoretical reasons.

The interest for quasi-contractions appeared also from the investigations of interpolation problems. Here the functions taken into consideration are meromorphic and the connection with the quadratic form (1.1) is justified by certain kernels as the Carathéodory kernel, the Schur kernel and the Nevanlinna kernel (see M.G.Kreĭn and H.Langer [13]). The study of quasi-contractions has been performed until now mainly with respect to lifting and extension problems of these operators (see [5],[6]).

The aim of this paper is to investigate quasi-contractive operators from both spectral and geometrical aspects, following closely the questions considered in [14] and [15]. The first problems we are dealing with are concerning the action of quasi-contractions on negative subspaces and maximal negative subspaces. Some of the proofs of the original results can be adapted to the case of quasi–contractions but it is more convenient to try first to reduce the quasi–contractive case to the contractive one. In Lemma 3.3 we show that any quasi–contraction is the restriction of a contraction, where the added space has a certain minimality property. The main result of Section 3 is Theorem 3.5 which proves that the compression of a quasi–contraction T to maximal uniformly negative subspaces is a semi-Fredholm operator whose index is equal to $\kappa^-(I - T^\sharp T) - \kappa^-(I - TT^\sharp)$. This result reflects a new phenomenon in the theory of operators in Kreĭn spaces which was hidden in the case of contractions [14] and it enables us to add some new insight on the geometry of contractive operators, too.

As in the case of contractions, the relation with the adjoint operator is complicated. It is natural to introduce the concept of double quasi–contractions (that is, both of the operator and its adjoint are contractions) and to investigate their properties and their connection with double contractions. These problems are considered in Section 4.

For the case of contractions, the relation with the adjoint operator is reflected also by a recent result of M. A. Dritschel and J. Rovnyak [7] which shows that discarding a uniformly negative subspace of the image space, a contractive operator can be reduced to a double contraction. This result has a counter-part in the framework of quasi–contractions, see Theorem 4.3.

The main result of Section 4 is a spectral characterization of double quasi–contractions (see Theorem 4.6) which generalizes the spectral characterization obtained by M. G. Kreĭn and Yu. L. Shmulyan in [15] in the framework of their theory of polar decompositions.

The technique we use in this paper is a combination of geometric methods (extensions of operators in Kreĭn spaces inspired by the technique used in [5], [6]) and the spectral theory of definitizable operators developed by M. G. Kreĭn and H .Langer [12] and H. Langer [16], [17]. We review the necessary results and notation in Section 2.

For the foundations of operator theory in Kreĭn spaces that we use here we recommend [3], [4]. Also, for survey variants of the theory of contractions in Kreĭn spaces see [1], [7].

The results we have obtained in this paper open some ways towards investigation of other problems on quasi–contractions and double quasi–contractions. In this respect, we can consider now the Ginzburg–Potapov transform of quasi–contractions as well as the fractional linear transformations associated with these operators. All these problems are the subject of our current research and will be published elsewhere.

We thank Aad Dijksma for carefully reading a preliminary version of this paper, pointing out some errors and for making useful suggestions which improved the presentation of this article.

2. Notation and Preliminary Results

2.1 Signatures. Let $(\mathcal{K}_i, [.,.])$ $(i = 1, 2)$ be *Kreĭn space s* and let $T \in \mathcal{L}(\mathcal{K}_1, \mathcal{K}_2)$ be a bounded linear operator. Then $T^\sharp \in \mathcal{L}(\mathcal{K}_2, \mathcal{K}_1)$ will denote its *adjoint*, i.e.

$$[Tx, y] = [x, T^\sharp y], \quad x \in \mathcal{K}_1, \ y \in \mathcal{K}_2.$$

In case we fix *fundamental symmetries* (briefly, f.s.) J_1 on $\mathcal{K}_1$ and J_2 on $\mathcal{K}_2$, i.e. we fix Hilbert spaces $(\mathcal{K}_i, (.,.)_{J_i})$, then $T^* \in \mathcal{L}(\mathcal{K}_2, \mathcal{K}_1)$ will denote the Hilbert space adjoint of T, that is

$$(Tx, y)_{J_1} = (x, T^*y)_{J_2}, \quad x \in \mathcal{K}_1, \ y \in \mathcal{K}_2.$$

In particular we have

$$T^\sharp = J_1 T^* J_2.$$

Whenever a Krein space $\mathcal{K}$ is considered we can choose a *fundamental decomposition* (in brief f.s.) $\mathcal{K} = \mathcal{K}^+[+]\mathcal{K}^-$, where $(\mathcal{K}^+, [\cdot, \cdot])$ and $(\mathcal{K}^-, -[\cdot, \cdot])$ are Hilbert spaces and $\mathcal{K}^+ \perp \mathcal{K}^-$.

Let $\mathcal{K}$ be a Kreĭn space and $A \in \mathcal{L}(\mathcal{K})$ be *selfadjoint*, i.e. $A = A^\sharp$. We consider another inner product on $\mathcal{K}$ defined by

$$[x, y]_A = [Ax, y], \quad x, y \in \mathcal{K},$$

and denote by $\kappa^-(A)$ the supremum of $\dim \mathcal{L}$ where $\mathcal{L}$ runs through the set of all finite dimensional subspaces of $\mathcal{K}$ such that

$$[x, x]_A < 0, \quad x \in \mathcal{L} \setminus \{0\}.$$

By definition $\kappa^-(A)$ is the *negative signature* of A and it is either a positive integer or the symbol ∞. Similarly one defines $\kappa^+(A)$, the *positive signature* of A. Also, the *null signature* of A is by definition $\kappa^0(A) = \dim \ker A$.

We also make the remark that $\kappa^-(A)$ equals the greatest number of negative eigenvalues (counted with their multiplicities) of the matrices $([x_i, x_j]_A)_{i,j=1}^n$, where $\{x_i\}_{i=1}^n \subset \mathcal{K}$ and $n \geq 0$.

2.2 Pseudo-regular subspaces. In this paper a *subspace* is always a closed linear manifold. For two subspaces $\mathcal{A}$ and $\mathcal{B}$ of the Kreĭn space $\mathcal{K}$, we use the symbol $\mathcal{A}[+]\mathcal{B}$ to denote the sum space if $\mathcal{A} \perp \mathcal{B}$, $\mathcal{A} + \mathcal{B}$ is closed, and $\mathcal{A} \cap \mathcal{B} = 0$.

A subspace $\mathcal{L}$ of the Kreĭn space is called *pseudo-regular* if $\mathcal{L} + \mathcal{L}^\perp$ is closed. The subspace $\mathcal{L}$ is pseudo-regular if and only if there exists a *regular* subspace $\mathcal{R}$ of $\mathcal{K}$ such that

$$\mathcal{L} = \mathcal{R}[+]\mathcal{L}^0,$$

where $\mathcal{L}^0 = \mathcal{L} \cap \mathcal{L}^\perp$ denotes the *isotropic subspace* of $\mathcal{L}$ (see e.g. [8]).

2.3 Definitizable Operators. A selfadjoint operator $A \in \mathcal{L}(\mathcal{K})$, $\mathcal{K}$ being a Kreĭn space, is called *definitizable* if there exists a non-trivial polynomial p such that

$$[p(A)x, x] \geq 0, \quad x \in \mathcal{K}.$$

If A is definitizable then $\sigma_0(A) = \sigma(A) \setminus \mathbf{R}$ (we denote the spectrum of A by $\sigma(A)$) is a finite set and there exists a finite set $c(A) \subset \mathbf{R}$, called the set of *critical points* of A, such that on $\mathcal{R}_A$, the Boolean algebra generated by intervals Δ of $\mathbf{R}$ with $\partial\Delta \cap c(A) = \emptyset$, there exists a mapping

$$E: \mathcal{R}_A \to \mathcal{L}(\mathcal{K})$$

(called the *spectral function* of A) with the following properties:

(i) $E(\Delta_1 \cap \Delta_2) = E(\Delta_1)E(\Delta_2), \quad \Delta_1, \Delta_2 \in \mathcal{R}_A$.

(ii) $E(\Delta_1 \cup \Delta_2) = E(\Delta_1) + E(\Delta_2) - E(\Delta_1 \cap \Delta_2), \quad \Delta_1, \Delta_2 \in \mathcal{R}_A$.

(iii) $\sigma(A|E(\Delta)\mathcal{K}) \subseteq \bar{\Delta}, \quad \Delta \in \mathcal{R}_A$.

(iv) $E(\Delta)$ is positive if $p|\Delta > 0$ and $E(\Delta)$ is negative if $p|\Delta < 0$.

For a detailed investigation of definitizable operators see [16] and [17].

We will need in the following a formula to compute the negative signature of a definitizable operator. Whenever σ is a spectral set of the operator A we use $E(\sigma; A)$ to denote the corresponding spectral projection obtained by the Riesz–Dunford functional calculus. If λ is isolated in the spectrum of A then $E(\lambda; A) = E(\{\lambda\}; A)$.

Lemma 2.1 *Let $A \in \mathcal{L}(\mathcal{K})$ be a selfadjoint definitizable operator and denote by E its spectral function. Then, for $\varepsilon > 0$ sufficiently small such that $c(A) \cap [-\varepsilon, \varepsilon] \subseteq \{0\}$, we have*

$$\begin{aligned}
\kappa^-(A) &= \sum_{\mathrm{Im}\,\lambda > 0} \mathrm{rank}\, E(\lambda; A) + \kappa^-(A|E((-\varepsilon, \varepsilon))\mathcal{K}) \\
&\quad + \kappa^-(E([\varepsilon, +\infty))) + \kappa^+(E((-\infty, \varepsilon])).
\end{aligned} \tag{2.1}$$

Proof. Indeed, $\sigma_0(A) = \sigma(A) \setminus \mathbf{R}$ is a finite set hence it is a spectral set of A. Denoting

$$\sigma_0^+(A) = \{\lambda \in \sigma_0(A)| \ \mathrm{Im}\,\lambda > 0\}, \quad \sigma_0^-(A) = \{\lambda \in \sigma_0(A)| \ \mathrm{Im}\,\lambda < 0\},$$

we also have

$$\sigma_0^+(A) = \{\bar{\lambda}| \ \lambda \in \sigma_0^-(A)\},$$

and the spectral subspaces $E(\sigma_0^+(A); A)\mathcal{K}$ and $E(\sigma_0^-(A); A)\mathcal{K}$ are unitary equivalent (as Hilbert spaces). Moreover, identifying both of these spaces with the same Hilbert space $\mathcal{H}$ then

$$E(\sigma_0(A); A)\mathcal{K} = \mathcal{H} \oplus \mathcal{H},$$

where the f.s. is given by $J(x \oplus y) = y \oplus x$, $x, y \in \mathcal{H}$. Then, with respect to this representation

$$A|E(\sigma_0(A); A)\mathcal{K} = \begin{bmatrix} B & 0 \\ 0 & B^* \end{bmatrix},$$

where $B \in \mathcal{L}(\mathcal{H})$ is invertible. This shows that

$$\kappa^-(A|E(\sigma_0(A); A)\mathcal{K}) = \dim \mathcal{H}$$
$$= \dim E(\sigma_0^+(A); A)\mathcal{K} = \sum_{\mathrm{Im}\,\lambda > 0} \mathrm{rank}\, E(\lambda; A). \tag{2.2}$$

Let now $\varepsilon > 0$ be sufficiently small such that $c(A) \cap [-\varepsilon, \varepsilon] \subset \{0\}$. Since

$$\sigma(A|E([\varepsilon, +\infty))\mathcal{K}) \subset [\varepsilon, +\infty),$$

it follows that there exists an operator $R \in \mathcal{L}(E([\varepsilon, +\infty))\mathcal{K})$, $R = R^\sharp$ and $\sigma(R) \subset [\varepsilon^{1/2}, +\infty)$ (in particular R is invertible) such that

$$A|E([\varepsilon, +\infty))\mathcal{K} = R^2.$$

This means

$$[Ax, x] = [Rx, Rx], \quad x \in E([\varepsilon, +\infty))\mathcal{K}.$$

Since R is invertible we obtain

$$\kappa^-(A|E([\varepsilon, +\infty))\mathcal{K}) = \kappa^-(E([\varepsilon, +\infty))). \tag{2.3}$$

In the same way we obtain

$$\kappa^-(A|E((-\infty, -\varepsilon])\mathcal{K}) = \kappa^+(E((-\infty, -\varepsilon])). \tag{2.4}$$

The formula (2.1) follows now from (2.2)-(2.4). ∎

3. Quasi-contractive operators

Let $\mathcal{K}_1$ and $\mathcal{K}_2$ be Kreĭn spaces. An operator $T \in \mathcal{L}(\mathcal{K}_1, \mathcal{K}_2)$ is *quasi-contractive*, by definition, if $\kappa^-(I - T^\sharp T)$ is finite. If $\kappa^-(I - T^\sharp T) = 0$ (i.e. $I - T^\sharp T$ is positive) then T is called *contractive*.

Proposition 3.1 *Let $T \in \mathcal{L}(\mathcal{K}_1, \mathcal{K}_2)$ be quasi–contractive and $\kappa = \kappa^-(I - T^\sharp T) < \infty$. Then $\ker T$ is a pseudo-regular subspace of $\mathcal{K}_1$ such that*

$$\kappa^0(\ker T) + \kappa^-(\ker T) \le \kappa. \tag{3.1}$$

Proof. Let us denote $A = I - T^\sharp T$. A is a selfadjoint operator in $\mathcal{K}_1$ and since $\kappa^-(A) = \kappa < \infty$ it follows (see [17]) that A is definitizable. If E denotes the spectral function of A then from Lemma 2.1 we obtain $\kappa^-(E(0, +\infty)) \le \kappa$, in particular $E(0, +\infty)\mathcal{K}_1$ is a Pontryagin space. Furthermore, since $\ker T \subseteq \ker T^\sharp T = \ker(I - A) \subseteq E(0, +\infty)\mathcal{K}_1$ then $\ker T$ is pseudo-regular and the inequality (3.1) holds. ∎

An operator $T \in \mathcal{L}(\mathcal{K}_1, \mathcal{K}_2)$ is *expansive (strictly expansive)* on a subspace $\mathcal{L} \in \mathcal{K}_1$ if $[Tx, Tx] \ge [x, x]$ ($[Tx, Tx] > [x, x]$) for all $x \in \mathcal{L}$ ($x \in \mathcal{L} \setminus \{0\}$). By definition, a quasi–contraction is strictly expansive on a subspace of finite maximal dimension $\kappa^-(I - T^\sharp T)$. The next result shows that we can choose this subspace to be regular.

Proposition 3.2 *Let $T \in \mathcal{L}(\mathcal{K}_1, \mathcal{K}_2)$ be quasi–contractive and denote $\kappa = \kappa^-(I - T^\sharp T)$. Then there exists a regular subspace $\mathcal{L}$ in $\mathcal{K}_1$, such that $\dim \mathcal{L} = \kappa$ and $T|\mathcal{L}$ is strictly expansive.*

Proof. Since $\kappa^-(I - T^\sharp T) = \kappa$ it follows that there exists a subspace $\mathcal{L}$ of $\mathcal{K}_1$, $\dim \mathcal{L} = \kappa$ such that

$$[Tx, Tx] > [x, x], \quad x \in \mathcal{L} \setminus \{0\}. \tag{3.2}$$

Since $\mathcal{L}$ is finite dimensional it is pseudo-regular hence

$$\mathcal{L} = \mathcal{L}^0[+]\mathcal{R},$$

with $\mathcal{R}$ a regular subspace of $\mathcal{K}_1$. To be more specific, we fix a f.s. J_i on $\mathcal{K}_1$ and take $\mathcal{R} = \mathcal{L} \ominus \mathcal{L}^0$.

Let us assume that $\mathcal{L}^0 \ne 0$, i.e. $\mathcal{L}$ is degenerate, and let $\{f_i\}_{i=1}^\kappa$ be an orthogonal basis of $\mathcal{L}$, i.e. it is a basis such that for all $i \ne j$ we have $[f_i, f_j] = 0$. Such an orthogonal basis can be obtained as the union of a J_1-orthonormal basis of $\mathcal{R}$ and a J_1-orthonormal basis of $\mathcal{L}^0$. Then $\mathrm{rank}\, ([f_i, f_j])_{i,j=1}^\kappa = \dim(\mathcal{R}) = \kappa - \dim(\mathcal{L}^0)$. Let us now fix the unitary norm $\| \cdot \|$ on $\mathcal{K}_1$, which is the norm associated to the f.s. J_1 on $\mathcal{K}_1$. We claim that for any $\epsilon > 0$ there exists a system of vectors $\{g_i\}_{i=1}^\kappa$ in $\mathcal{K}_1$ such that for all $i \in \{1, 2, \ldots .\kappa\}$ we have $\|f_i - g_i\| < \epsilon$ and $\mathrm{rank}\, ([g_i, g_j])_{i,j=1}^\kappa = \kappa$. The claim is easy to establish by perturbing only the vectors f_i in $\mathcal{L}^0$ with the vectors $g_i = f_i + \epsilon J_1 f_i$, for the vectors $f_j \in \mathcal{R}$ letting $g_j = f_j$.

Let us now consider $\{\lambda_i\}_{i=1}^{\kappa} \in \mathbf{C}$ be such that $\sum_{i=1}^{\kappa} |\lambda_i|^2 = 1$ and denote

$$x = \sum_{i=1}^{\kappa} \lambda_i f_i, \quad y = \sum_{i=1}^{\kappa} \lambda_i g_i. \tag{3.3}$$

Then

$$\|x - y\| = \|\sum_{i=1}^{\kappa} \lambda_i(f_i - g_i)\| \leq \sum_{i=1}^{\kappa} |\lambda_i| \cdot \|f_i - g_i\|$$

$$\leq \left(\sum_{i=1}^{\kappa} |\lambda_i|^2\right)^{1/2} \cdot \left(\sum_{i=1}^{\kappa} \|f_i - g_i\|^2\right)^{1/2} \leq \sqrt{\kappa} \cdot \epsilon.$$

Using the boundedness of T and the normic joint continuity of the inner products, similar majorizations as before show that for any $\delta > 0$ there exists $\epsilon > 0$, independent of $\{\lambda_i\}_{i=1}^{\kappa}$, such that if x and y are as in (3.3) and $\|f_i - g_i\| < \epsilon$ then

$$|([y, y] - [Ty, Ty]) - ([x, x] - [Tx, Tx])| < \delta.$$

Letting δ be sufficiently small, from (3.2) it follows that we also have $[Ty, Ty] > [y, y]$, and since this inequality is homogenous of order two in y, we thus obtained

$$[Ty, Ty] > [y, y], \quad y \in \mathcal{L}_\delta \setminus \{0\},$$

where we let $\mathcal{L}_\delta$ be the space generated by the system of vectors $\{g_i\}_{i=1}^{\kappa}$. Also, since rank $([g_i, g_j])_{i,j=1}^{\kappa} = \kappa$ it follows that the space $\mathcal{L}_\delta$ is regular of dimension κ. We replace $\mathcal{L}$ by $\mathcal{L}_\delta$. $\blacksquare$

In the following, in order to prove certain results about quasi-contractions, we will use some related results about contractions. To do this we need first a notion of the proximity of quasi-contractive operators to contractive operators. One possible formulation is the following lemma (see also [5] for related results).

Lemma 3.3 *Let $T \in \mathcal{L}(\mathcal{K}_1, \mathcal{K}_2)$ be quasi-contractive and denote $\kappa = \kappa^-(I - T^\sharp T)$. Then, given a Kreĭn space $\mathcal{K}_2'$, there exists a contraction $\tilde{T} \in \mathcal{L}(\mathcal{K}_1, \mathcal{K}_2[+]\mathcal{K}_2')$ such that $P_{\mathcal{K}_2}\tilde{T} = T$ if and only if $\kappa^-(\mathcal{K}_2') \geq \kappa$.*

Proof. Let us fix f.s. J_1 on $\mathcal{K}_1$ and J_2 on $\mathcal{K}_2$. Since $\kappa = \kappa^-(I - T^\sharp T) < \infty$ it follows that the operator $J_1 - T^*J_2T$, considered as a selfadjoint operator in the Hilbert space $(\mathcal{K}_1, (.,.)_{J_1})$, has exactly κ negative eigenvalues. Denoting $J_T = \text{sgn}(J_1 - T^*J_2T)$ let

$$J_T = J_T^+ - J_T^-$$

be its Jordan decomposition. We have rank $J_T^- = \kappa$. Let us also denote

$$D_T = |J_1 - T^*J_2T|^{1/2}.$$

130   A. Gheondea

We consider now the operator

$$D_T^- = J_T^- D_T$$

and notice that

$$(D_T^-)^2 = (J_1 - T^* J_2 T)^-, \tag{3.4}$$

where

$$J_1 - T^* J_2 T = (J_1 - T^* J_2 T)^+ - (J_1 - T^* J_2 T)^-$$

is the Jordan decomposition. We consider now $\mathcal{D}_T^- = \mathcal{R}(D_T^-)$, which is a subspace of $\mathcal{K}_1$ of dimension κ, regarded as an anti-Hilbert space with the negative inner product inherited from $-(.,.)_{J_1}$. Then let us define

$$\tilde{T}_0 = \left[\begin{array}{c} T \\ D_T^- \end{array} \right] : \mathcal{K}_1 \to \mathcal{K}_2[+]\mathcal{D}_T^-. \tag{3.5}$$

$\tilde{T}_0$ is contraction. Indeed, $\tilde{J}_2 = J_2 \oplus -I$ is a f.s. of $\mathcal{K}_2[+]\mathcal{D}_T^-$ and using (3.4) we have

$$J_1 - \tilde{T}_0^* \tilde{J}_2 \tilde{T}_0 = J_1 - T^* J_2 T - (D_T^-)^2 = (J_1 - T^* J_2 T)^+ \geq 0.$$

If $\mathcal{K}_2^-$ is a Kreĭn space with $\kappa^-(\mathcal{K}_2') \geq \kappa$ then there exists a bounded isometric operator $V : \mathcal{D}_T^- \to \mathcal{K}_2'$ and denote

$$\tilde{T} = \left[\begin{array}{c} T \\ V D_T^- \end{array} \right] : \mathcal{K}_1 \to \mathcal{K}_2[+]\mathcal{K}_2^-.$$

Then $P_{\mathcal{K}_2} \tilde{T} = T$ and $\tilde{T}$ is contraction since $\tilde{T}_0$ is.

The converse implication is a direct consequence of [6]. ∎

As a first application we can investigate the action of a quasi–contraction on negative subspaces.

Corollary 3.4 *Let* $T \in \mathcal{L}(\mathcal{K}_1, \mathcal{K}_2)$ *be a quasi–contraction and let us also denote* $\kappa = \kappa^-(I - T^\sharp T)$. *Then:*

(i) *If* $\mathcal{L}$ *is a negative subspace of* $\mathcal{K}_1$ *then* $T\mathcal{L}$ *is closed and* $\kappa^+(T\mathcal{L}) \leq \kappa$.

(ii) *If* $\mathcal{L}$ *is a strictly negative subspace of* $\mathcal{K}_1$ *then* $T\mathcal{L}$ *is a subspace of* $\mathcal{K}_2$ *such that* $\kappa^0(T\mathcal{L}) + \kappa^+(T\mathcal{L}) \leq \kappa$.

(iii) *If* $\mathcal{L}$ *is a uniformly negative subspace of* $\mathcal{K}_1$ *then* $T\mathcal{L}$ *is a pseudo–regular subspace of* $\mathcal{K}_2$ *(such that* $\kappa^0(T\mathcal{L}) + \kappa^+(T\mathcal{L}) \leq \kappa$*).*

Proof.(i) Applying Lemma 3.3 we obtain a Kreĭn space $\mathcal{K}_2'$ such that $\kappa^-(\mathcal{K}_2') = \dim \mathcal{K}_2' = \kappa$ and a contraction $\tilde{T} \in \mathcal{L}(\mathcal{K}_1, \mathcal{K}_2[+]\mathcal{K}_2')$ such that $P_{\mathcal{K}_2}\tilde{T} = T$. We identify $\mathcal{K}_2$ with $\mathcal{K}_2[+]\{0\}$ as a subspace of $\mathcal{K}_2[+]\mathcal{K}_2'$. If $\mathcal{L}$ is a negative subspace of $\mathcal{K}_1$ then, using [14], Theorem 2.2, it follows that $\tilde{T}\mathcal{L}$ is a negative subspace in $\mathcal{K}_2[+]\mathcal{K}_2'$. We consider now the subspace

$$\mathcal{S} = \tilde{T}\mathcal{L} + \mathcal{K}_2'.$$

$\mathcal{S}$ is closed since $\tilde{T}\mathcal{L}$ is closed and $\mathcal{K}_2'$ is finite dimensional. Moreover, there exists a subspace $\mathcal{M}$ of $\mathcal{K}_2[+]\mathcal{K}_2'$ such that $\dim \mathcal{M} \le \kappa$ and

$$\mathcal{S} = \tilde{T}\mathcal{L}[+]\mathcal{M}. \tag{3.6}$$

Since $\tilde{T}\mathcal{L}$ is negative it follows $\kappa^+(\mathcal{S}) \le \kappa$. Using $P_{\mathcal{K}_2}\tilde{T} = T$ and the definition of $\mathcal{S}$ we obtain

$$\mathcal{S} = T\mathcal{L}[+]\mathcal{K}_2', \tag{3.7}$$

hence $T\mathcal{L}$ is closed and $\kappa^+(T\mathcal{L}) \le \kappa^+(\mathcal{S}) \le \kappa$.

(ii) Assuming that the subspace $\mathcal{L}$ is strictly negative then $\tilde{T}\mathcal{L}$ is the same [14]. Using (3.6) and (3.7) we obtain

$$\kappa^0(T\mathcal{L}) + \kappa^+(T\mathcal{L}) \le \kappa^0(\mathcal{S}) + \kappa^+(\mathcal{S}) \le \kappa.$$

(iii) If the subspace $\mathcal{L}$ is uniformly negative then $\tilde{T}\mathcal{L}$ is the same (also by [14]). From (3.6) and $\dim \mathcal{M} \le \kappa$ it follows that $\mathcal{S}$ is pseudo-regular and $\kappa^0(\mathcal{S})+\kappa^+(\mathcal{S}) \le \kappa$. Since these kind of pseudo-regular subspaces are hereditary (e.g. see [8]) and $T\mathcal{L} \subseteq \mathcal{S}$ then $T\mathcal{L}$ is also pseudo-regular. $\blacksquare$

The main result of this section is

Theorem 3.5 *Let $T \in \mathcal{L}(\mathcal{K}_1, \mathcal{K}_2)$ be quasi-contractive and consider for $i = 1,2$ the f.d. $\mathcal{K}_i = \mathcal{K}_i^+[+]\mathcal{K}_i^-$. Then $P_{\mathcal{K}_2^-}T|\mathcal{K}_1^- \in \mathcal{L}(\mathcal{K}_1^-, \mathcal{K}_2^-)$ is a semi-Fredholm operator, more precisely it has closed range and finite dimensional kernel, and its index is*

$$\mathrm{ind}\,(P_{\mathcal{K}_2^-}T|\mathcal{K}_1^-) = \kappa^-(I - T^\sharp T) - \kappa^-(I - TT^\sharp). \tag{3.8}$$

Proof. With respect to the f.d. $\mathcal{K}_i = \mathcal{K}_i^+[+]\mathcal{K}_i^-$, $i = 1,2$ we consider the matrix representation of T

$$T = \begin{bmatrix} T_{11} & T_{12} \\ T_{21} & T_{22} \end{bmatrix}, \tag{3.9}$$

in particular $P_{\mathcal{K}_2^-}T|\mathcal{K}_1^- = T_{22}$. With this notation, for any $x \in \mathcal{K}_1^-$ we have

$$[T_{22}x, T_{22}x] \le [T_{12}x, T_{12}x] + [T_{22}x, T_{22}x] = [Tx, Tx]$$

hence

$$[x,x] - [T_{22}x, T_{22}x] \geq [x,x] - [Tx, Tx].$$

Considering T_{22} as an operator acting between the negative definite Kreĭn spaces $\mathcal{K}_1^-$ and $\mathcal{K}_2^-$, the latter inequality yields

$\kappa^-(I - T_{22}^\sharp T_{22}) \leq \kappa^-(I - T^\sharp T)$ hence, since T is quasi–contractive, using Proposition 3.1 it follows that $\ker T_{22}$ is finite dimensional and then, from Corollary 3.4 it follows that $\mathcal{R}(T_{22}) = T_{22}\mathcal{K}_1^-$ is closed. We thus proved that T_{22} is a semi-Fredholm operator. In order to prove the formula (3.8) we proceed stepwise.

1. We first assume that $\kappa^-(I - T^\sharp T) = \kappa^-(I - TT^\sharp) = 0$. Then T_{22} is invertible (cf.[14], Theorem 3.1) hence the formula (3.8) is trivially satisfied in this case.

2. Assume now that T is contraction. We distinguish two subcases:

a) If $\kappa^-(I - TT^\sharp)$ is finite then, applying the same argument as in the proof of Lemma 3.3 for $T^\sharp$ instead of T, we obtain a negative definite Kreĭn space $\mathcal{D}_{T^\sharp}^-$ such that

$$\kappa^-(\mathcal{D}_{T^\sharp}^-) = \dim \mathcal{D}_{T^\sharp}^- = \kappa^-(I - TT^\sharp) < \infty$$

and an operator $\tilde{T} \in \mathcal{L}(\mathcal{K}_1[+]\mathcal{D}_{T^\sharp}^-)$ defined by

$$\tilde{T} = [T \quad D_{T^\sharp}^-]. \tag{3.10}$$

We claim that $\tilde{T}$ is a double contraction, i.e.

$$\kappa^-(I - \tilde{T}^\sharp \tilde{T}) = \kappa(I - \tilde{T}\tilde{T}^\sharp) = 0.$$

Indeed, $\tilde{T}^\sharp$ is contraction as the proof of Lemma 3.3 shows. In order to prove $\tilde{T}$ is contraction we use [5], Lemma 2.1 to produce a contractive extension of $\tilde{T}$ hence $\tilde{T}$ is also contractive. Consider now the f.d.

$$\mathcal{K}_1[+]\mathcal{D}_{T^\sharp}^- = \mathcal{K}_1^+[+](\mathcal{K}_1^-[+]\mathcal{D}_{T^\sharp}^-)$$

and with respect to this we represent

$$\tilde{T} = \begin{bmatrix} T_{11} & T_{12} & X_{12} \\ T_{21} & T_{22} & X_{22} \end{bmatrix}. \tag{3.11}$$

With our notation we have

$$\tilde{T}_{22} = [T_{22} \quad X_{22}]. \tag{3.12}$$

Using the fact proved at the first step, taking into account that X_{22} has finite rank and that the index of Fredholm operators is invariant under compact perturbation, we have

$$\begin{aligned} 0 &= \mathrm{ind}\,(\tilde{T}_{22}) = \mathrm{ind}\,([T_{22}\ 0]) \\ &= \mathrm{ind}\,(T_{22}) + \dim \mathcal{D}_{T^\sharp}^- \\ &= \mathrm{ind}\,(T_{22}) + \kappa^-(I - TT^\sharp), \end{aligned} \tag{3.13}$$

hence

$$\operatorname{ind}(T_{22}) = -\kappa^-(I - TT^\sharp),$$

and the formula (3.8) is proved in this case.

b) Assume that $\kappa^-(I - TT^\sharp)$ is infinite. We consider as before the negative definite Kreĭn space $\mathcal{D}_{T^\sharp}^-$ and the operator $\tilde{T}$ in (3.10). The proof that $\tilde{T}$ is double contractive is the same as before. The operator $\tilde{T}_{22}$ in (3.12) is invertible, hence

$$\mathcal{K}_2^- = \mathcal{R}(\tilde{T}_{22}) = \mathcal{R}(T_{22}) + \mathcal{R}(X_{22})$$

and since T_{22} itself is injective we have

$$\mathcal{R}(T_{22}) \cap \mathcal{R}(X_{22}) = 0,$$

hence

$$\mathcal{K}_2^- = \mathcal{R}(T_{22}) \dotplus \mathcal{R}(X_{22}). \tag{3.14}$$

We make now the remark that we can change the f.d. $\mathcal{K}_2 = \mathcal{K}_2^+[+]\mathcal{K}_2^-$ without restricting the generality. Indeed, making this changing we have to replace T_{22} by UT_{22} where U is a unitary operator mapping $\mathcal{K}_2^-$ onto some maximal uniformly negative subspace and

$$\operatorname{ind}(UT_{22}) = \operatorname{ind}(U) + \operatorname{ind}(T_{22}) = \operatorname{ind}(T_{22}).$$

On the other hand, since $\tilde{T}$ is double contraction it follows that $\tilde{T}(\mathcal{K}_1^-[+]\mathcal{D}_{T^\sharp}^-)$ is a maximal uniformly negative subspace of $\mathcal{K}_2$ (cf.[14], Theorem 3.2) hence, without restricting the generality we can assume

$$\tilde{T}(\mathcal{K}_1^-[+]\mathcal{D}_{T^\sharp}^-) = \mathcal{K}_2^-.$$

Then the operator X_{12} in (3.11) vanishes hence $X_{22} = \mathcal{D}_{T^\sharp}^-$ and this yields

$$\operatorname{rank}(X_{22}) = \dim \mathcal{D}_{T^\sharp}^- = \infty.$$

From (3.13) we infer now that $\mathcal{R}(T_{22})$ has infinite codimension in $\mathcal{K}_2^-$, in particular

$$\operatorname{ind}(T_{22}) = -\infty,$$

hence the formula (3.8) is again verified.

3. Let now T be a quasi-contraction. Using Lemma 2.1 in [5], we obtain a Kreĭn space $\mathcal{D}_T$ with

$$\kappa^-(\mathcal{D}_T) = \kappa^-(I - T^\sharp T) < \infty$$

and an operator $\tilde{T} \in \mathcal{L}(\mathcal{K}_1, \mathcal{K}_2[+]\mathcal{D}_T)$

$$\tilde{T} = \begin{bmatrix} T \\ D_T \end{bmatrix} : \mathcal{K}_1 \rightarrow \mathcal{K}_2[+]\mathcal{D}_T ,$$

such that

$$\kappa^-(I - \tilde{T}^\sharp \tilde{T}) = 0, \quad \kappa^-(I - \tilde{T}\tilde{T}^\sharp) = \kappa^-(I - TT^\sharp).$$

On $\mathcal{D}_T$ we fix the f.d. $\mathcal{D}_T = \mathcal{D}_T^+[+]\mathcal{D}_T^-$ œ(where $\mathcal{D}_T^-$ can be chosen as the space considered during the proof of Lemma 3.3) and, with our notation, we have

$$\tilde{T}_{22} = \left[\begin{array}{c} T_{22} \\ X_{22} \end{array} \right] : \mathcal{K}_1^- \to \mathcal{K}_2^-[+]\mathcal{D}_T^-.$$

Using the result obtained at the previous step we have

$$-\kappa^-(I - TT^\sharp) = \mathrm{ind}\,(\tilde{T}_{22}) = \mathrm{ind}\,\left(\left[\begin{array}{c} T_{22} \\ 0 \end{array} \right] \right)$$

$$= \mathrm{ind}\,(T_{22}) - \dim \mathcal{D}_T^- = \mathrm{ind}\,(T_{22}) \quad -\kappa^-(I - T^\sharp T),$$

hence

$$\mathrm{ind}\,(T_{22}) = \kappa^-(I - T^\sharp T) - \kappa^-(I - TT^\sharp),$$

which proves the formula (3.8). ∎

Remark 3.6 Let T be in $\mathcal{L}(\mathcal{K}_1, \mathcal{K}_2)$. According to [2], Proposition 3.1, the following equality holds

$$\kappa^-(I - T^\sharp T) + \kappa^-(\mathcal{K}_2) = \kappa^-(I - TT^\sharp) + \kappa^-(\mathcal{K}_1). \tag{3.15}$$

This shows that in case either $\mathcal{K}_1$ or $\mathcal{K}_2$ are Pontryagin spaces (of finite negative signatures) then the formula (3.8) can be obtained at once.

Corollary 3.7 *Let $T \in \mathcal{L}(K_1, \mathcal{K}_2)$ be quasi–contractive, fix f.d. $\mathcal{K}_i = \mathcal{K}_i^+[+]\mathcal{K}_i^-$, $i = 1, 2$ and consider the representation (3.9). Then for any (Hilbert space) contraction $K \in \mathcal{L}(\mathcal{K}_1^-, \mathcal{K}_1^+)$, the operator $T_{21}K + T_{22}$ is semi–Fredholm, more precisely it has closed range and finite dimensional kernel, and its index is*

$$\mathrm{ind}\,(T_{21}K + T_{22}) = \kappa^-(I - T^\sharp T) - \kappa^-(I - TT^\sharp). \tag{3.16}$$

Proof. Let $K \in \mathcal{L}(\mathcal{K}_1^-, \mathcal{K}_2^+)$ be a (Hilbert space) contraction and denote

$$\mathcal{M} = G(K) = \{Kx + x \mid x \in \mathcal{K}_1^-\}.$$

$\mathcal{M}$ is a negative subspace of the Kreĭn space $\mathcal{K}_1$ and

$$T\mathcal{M} = \{(T_{11}Kx + T_{12}x) + (T_{21}Kx + T_{22}x) \mid x \in \mathcal{K}_1^-\}.$$

is a (closed) subspace of $\mathcal{K}_2$ such that $\kappa^+(T\mathcal{M}) \leq \kappa^-(I - T^\sharp T)$ (cf. Corollary 3.4). It follows that

$$\mathcal{R}(T_{21}K + T_{22}) = J_2^- T\mathcal{M}.$$

Using a similar argument as in the proof of Corollary 3.4 we show that $T_{21}K + T_{22}$ has closed range. Let us consider the subspace of $\mathcal{K}_1$

$$\mathcal{L} = \{Kx + x \mid x \in \ker(T_{21}K + T_{22})\}.$$

$\mathcal{L}$ is a subspace of $\mathcal{M}$ hence it is negative. From Corollary 3.4 $T\mathcal{L}$ is closed. Since $T\mathcal{L} \subseteq \mathcal{K}_2^+$ it follows

$$\dim(T\mathcal{L}) = \kappa^+(T\mathcal{L}) \le \kappa^+(T\mathcal{M}) \le \kappa^-(I - T^\sharp T) < \infty.$$

On the other hand, from Proposition 3.1 and the negativity of $\mathcal{L}$ we obtain that $\ker T \cap \mathcal{L}$ is finite dimensional hence, using the preceding inequality, it follows that $\mathcal{L}$ itself is finite dimensional. Since $\dim \mathcal{L} = \dim(\ker(T_{21}K + T_{22}))$ we thus proved that $\ker(T_{21}K + T_{22})$ has finite dimension. We proved that the operator $T_{21}K + T_{22}$ is semi-Fredholm.

The same argument as before shows that for any $\lambda \in [0,1]$ the operator $\lambda T_{21}K + T_{22}$ is semi-Fredholm. Using the local stability of the index of semi-Fredholm operator (e.g. see [11]) it follows that $\mathrm{ind}\,(\lambda T_{21}K + T_{22})$ is independent of $\lambda \in [0,1]$ hence, using Theorem 3.5,

$$\mathrm{ind}\,(T_{21}K + T_{22}) = \mathrm{ind}\,(T_{22}) = \kappa^-(I - T^\sharp T) - \kappa^-(I - TT^\sharp).$$

The proof is complete. ∎

The preceding result enables us to add some new insight on the action of contraction operators on maximal negative subspaces (compare with [14], Theorem 2.3).

Corollary 3.8 *Let $T \in \mathcal{L}(K_1, K_2)$ be a contraction and let $\mathcal{L}$ be a maximal negative subspace in $\mathcal{K}_1$. Then $T\mathcal{L}$ is a negative subspace of $\mathcal{K}_2$ such that*

$$\kappa^-((T\mathcal{L})^\perp) = \kappa^-(I - TT^\sharp).$$

Proof. Fix f.d. $\mathcal{K}_i = \mathcal{K}_i^+[+]\mathcal{K}_i^-$, $i = 1,2$ and let K denote the angular operator of $\mathcal{L}$. We consider the representation (3.9). Since T is contractive, the operator $T_{21}K + T_{22}$ is injective hence using (3.16) we get

$$\kappa^-(I - TT^\sharp) = -\mathrm{ind}\,(T_{21}K + T_{22}) = \mathrm{codim}_{\mathcal{K}_2^-}(\mathcal{R}(T_{21}K + T_{22})) = \kappa^-((T\mathcal{L})^\perp).$$

∎

We conclude this section with a result concerning the product of quasi–contractions. This represent the answer to a question raised by Aad Dijksma.

Corollary 3.9 *Let $T \in \mathcal{L}(\mathcal{K}_1, \mathcal{K}_2)$ and $S \in \mathcal{L}(\mathcal{K}_2, \mathcal{K}_3)$ be quasi-contractions. Then ST is also a quasi-contraction such that*

$$\kappa^-(I - (ST)^\sharp(ST)) \le \kappa^-(I - T^\sharp T) + \kappa^-(I - S^\sharp S)$$

and the following formula holds

$$\kappa^-(I - (ST)^\sharp(ST)) - \kappa^-(I - (ST)(ST)^\sharp)$$

$$= \kappa^-(I - T^\sharp T) - \kappa^-(I - TT^\sharp) + \kappa^-(I - S^\sharp S) - \kappa^-(I - SS^\sharp). \tag{3.17}$$

Proof. The fact that ST is also quasi-contractive follows from

$$I - (ST)^\sharp ST = I - T^\sharp T + T^\sharp(I - S^\sharp S)T.$$

This proves also the inequality.

Let us fix now the f.d. $\mathcal{K}_i = \mathcal{K}_i^+[+]\mathcal{K}_i^-$, for $i = 1, 2, 3$, and consider the matrix representations of S and T

$$T = \begin{bmatrix} T_{11} & T_{12} \\ T_{21} & T_{22} \end{bmatrix}, \quad T = \begin{bmatrix} S_{11} & S_{12} \\ S_{21} & S_{22} \end{bmatrix}.$$

We now consider the same kind of matrix representation of ST and notice that

$$(ST)_{22} = S_{21}T_{12} + S_{22}T_{22}. \tag{3.18}$$

For $\delta \in [0, 1]$ we consider the contractions $C_\delta \in \mathcal{L}(\mathcal{K}_2)$ defined by

$$C_\delta = \begin{bmatrix} \sqrt{1 - \delta^2} & 0 \\ 0 & \sqrt{1 + \delta^2} \end{bmatrix},$$

where the matrix of C_δ has to be understood with respect to the f.d. $\mathcal{K}_2 = \mathcal{K}_2^+[+]\mathcal{K}_2^-$. Then $T_\delta = C_\delta T$ is quasi-contractive and the same is ST_δ. Applying Theorem 3.5 to the quasi-contractions ST_δ it follows that for all $\delta \in [0, 1]$ the operators

$$F_\delta = \sqrt{1 - \delta^2}\, S_{21}T_{12} + \sqrt{1 + \delta^2}\, S_{22}T_{22}$$

are semi-Fredholm and then using the local stability of the index of semi-Fredholm operators we obtain

$$\operatorname{ind} F_0 = \operatorname{ind} F_1,$$

hence, taking into account of (3.18) we obtain

$$\operatorname{ind}((ST)_{22}) = \operatorname{ind}(S_{21}T_{12} + S_{22}T_{22}) = \operatorname{ind}(F_0)$$

$$= \operatorname{ind}(F_1) = \operatorname{ind}(S_{22}T_{22}) = \operatorname{ind}(S_{22}) + \operatorname{ind}(T_{22}).$$

Applying now the formula (3.8) for T, S, and ST we obtain the equation (3.17). ∎

Remark 3.10 It turns out that the index formula obtained in Theorem 3.5 enables us to add some new insight in the lifting theory of quasi-contractions. More precisely, let $T \in \mathcal{L}(\mathcal{K}_1, \mathcal{K}_2)$ be a quasi-contraction and let $\mathcal{K}_1'$ and $\mathcal{K}_2'$ be Kreĭn spaces. If $\mathcal{K}_1'$ are Pontryagin spaces then for any quasi-contractive lifting $\tilde{T} \in \mathcal{L}(\mathcal{K}_1[+]\mathcal{K}_1', \mathcal{K}_2[+]\mathcal{K}_2')$, i.e. $P_{\mathcal{K}_2}\tilde{T}|\mathcal{K}_1 = T$, we must have the equality

$$\kappa^-(I - \tilde{T}^\sharp\tilde{T}) - \kappa^-(I - \tilde{T}\tilde{T}^\sharp)$$

$$= \kappa^-(I - T^\sharp T) - \kappa^-(I - TT^\sharp) + \kappa^-(\mathcal{K}_1') - \kappa^-(\mathcal{K}_2').$$

(Compare with [5] and [6] where this kind of identities were obtained using the desription of the lifted operators.)

4. Doubly quasi–contractive operators

As we already mentioned, for any operator T acting between Kreĭn spaces $\mathcal{K}_1$ and $\mathcal{K}_2$, the following relation holds

$$\kappa^-(I - T^\sharp T) + \kappa^-(\mathcal{K}_2) = \kappa^-(I - TT^\sharp) + \kappa^-(\mathcal{K}_1),$$

(see Remark (3.6)). Thus, in case $\mathcal{K}_1$ and $\mathcal{K}_2$ are Pontryagin spaces (of finite negative signatures), this shows that the numbers $\kappa^-(I - T^\sharp T)$ and $\kappa^-(I - TT^\sharp)$ are simultaneously determined, in particular they are simultaneously finite or not. If $\kappa^-(\mathcal{K}_1)$ and $\kappa^-(\mathcal{K}_2)$ are infinite then it can be shown by examples that the numbers $\kappa^-(I - T^\sharp T)$ and $\kappa^-(I - TT^\sharp)$ are fairly arbitrary.

An operator $T \in \mathcal{L}(\mathcal{K}_1, \mathcal{K}_2)$ is called a *double quasi-contraction* if both of T and $T^\sharp$ are quasi–contractions, i.e. both of the numbers $\kappa^-(I - T^\sharp T)$ and $\kappa^-(I - TT^\sharp)$ are finite. As a first task we investigate conditions assuring the double quasi–contractivity of quasi-contractions.

Theorem 4.1 *Let $T \in \mathcal{L}(, \mathcal{K}_1, \mathcal{K}_2)$ be quasi-contractive. The following statements are equivalent:*

(i) *T is double quasi-contraction.*

(ii) *For some (equivalently, for any) f.d. $\mathcal{K}_i = \mathcal{K}_i^+[+]\mathcal{K}_i^-$, $i = 1, 2$, the operator $P_{\mathcal{K}_2^-}T|\mathcal{K}_1^-$ in $\mathcal{L}(\mathcal{K}_1^-, \mathcal{K}_2^-)$ is Fredholm .*

(iii) *For some (equivalently, for any) maximal negative subspace $\mathcal{M}$ in $\mathcal{K}_1$, the number $\kappa^-((T\mathcal{M})^\perp)$ is finite.*

(iv) *For some (equivalently, for any) maximal negative subspace $\mathcal{L}$ in $\mathcal{K}_2$ the linear manifold $T^\sharp\mathcal{L}$ is closed and the numbers $\kappa^+(T^\sharp\mathcal{L})$ and $\dim(\ker T^\sharp \cap \mathcal{L})$ are finite.*

Proof. $(i) \Leftrightarrow (ii)$ Since T is quasi–contractive and f.d. $\mathcal{K}_i = \mathcal{K}_i^+[+]\mathcal{K}_i^-$, $i = 1, 2$ are fixed, from Theorem 3.5 it follows that $\kappa^-(I - TT^\sharp)$ is finite if and only if the operator $P_{\mathcal{K}_2^-}T|\mathcal{K}_1^-$ is Fredholm.

$(i) \Rightarrow (iii)$ Let $\mathcal{M}$ be a maximal negative subspace in $\mathcal{K}_1$ and $f \in (T\mathcal{M})^\perp$. Then $T^\sharp f \in \mathcal{M}^\perp$ hence $[T^\sharp f, T^\sharp f] \geq 0$ and this yields

$$[f, f] - [T^\sharp f, T^\sharp f] < 0,$$

provided f is assumed negative. This shows that

$$\kappa^-((T\mathcal{M})^\perp) \leq \kappa^-(I - TT^\sharp),$$

hence $\kappa^-((T\mathcal{M})^\perp)$ is finite.

(iii)$\Rightarrow$ (ii) Let $\mathcal{M}$ be a maximal negative subspace of $\mathcal{L}_1$ such that $\kappa^-((T\mathcal{M})^\perp)$ is finite. Fix f.d. $\mathcal{K}_i = \mathcal{K}_i^+[+]\mathcal{K}_i^-$, $i = 1, 2$ and consider the representation (3.9) of T. Let $K \in \mathcal{L}(\mathcal{K}_1^-, \mathcal{K}_1^+)$ be the angular operator of $\mathcal{M}$. Since $\mathcal{K}_2^- \ominus \mathcal{R}(T_{21}K + T_{22})$ is a subspace in $(T\mathcal{M})^\perp$ it follows that $\mathrm{codim}_{\mathcal{K}_2^-}(\mathcal{R}(T_{21}K + T_{22}))$ is finite hence, using Corollary 3.7, the operator $T_{21}K + T_{22}$ is Fredholm. In particular, this means that $\mathrm{ind}\,(T_{21}K + T_{22})$ is finite. From (3.15) we obtain that $\kappa^-(I - TT^\sharp)$ is also finite.

(i)$\Rightarrow$ (iv) Let $\mathcal{L}$ be a maximal negative subspace of $\mathcal{K}_2$. Since $T^\sharp$ is quasi-contraction, from Proposition 3.1 it follows that $\ker T^\sharp \cap \mathcal{L}$ is finite dimensional. Moreover, using Corollary 3.4 we obtain that $T^\sharp \mathcal{L}$ is closed and $\kappa^+(T^\sharp\mathcal{L}) \leq \kappa^-(I - TT^\sharp) < \infty$. The fact that $\kappa^-((T^\sharp\mathcal{L})^\perp)$ is also finite follows from the proof of (i)$\Rightarrow$ (iii) replacing T by $T^\sharp$ and $\mathcal{M}$ by $\mathcal{L}$.

(iv)$\Rightarrow$ (ii) Let $\mathcal{L}$ be as stated at (iv). Fix a f.d. $\mathcal{K}_1 = \mathcal{K}_1^+[+]\mathcal{K}_1^-$. Since T is quasi-contractive $T\mathcal{K}_1^-$ is a pseudo-regular subspace of $\mathcal{K}_2$ such that $\kappa^0(T\mathcal{K}_1^-) + \kappa^+(T\mathcal{K}_1^-) \leq \kappa^-(I - T^\sharp T) < \infty$, hence

$$T\mathcal{K}_1^- = \mathcal{S}^-[+]\mathcal{S}^0[+]\mathcal{S}^+,$$

where $\mathcal{S}^-$ is uniformly negative and $\dim \mathcal{S}^0 + \dim \mathcal{S}^+ \leq \kappa^-(I - T^\sharp T)$. Choose f.d. $\mathcal{K}_2 = \mathcal{K}_2^+[+]\mathcal{K}_2^-$ such that $\mathcal{S}^- \subseteq \mathcal{K}_2^-$ and $\mathcal{S}^+ \subseteq \mathcal{K}_2^+$ (this is possible by the Phillips Extension Theorem) and represent T as in (3.9) with respect to these f.d.. Then

$$\mathrm{rank}\,(T_{12}) \leq \dim \mathcal{S}^0 + \dim \mathcal{S}^+ \leq \kappa^-(I - T^\sharp T) < \infty.$$

We consider now the angular operator L of $\mathcal{L}$

$$\mathcal{L} = \{x + Lx \mid x \in \mathcal{K}_2^-\}$$

and notice that $T^\sharp$ has the representation

$$T^\sharp = \begin{bmatrix} T_{11}^* & -T_{21}^* \\ -T_{12}^* & T_{22}^* \end{bmatrix},$$

hence

$$T^\sharp \mathcal{L} = \{(T_{11}^* Lx - T_{21}^*) + (-T_{12}^* Lx + T_{22}^* x)|\ x \in \mathcal{K}_2^-\}.$$

Since $T^\sharp \mathcal{L}$ is closed and $\kappa^+(T^\sharp \mathcal{L})$ is finite, from here we infer (e.g. as in the proof of Corollary 3.4) that $\mathcal{R}(-T_{12}^* L + T_{22}^*)$ is closed. Since $\kappa^+(T^\sharp \mathcal{L})$ and $\dim(\ker T^\sharp \cap \mathcal{L})$ are finite it follows that $\ker(-T_{12}^* L + T_{22}^*)$ is finite. We have thus proved that $-T_{12}^* L + T_{22}^*$ is semi-Fredholm and

$$\operatorname{ind} T_{22}^* = \operatorname{ind}(-T_{12}^* L + T_{22}^*) < \infty,$$

where we took into account that T_{12} has finite rank. Since we already know that $\operatorname{ind} T_{22} < \infty$ (from Theorem 3.5) it follows that T_{22} has finite index, hence it is a Fredholm operator. $\blacksquare$

The equivalence of (i) and (iv) in Theorem 4.1 has as consequence a characterization of double contractions obtained by M. A. Dritschel and J. Rovnyak by a completely different method (cf. [7], Theorem 2.1.5).

Corollary 4.2 *Let $T \in \mathcal{L}(\mathcal{K}_1, \mathcal{K}_2)$ be contraction. Then T is a double contraction if and only if $T^\sharp$ maps some maximal negative subspace of $\mathcal{K}_2$ in a one-to-one way into a negative subspace of $\mathcal{K}_1$.*

Proof. One implication is clear. Conversely, assume that $\mathcal{L}$ is a maximal negative subspace of $\mathcal{K}_2$ which is mapped by $T^\sharp$ in a one-to-one way into a negative subspace of $\mathcal{K}_1$. This means that $T^\sharp \mathcal{L}$ is closed and $\kappa^+(T^\sharp \mathcal{L}) = \dim(\ker T^\sharp \cap \mathcal{L}) = 0$. From Theorem 4.1 it follows that $\kappa^-(I - TT^\sharp)$ is finite. Thus

$$\operatorname{ind} T_{22} = -\kappa^-(I - TT^\sharp) \le 0.$$

On the other hand, with the notation as in the proof of $(iv) \Rightarrow (ii)$ of Theorem 4.1, from $\kappa^+(T^\sharp \mathcal{L}) = \dim(\ker T^\sharp \cap \mathcal{L}) = 0$ it follows $\ker(-T_{12}^* L + T_{22}^*) = 0$ hence

$$\operatorname{ind} T_{22} = -\operatorname{ind} T_{22}^* = -\operatorname{ind}(-T_{12}^* L + T_{22}^*) \ge 0.$$

From these two inequalities we obtain $\kappa^-(I - TT^\sharp) = 0$. $\blacksquare$

We are concerned now with the question of estimating the gap between quasi-contractions and double quasi-contractions.

Theorem 4.3 *Let $T \in \mathcal{L}(\mathcal{K}_1, \mathcal{K}_2)$ be a quasi-contraction and denote $\kappa = \kappa^-(I - T^\sharp T)$. Then there exists a regular subspace $\mathcal{K}$ of $\mathcal{K}_2$ such that $\kappa^+(\mathcal{K}^\perp) \le \kappa$, the operator $\hat{T} = P_\mathcal{K} T \in \mathcal{L}(\mathcal{K}_1, \mathcal{K})$ is double quasi-contraction,*

$$\kappa^-(I - \hat{T}^\sharp \hat{T}) \le \kappa^-(I - \hat{T}\hat{T}^\sharp) \le 2\kappa$$

and

$$\kappa^-(I - \hat{T}\hat{T}^\sharp) - \kappa^-(I - \hat{T}^\sharp \hat{T}) \le \kappa.$$

Proof. We use Lemma 3.3 to produce a contractive operator $\tilde{T}\colon \mathcal{K}_1 \to \mathcal{K}_2[+]\mathcal{K}_2'$, where $\dim \mathcal{K}_2' = \kappa^-(\mathcal{K}_2') = \kappa$ and $P_{\mathcal{K}_2}\tilde{T} = T$. Then we apply [7], Theorem 2.1.4 to produce a uniformly negative subspace $\mathcal{L}$ of $\mathcal{K}_2[+]\mathcal{K}_2'$ such that denoting $\mathcal{R} = \mathcal{L}^\perp$ the operator $P_{\mathcal{R}}\tilde{T}\colon \mathcal{K}_1 \to \mathcal{R}$ is double contractive.

We claim that $\mathcal{R}' = \mathcal{R} + \mathcal{K}_2'$ is a regular subspace with uniformly negative orthogonal companion. Indeed, $\mathcal{R}'$ is closed since $\mathcal{K}_2'$ is finite dimensional and since $\mathcal{R}'^\perp = \mathcal{R}^\perp \cap \mathcal{K}_2'^\perp \subset \mathcal{L}$ it follows that $\mathcal{R}'^\perp$ is uniformly negative, in particular it is regular, hence $\mathcal{R}'$ is also regular.

Letting now $\check{T} = P_{\mathcal{R}'}\tilde{T}\colon \mathcal{K}_1 \to \mathcal{R}'$ we obtain a double quasi–contraction such that

$$\kappa^-(I - \check{T}^\sharp\check{T}) \leq \kappa^-(I - \check{T}\check{T}^\sharp) \leq \kappa,$$

and

$$\kappa^-(I - \check{T}\check{T}^\sharp) - \kappa^-(I - \check{T}^\sharp\check{T}) \leq \kappa.$$

These follow from the Remark 3.6 and the index formula obtained in Theorem 3.5.

We define now $\mathcal{K}$ as the orthogonal complement of $\mathcal{K}_2'$ with respect to $\mathcal{R}'$, i.e. $\mathcal{R}' = \mathcal{K}[+]\mathcal{K}_2'$. Then $\kappa^+(\mathcal{K}^\perp) \leq \kappa$. Denoting $\hat{T} = P_{\mathcal{K}}\check{T}$ it follows $\hat{T} = P_{\mathcal{K}}\tilde{T} = P_{\mathcal{K}}T$ and the formulæ and the inequalities for the signatures of defect of $\hat{T}$ follow immediately using again Remark 3.6, Theorem 3.5 and the preceding inequalities. ∎

Other results concerning extensions to double quasi–contractions can be proved (e.g. see the technique developed in [2], [5], [6], and [7]). Here we only record a consequence of the existence of the elementary rotation which gives an estimation of the gap between double quasi–contractions and double contractions.

Theorem 4.4 *Let $T \in \mathcal{L}(\mathcal{K}_1, \mathcal{K}_2)$ be a double quasi–contraction and let also $\mathcal{K}_1'$ and $\mathcal{K}_2'$ be Kreĭn spaces. Then there exists a double contraction $\tilde{T} \in \mathcal{L}(\mathcal{K}_1[+]\mathcal{K}_1', \mathcal{K}_2[+]\mathcal{K}_2')$ such that*

$$P_{\mathcal{K}_2}\tilde{T}|\mathcal{K}_1 = T \tag{4.1}$$

if and only if

$$\kappa^-(I - TT^\sharp) \leq \kappa^-(\mathcal{K}_1'), \quad \kappa^-(I - T^\sharp T) \leq \kappa^-(\mathcal{K}_2'), \tag{4.2}$$

and

$$\kappa^-(I - T^\sharp T) + \kappa^-(\mathcal{K}_1') = \kappa^-(I - TT^\sharp) + \kappa^-(\mathcal{K}_2'). \tag{4.3}$$

Proof. Let $\tilde{T} \in \mathcal{L}(\mathcal{K}_1[+]\mathcal{K}_1', \mathcal{K}_2[+]\mathcal{K}_2')$ be a double contraction such that (4.1) holds. Then, the inequalities (4.2) are consequences of [5], Theorem 2.3. In order to prove (4.3) we fix f.d. on $\mathcal{K}_1, \mathcal{K}_2, \mathcal{K}_1'$ and $\mathcal{K}_2'$. Then consider the operators

$$\tilde{T}_{22} = P_{\mathcal{K}_2^-[+]\mathcal{K}_2'^-}\tilde{T}|\mathcal{K}_1^-[+]\mathcal{K}_1'^-, \quad T_{22} = P_{\mathcal{K}_2^-}T|\mathcal{K}_1^-.$$

From (4.1) we have, with respect to the decompositions $\mathcal{K}_1^-[+]\mathcal{K}_1'^-$ and $\mathcal{K}_2^-[+]\mathcal{K}_2'^-$, the following representation

$$\tilde{T}_{22} = \begin{bmatrix} T_{22} & X \\ Y & Z \end{bmatrix}.$$

Since $\tilde{T}_{22}$ is invertible, in particular it is a Fredholm operator of index 0, and taking into account that T_{22} is a Fredholm operator of index

$$\operatorname{ind}(T_{22}) = \kappa^-(I - T^\sharp T) - \kappa^-(I - TT^\sharp),$$

we easily obtain the formula (4.3).

In order to prove the converse implication we first produce a certain doubly contractive extension of T. To this end let $R(T) \in \mathcal{L}(\mathcal{K}_1[+]\mathcal{D}_{T^*}, \mathcal{K}_2[+]\mathcal{D}_T)$ be the elementary rotation of T (see [2]). $R(T)$ is a unitary extension of T, in particular it is double contraction.

Let now $\mathcal{K}_1'$ and $\mathcal{K}_2'$ be Kreĭn spaces such that (4.2) and (4.3) hold. Since (4.2) holds there, exists $\Gamma_2 \in \mathcal{L}(\mathcal{D}_T, \mathcal{K}_2')$ contraction and also there exists $\Gamma_1 \in \mathcal{L}(\mathcal{K}_1', \mathcal{D}_{T^*})$ such that $\Gamma_1^\sharp$ is a contraction. Then (see (3.14)) we have

$$\kappa^-(I - TT^\sharp) + \kappa^-(I - \Gamma_1^\sharp\Gamma_1) = \kappa^-(\mathcal{K}_1'),$$

and

$$\kappa^-(I - T^\sharp T) + \kappa^-(I - \Gamma_2\Gamma_2^\sharp) = \kappa^-(\mathcal{K}_2'),$$

hence, using (4.3) it follows that there exists a double contraction $\Gamma \in \mathcal{L}(\mathcal{D}_{\Gamma_1}, \mathcal{D}_{\Gamma_2^*})$. We consider now the operators

$$\hat{\Gamma}_1 \in \mathcal{L}(\mathcal{K}_1[+]\mathcal{K}_1', \mathcal{K}_1[+]\mathcal{D}_{T^*}), \quad \hat{\Gamma}_1 = \begin{bmatrix} I & 0 \\ 0 & \Gamma_1 \end{bmatrix},$$

$$\hat{\Gamma}_2 \in \mathcal{L}(\mathcal{K}_2[+]\mathcal{D}_T, \mathcal{K}_2[+]\mathcal{K}_2'), \quad \hat{\Gamma}_2 = \begin{bmatrix} I & 0 \\ 0 & \Gamma_2 \end{bmatrix},$$

and

$$\hat{\Gamma} \in \mathcal{L}(\mathcal{K}_1[+]\mathcal{D}_{\Gamma_1}, \mathcal{K}_2[+]\mathcal{D}_{\Gamma_2^*}), \quad \hat{\Gamma} = \begin{bmatrix} 0 & 0 \\ 0 & \Gamma \end{bmatrix}.$$

Using these we define

$$\tilde{T} = \hat{\Gamma}_2 R(T)\hat{\Gamma}_1 + D_{\hat{\Gamma}_2^*}\hat{\Gamma}D_{\hat{\Gamma}_1}.$$

Using the operators $U(\tilde{T})$ and $U_*(\tilde{T})$ obtained in [5], Theorem 2.3, it follows that $\tilde{T}$ is double contraction and clearly it satisfies the condition (4.1). ∎

In the remaining part of this section we will concentrate on a spectral characterization of double quasi–contractions, more precisely we will generalize the spectral characterization of double contractions obtained by M.G.Kreĭn and Yu.L.Shmulyan in [15], Theorem 5.3 to the case of double quasi–contractions.

Lemma 4.5 *Let $T \in \mathcal{L}(\mathcal{K}_1, \mathcal{K}_2)$ be a double quasi-contraction. Then:*

(i) $\sigma(T^\sharp T) \setminus \mathbf{R}_+$ *is a finite set of points, all of them eigenvalues of finite multiplicities (i.e. the corresponding root subspaces have finite dimension).*

(ii) *Denoting*
$$\kappa_1 = \kappa^-(I - T^\sharp T), \;\; \kappa_2 = \kappa^-(I - TT^\sharp),$$
then:

$$\operatorname{rank} E(\{\lambda \in \sigma(T^\sharp T) \setminus \mathbf{R}_+ \mid \operatorname{Im}\lambda \geq 0\}; T^\sharp T) \leq 2(\kappa_1 + 1)\kappa_1 + \kappa_2. \tag{4.4}$$

Proof. According to [16] the operator $T^\sharp T$ is definitizable and denoting by E its spectral function then from Lemma 2.1 it follows that for $\varepsilon > 0$ sufficiently small we have

$$\kappa^-(I - T^\sharp T) = \kappa^+(E(1 + \varepsilon, +\infty)) + \kappa^-(E(-\infty, 1 - \varepsilon))$$
$$+\kappa^-((I - T^\sharp T)|E(1 - \varepsilon, 1 + \varepsilon)\mathcal{K}_1) + \sum_{\operatorname{Im}\lambda > 0} \operatorname{rank}(E(\lambda; T^\sharp T)). \tag{4.5}$$

From here we already know that the non-real part of $\sigma(T^\sharp T)$ is finite and contains only eigenvalues of finite multiplicity. Let $\Delta \in \mathcal{R}_{T^\sharp T}$ be a set contained in $(-\infty, 0)$ and such that $c(T^\sharp T) \cap \Delta = \emptyset$ and $E(\Delta)$ is positive. Then, for any $f \in E(\Delta)\mathcal{K}_1 \setminus \{0\}$ we have (see [16])

$$[T^\sharp T f, f] = \int_\Delta \lambda d[E(\lambda) f, f] < 0 \tag{4.6}$$

and

$$[(T^\sharp T)^2 f, f] = \int_\Delta \lambda^2 d[E(\lambda) f, f] > 0. \tag{4.7}$$

Since T is one-to-one on the subspace $E(\Delta)\mathcal{K}_1$, taking into account that

$$[Tf, Tf] - [T^\sharp T f, T^\sharp T f] = [T^\sharp T f, f] - [(T^\sharp T)^2 f, f] \tag{4.8}$$

and using (4.6) and (4.7), it follows that

$$\operatorname{rank}(E(\Delta)) \leq \kappa^-(I - TT^\sharp). \tag{4.9}$$

In particular this and (4.5) show that the set $\sigma(T^\sharp T) \setminus \mathbf{R}_+$ is finite.

Let $\alpha \in c(T^\sharp T)$ be negative. Then α is an isolated eigenvalue of $T^\sharp T$. We denote by k_α the geometric multiplicity of α. Then $\ker(\alpha I - T^\sharp T)^{k_\alpha}$ is the root subspace coresponding to α and the following decomposition holds

$$\ker(\alpha I - T^\sharp T)^{k_\alpha} = (\mathcal{S}_+[+]\mathcal{S}_-[+]\mathcal{S}_0)\dot{+} \sum_{k=1}^{k_\alpha - 1} \mathcal{S}_k, \tag{4.10}$$

where $\mathcal{S}_+$ is a positive subspace, $\mathcal{S}_-$ a negative subspace and $\mathcal{S}_0$ a neutral subspace such that

$$\ker(\alpha I - T^\sharp T) = \mathcal{S}_+[+]\mathcal{S}_-[+]\mathcal{S}_0$$

and for any $0 \leq k \leq k_\alpha - 1$ we have

$$\ker(\alpha I - T^\sharp T)^{k+1} = \ker(\alpha I - T^\sharp T)\dot{+}\sum_{j=1}^{k}\mathcal{S}_j.$$

Using now an argument as in the proof of [10], Theorem 7.2 it follows that each subspace $\mathcal{S}_k$ is isomorphic with some neutral subspace of $\ker(\alpha I - T^\sharp T)$ hence

$$\dim\mathcal{S}_k \leq \kappa^-(I - T^\sharp T), \quad 0 \leq k \leq k_\alpha - 1.$$

It follows

$$\dim(\sum_{k=1}^{k_\alpha-1}\mathcal{S}_k) \leq (k_\alpha - 1)\kappa_1. \tag{4.11}$$

Since $\sigma(T^\sharp T)\setminus\mathbf{R}_+$ is finite it follows that this is a spectral set of $T^\sharp T$ and also (see [17]) all the elements of $\sigma(T^\sharp T)\setminus\mathbf{R}_+$ are eigenvalues. In particular, the same holds for the set $\sigma = \{\lambda \in \sigma(T^\sharp T)\setminus\mathbf{R}_+|\ \operatorname{Im}\lambda \geq 0\}$. We decompose σ into the disjoint union

$$\sigma = \sigma_+ \sqcup \sigma_- \sqcup (c(T^\sharp T)\cap\sigma)\sqcup\sigma_0, \tag{4.12}$$

where

$$\sigma_+ = \{\lambda < 0|\ E(\lambda)\ \text{positive}\ \},$$

$$\sigma_- = \{\lambda < 0|\ E(\lambda)\ \text{negative}\ \},$$

$$\sigma_0 = \{\lambda \in \sigma|\operatorname{Im}\lambda > 0\}.$$

All these are spectral sets for $T^\sharp T$. From (4.5) it follows

$$\dim(E(\sigma_0\cup\sigma_-;T^\sharp T)\mathcal{K}_1[+]\sum_{\alpha\in c(T^\sharp T)\cap\sigma}(\mathcal{S}_-(\alpha)[+]\mathcal{S}_0(\alpha))) \leq \kappa_1. \tag{4.13}$$

On the other hand, a slight variation of the argument used to prove (4.9) shows that

$$\dim(E(\sigma_+)\mathcal{K}_1[+]\sum_{\alpha\in c(T^\sharp T)\cap\sigma}\mathcal{S}_+(\alpha)) \leq \kappa_2. \tag{4.14}$$

Also, since $T^\sharp T$ has a definitizing polynomial of degree $2\kappa_1 + 1$ it follows (see [17]) that

$$\sum_{\alpha\in c(T^\sharp T)}(k_\alpha - 1) \leq 2\kappa_1 + 1,$$

144 A. Gheondea

hence, using (4.11), we obtain

$$\dim(\sum_{\alpha \in c(T^\sharp T) \cap \sigma} \sum_{k=1}^{k_\alpha - 1} \mathcal{S}_k(\alpha)) \le (2\kappa_1 + 1)\kappa_1. \tag{4.15}$$

From (4.13)-(4.15) and taking into account of the decompositions (4.12) and (4.10) we obtain the estimation given in (4.4). In particular, this shows that the root subspace of any eigenvalue in $\sigma(T^\sharp T) \setminus \mathbf{R}_+$ is finite dimensional. $\blacksquare$

Theorem 4.6 *Let $T \in \mathcal{L}(\mathcal{K}_1, \mathcal{K}_1)$ be quasi–contractive. Then, in order that T be doubly quasi–contractive, it is necessary and sufficient that T satisfy the following conditions:*

(i) *$\sigma(T^\sharp T) \setminus \mathbf{R}_+$ is a finite set of points, all of them eigenvalues of finite multiplicities.*

(ii) *$\ker T^\sharp$ is a pseudo–regular subspace such that both $\kappa^-(\ker T^\sharp)$ and $\kappa^0(\ker T^\sharp)$ are finite.*

Proof. Assuming that T is doubly quasi–contractive, the assertion (i) was already proved in Lemma 4.5 while the assertion (ii) is a consequence of Proposition 3.1.

Conversely, let us assume that the quasi–contraction T satisfies both of the conditions (i) and (ii). In order to prove that $T^\sharp$ is also a quasi–contraction, we adapt the proof of [15], Theorem 5.2, i.e. discarding some finite dimensional subspaces, we obtain a certain left polar decomposition of T. We devide the proof into several steps.

1. *Without restricting the generality, we can assume that there exists $R \in \mathcal{L}(\mathcal{K}_1)$, $R^\sharp = R$ such that*

(a) *$T^\sharp T = R^2$,*

(b) *$\sigma(R) \subset [0, +\infty)$,*

(c) *$\ker R = \ker(T^\sharp T)$ is uniformly positive.*

Indeed, let E_0 be the spectral projection of $T^\sharp T$ corresponding $\sigma(T^\sharp T) \setminus \mathbf{R}_+$. Then E_0 has finite rank. Let us denote by $A \in \mathcal{L}((I - E_0)\mathcal{K}_1)$ the operator

$$A = T^\sharp T | (I - E_0)\mathcal{K}_1.$$

Then $\sigma(A) \subset \mathbf{R}_+$ and since $T^\sharp T$ is definitizable, so is A. Letting E denote the spectral function of A we choose $\varepsilon > 0$ sufficiently small such that the closed interval $[0, \varepsilon]$ contains no critical points of A, except (possibly) 0. We denote

$$A_1 = E([0, \varepsilon])A, \quad A_2 = E((\varepsilon, +\infty))A,$$

as operators acting on appropriate spectral subspaces.

The operator A_2 is selfadjoint, $\sigma(A_2) \subset [\varepsilon, +\infty)$ hence, by Riesz–Dunford functional calculus, there exists its square root, $R_2 = A_2^{1/2}$, $R_2 = R_2^\sharp$ and $\sigma(R_2) \subset [\varepsilon^{1/2}, +\infty)$.

The operator A_1 is selfadjoint definitizable, $\sigma(A_1) \subseteq [0, \varepsilon]$ and $c(A_1) \subseteq \{0\}$. Let $\mathcal{N}$ be a neutral subspace in $E([0, \varepsilon])$ which is maximal A_1-invariant. Then $\mathcal{N}$ is finite dimensional since $\kappa^-(E([0, \varepsilon]))$ is finite. For a fixed f.s. J on the Kreĭn space $E([0, \varepsilon])\mathcal{K}_1$, the subspace $\mathcal{N}\dot{+}J\mathcal{N}$ is regular hence there exists a regular subspace $\hat{\mathcal{K}}_1$ such that

$$E([0, \varepsilon])\mathcal{K}_1 = \mathcal{N}\dot{+}\hat{\mathcal{K}}_1\dot{+}J\mathcal{N}$$

and, with respect to this decomposition, A_1 has the representation

$$A_1 = \begin{bmatrix} * & * & * \\ 0 & \hat{A}_1 & * \\ 0 & 0 & * \end{bmatrix}$$

where $\hat{A}_1 \in \mathcal{L}(\hat{\mathcal{K}}_1)$ is selfadjoint definitizable with no critical points at all. It follows that $\hat{A}_1$ is similar with a selfadjoint operator on a Hilbert space, hence there exists its square root $R_1 = \hat{A}_1^{1/2}$, R_1 is selfadjoint and $\sigma(R_1) \subseteq [0, \varepsilon^{1/2}]$. Let now P denote the orthogonal projection of $\mathcal{K}_1$ onto the regular subspace

$$E(\sigma(T^\sharp T) \setminus \mathbf{R}_+; T^\sharp T)\mathcal{K}_1[+](\mathcal{N}\dot{+}J\mathcal{N})[+]\mathcal{S},$$

where $\mathcal{S}$ is a maximal uniformly negative subspace of $\ker \hat{A}_1$. Considering now the operator

$$R = R_1|\mathcal{S}^\perp[+]R_2,$$

we verify immediately that R satisfies the required conditions (a), (b), and (c), for $T|(I-P)\mathcal{K}_1$ instead of T. Since P has finite rank, $T|(I-P)\mathcal{K}_1$ is also quasi-contractive and T and $T|(I - P)\mathcal{K}_1$ are simultaneously doubly quasi-contractive.

2. *There exists a bounded isometry* $W : \overline{\mathcal{R}(T)} \rightarrow \mathcal{K}_1$ *such that*

$$R = WT. \tag{4.16}$$

Indeed, since $\ker T \subseteq \ker T^\sharp T = \ker R$, (4.16) uniquely determines an operator $W : R(T) \rightarrow \mathcal{K}_1$. Since $R^2 = T^\sharp T$ it follows that W is isometric. We now make the remark that the properties (a), (b), and (c) of R enable the same reasoning as in the proof of [15], Theorem 5.1 work, thus proving that W is bounded.

3. *There exists a finite rank orthogonal projection* $Q \in \mathcal{L}(\mathcal{K}_2)$ *and a partial isometry* $V \in \mathcal{L}(\mathcal{K}_1, \mathcal{K}_2)$ *such that* $V^\sharp$ *is quasi-contractive and*

$$VR = (I - Q)T. \tag{4.17}$$

146   A. Gheondea

Indeed, we consider the factorization (4.16) and notice that $\overline{\mathcal{R}(T)} = (\ker T^\sharp)^\perp$ is pseudo-regular hence we have a decomposition

$$\overline{\mathcal{R}(T)} = \mathcal{R}[+]\mathcal{P},$$

where $\mathcal{R}$ is a regular subspace of $\mathcal{K}_2$ and $\mathcal{P}$ is a neutral subspace of $\mathcal{K}_2$ such that $\dim \mathcal{P} = \kappa^0(\overline{\mathcal{R}(T)}) = \kappa^0(\ker T^\sharp) < \infty$. Moreover, there exists a regular subspace $\mathcal{S} \subseteq \mathcal{R}^\perp$ such that $\mathcal{P} \subseteq \mathcal{S}$ and $\dim \mathcal{S} = 2 \dim \mathcal{P} < \infty$. We let Q denote the orthogonal projection onto $\mathcal{S}$. Since $\mathcal{R}$ is regular and W is isometric, it follows that $W|\mathcal{R}$ is one-to-one and $W\mathcal{R}$ is a regular subspace, too. Let $V = (W|\mathcal{R})^{-1} : W\mathcal{R} \to \mathcal{R}$ and extend it to a partial isometry $V \in \mathcal{L}(\mathcal{K}_1, \mathcal{K}_2)$ such that the left support of V is $\mathcal{R}$ and its right support is $W\mathcal{R}$. Then the factorization (4.17) follows. In addition, $V^\sharp$ is quasi–contractive since

$$\kappa^-(I - VV^\sharp) = \kappa^-(\mathcal{R}^\perp) = \kappa^-(\ker T^\sharp) + \kappa^0(\ker T^\sharp) < \infty.$$

4. $T^\sharp$ *is quasi–contractive.* Indeed, from (4.17) we have

$$T^\sharp(I - Q) = RV^\sharp.$$

Since $R^2 = T^\sharp T$ and T is quasi–contractive, so is R. $V^\sharp$ is also quasi–contractive; hence the same is true of $T^\sharp(I - Q)$. Since Q has finite rank, from here we infer that $T^\sharp$ is quasi–contractive. $\blacksquare$

Remark 4.7 The proof of Theorem 4.6 gives also an estimation for $\kappa^-(I - TT^\sharp)$. However, we think that this estimate can be considerably improved.

References

[1] T. Ando: *Linear Operators in Kreĭn Spaces*, Lecture Notes, Hokkaido University 1979.

[2] Gr. Arsene, T. Constantinescu and A. Gheondea: Lifting of operators and prescribed numbers of negative squares, *Michigan Math. J.*, **34**(1987), 201–216.

[3] T. Ya. Azizov and I. S. Iokhvidov: *Foundations of the Theory of Linear Operators in Spaces with Indefinite Metric* [Russian], Nauka, Moscow 1986.

[4] J. Bognár: *Indefinite Inner Product Spaces*, Springer-Verlag, Berlin 1974.

[5] T. Constantinescu and A. Gheondea: Minimal signature in lifting of operators, I, *J. Operator Theory*, **22**(1989), 345–367.

[6] T. Constantinescu and A. Gheondea: Minimal signature in lifting of operators, II, *J. Funct. Anal.*, **103**(1992), 317–351.

[7] M. A. Dritschel and J. Rovnyak: Extension theorems for contraction operators on Kreĭn spaces, in *Operator Theory: Advances and Applications*, Vol. 47, Birkhäuser-Verlag, Basel-Boston 1990, pp. 221–305.

[8] A. Gheondea: On the geometry of pseudo-regular subspaces of a Kreĭn space, in *Operator Theory: Advances and Applications*, Vol. 14, Birkhäuser Verlag, Basel-Boston 1984, pp. 141–156.

[9] Yu. P. Ginzburg: *J-nonexpansive analytic operator functions* [Russian], Candidate's Dissertation, Odessa 1958.

[10] I. S. Iokhvidov; M. G. Kreĭn and H. Langer: *Introduction to the Spectral Theory of Operators in Spaces with an Indefinite Metric*, Akademie-Verlag, Berlin 1983.

[11] T. Kato: *Perturbation Theory of Linear Operators*, Springer–Verlag, Berlin-Heidelberg–New York 1966.

[12] M. G. Kreĭn; H. Langer: On the spectral function of a selfadjoint operator in a space with indefinite metric [Russian], *Dokl. Akad. Nauk SSSR*, **152**(1963), 39–42.

[13] M. G. Kreĭn and H. Langer: Über einige Fortsetzunsprobleme, die eng mit der Theorie Hermitescher Operatoren in Raume Π_κ zusammenhängen. I. Einige Funktionenklassen und ihre Darstellungen, *Math. Nach.*, **77**(1977), 187–236.

[14] M. G. Kreĭn and Yu. L. Shmulyan: Plus-operators in spaces with indefinite metric [Russian], *Mat. Issled*, **1**(1966),131–161.

[15] M. G. Kreĭn and Yu. L. Shmulyan: *J*-polar representations of plus-operators [Russian], *Mat. Issled.* **1**(1966), 172-210.

[16] H. Langer: *Spektraltheorie linearer Operatoren in J-Räumen und einige Anwendungen auf den Schar $L(\lambda) = \lambda^2 + \lambda B + C$*, Habilitationsschrift, Dresden 1965.

[17] H. Langer: Spectral functions of definitizable operators in Kreĭn spaces, in *Lecture Notes in Mathematics*, Vol. **948**, Springer-Verlag, Berlin-Heidelberg-New York 1983.

[18] V. P. Potapov: The multiplicative structure of J-contractive matrix functions [Russian], *Trudy Moskov. Mat. Obshch.* **4**(1955), 125–236.

A. Gheondea

Institutul de Matematică
al Academiei Române
C.P. 1-764, Bucureşti 70700
România

Operator Theory:
Advances and Applications, Vol. 61
© 1993 Birkhäuser Verlag Basel

Antitonicity of the Inverse and J–Contractivity

Seppo Hassi[*] and Kenneth Nordstrœm[†]

Abstract. A well–known result in operator theory states that the inverse operator function is antitone, relative to the usual ordering by positivity, on the set of positive invertible operators, i.e., given two such operators A and B, $(\star) A \leq B \implies A^{-1} \geq B^{-1}$. In this paper recent work by Shmul'yan [S] and the authors [HN] on the extension of $(\star)$ to the case of selfadjoint invertible A and B is reviewed. Some of the results in [S] and [HN] are here given different proofs. In particular, a criterion for $(\star)$ in terms of the spectrum of $B^{-1}A$ is proved without relying on results from the theory of indefinite inner product spaces. The geometrical aspects of $(\star)$ are explored, and are shown to lead naturally to the study of convex sets of invertible selfadjoint operators. In particular, criteria for a set of operators to be coverable by a convex set of invertible selfadjoint operators are derived. These results are used to obtain additional (geometrical) characterizations of $(\star)$ as well as further criteria for the J-bicontractivity of J-contractions in a Kreĭn space.

1.   Operator inequalities, inertia, and J-contractivity

Throughout this paper we shall be concerned with everywhere defined linear operators acting on a complex Hilbert space $\mathcal{H}$. Unless otherwise stated, all the operators are assumed to be bounded [i.e., elements of $\mathcal{B}(\mathcal{H})$], and by an invertible operator we understand a bounded operator whose inverse is bounded as well.

According to a well-known result in operator theory, the inverse operator function is antitone, relative to the usual ordering by positivity, on the set of positive invertible operators, i.e., given positive invertible operators A and B, we have the implication

$$A \leq B \implies A^{-1} \geq B^{-1}. \tag{1.1}$$

Below in this section we review recent work by Shmul'yan and, unaware of this earlier work, by the authors (cf. [S] and [HN]) on the extension of (1.1) from the class of positive invertible operators to the larger class of selfadjoint invertible operators.

[*]Research carried out in part while visiting TU Wien.
[†]Research carried out in part while visiting Institut für Mathematik, Universität Augsburg, and supported by grants of the Deutsche Forschungsgemeinschaft and the Academy of Finland.

In Section 2 some of the results in [S] and [HN] are proved afresh, using different methods. In particular, a spectral characterization of (1.1), given in both papers, is proved here without relying on results from the theory of indefinite inner product spaces.

The geometrical aspects of (1.1), which were briefly considered at the end of [HN], are explored more fully. In the final section it is shown that the antitonicity problem leads naturally to the study of convexity of sets of invertible operators. In particular, criteria for a set of operators to be coverable by a convex set of invertible selfadjoint operators are derived. In Theorem 6 such criteria are given in terms of the spectrum of pairs of selfadjoint operators, and in terms of the existence of a covering convex set of selfadjoint operators with constant inertia. This result is used to obtain further characterizations of the antitonicity relation (1.1), and yields also additional criteria for the J-bicontractivity of J-contractions in a Kreĭn space.

Suppose that $\dim \mathcal{H} < \infty$, and define the inertia of a Hermitian matrix H by $\text{In}(H) = (i_+(H), i_-(H), i_0(H))$, where $i_+(H)$, $i_-(H)$, and $i_0(H)$ denote the number of positive, negative, and zero eigenvalues of H, respectively, counting multiplicities. For Hermitian nonsingular matrices A and B, we then have the following result

$$A \leq B \implies A^{-1} \geq B^{-1} \quad \text{iff} \quad \text{In}(A) = \text{In}(B), \tag{1.2}$$

cf. [N, p. 4474] or [S, Theorem 5].

Now consider the general case $\dim \mathcal{H} = \infty$. For a selfadjoint operator H, let

$$H = \int_{-\|H\|}^{\|H\|} \lambda \, dE(\lambda)$$

be its spectral representation, and let

$$P_+ = \int_0^{\|H\|} dE(\lambda), \quad P_- = \int_{-\|H\|}^{-0} dE(\lambda), \quad \text{and} \quad P_0 = E(0) - E(-0),$$

with corresponding ranges $\mathcal{H}_+ = \mathcal{R}(P_+)$, $\mathcal{H}_- = \mathcal{R}(P_-)$, and $\mathcal{H}_0 = \mathcal{R}(P_0)$. The inertia of H is defined as the ordered triple

$$\text{In}(H) = (i_+(H), i_-(H), i_0(H)), \tag{1.3}$$

where

$$i_+(H) = \dim \mathcal{H}_+, \quad i_-(H) = \dim \mathcal{H}_-, \quad \text{and} \quad i_0(H) = \dim \mathcal{H}_0. \tag{1.4}$$

The following example shows that the problem of extending the finite-dimensional result (1.2) to the Hilbert space-setting is non-trivial.

Example 1.1 Take $\mathcal{H} = l^2(\mathbb{Z}) = \{(x_i)_{i=-\infty}^{\infty} : \sum_{-\infty}^{\infty} |x_i|^2 < \infty\}$, and define $J : \mathcal{H} \to \mathcal{H}$, $x = (x_i) \in \mathcal{H}$, by

$$(Jx)_i = \begin{cases} x_i, & \text{if } i \leq 0, \\ -x_i, & \text{if } i > 0. \end{cases}$$

Then $J = J^* = J^{-1}$ is a canonical symmetry on $\mathcal{H}$, and $(\mathcal{H}, (J\cdot, \cdot))$ defines a Kreĭn space. Consider the bilateral shift operator U on $\mathcal{H}$, defined by

$$(Ux)_i = x_{i+1}, \ i \in \mathbb{Z},$$

and choose $A = U^*JU$ and $B = J$. Then A and B are invertible selfadjoint operators satisfying $A \leq B$ and $\text{In}(A) = \text{In}(B)$, but $A^{-1} \not\geq B^{-1}$ (in fact $A^{-1} \leq B^{-1}$).

The above example shows that (1.2) does not hold for invertible selfadjoint operators A and B on a general Hilbert space, and suggests the following problems:

(a) Determine the precise extent to which the finite-dimensional result (1.2) generalizes as such.

(b) Derive criteria for (1.1) to be valid when A and B are invertible selfadjoint operators.

Before addressing problem (b), it is instructive to consider first the proof of (1.1) for positive A and B. Among the various possible proofs, a particularly illuminating one goes as follows:

$$\begin{aligned} A \leq B \quad &\Longleftrightarrow \quad B^{-1/2}AB^{-1/2} \leq I \\ &\Longleftrightarrow \quad (B^{-1/2}A^{1/2})(A^{1/2}B^{-1/2}) \leq I \\ &\Longleftrightarrow \quad A^{1/2}B^{-1}A^{1/2} \leq I \\ &\Longleftrightarrow \quad B^{-1} \leq A^{-1}, \end{aligned}$$

i.e., along with $U = A^{1/2}B^{-1/2}$, U^* is also a contraction.

In the general case of selfadjoint A and B only, the square-root technique is not available. Suppose that $A \leq B$ and $A^{-1} \geq B^{-1}$. Then, as noted in [S, Theorem 1] and [HN, Lemma 2], we have $\text{In}(A) = \text{In}(B)$. Thus equality of inertia remains necessary for (1.1).

To obtain sufficient conditions recall a result by Köthe [K, Satz 1.2], according to which, for two invertible selfadjoint operators G and H,

$$\text{In}(G) = \text{In}(H) \iff G \simeq H. \tag{1.5}$$

The right-hand side of (1.5) denotes congruence of G and H, i.e., there exists an invertible operator S such that $G = S^*HS$. Suppose now that $A \leq B$ and $\text{In}(A) = \text{In}(B)$. Then, for a canonical symmetry J, we have $A \simeq B \simeq J$. Using (1.5), it follows that

$$A \leq B \iff S_1^*JS_1 \leq S_2^*JS_2. \tag{1.6}$$

Defining $U = S_1 S_2^{-1}$, (1.6) takes the form

$$A \le B \iff U^* J U \le J,$$

i.e., U is J-contractive. On the other hand,

$$A^{-1} \ge B^{-1} \iff U J U^* \le J,$$

that is, U^* is J-contractive.

To formulate a theorem, which gives criteria as called for in problem (b), we need to introduce some concepts and notation. Let J be a canonical symmetry on $\mathcal{H}$. Then $J = P_+ - P_-$, with $P_\pm^2 = P_\pm$ and $P_+ + P_- = I$ [take $P_\pm = (I \pm J)/2$]. For any J-nonpositive subspace $\mathcal{L} \subset \mathcal{H}$, define a linear operator $K_\mathcal{L}$ by

$$K_\mathcal{L} = P_+(P_- \mid \mathcal{L})^{-1} : P_-\mathcal{L} \to \mathcal{R}(P_+),$$

the so-called angular operator of the J-nonpositive subspace $\mathcal{L}$. Further denote by $\mathcal{K}^-$ the operator ball

$$\mathcal{K}^- = \{K_\mathcal{L} : \mathcal{L} \in \mathcal{M}^-\},$$

where $\mathcal{M}^-$ is the set of all maximal J-nonpositive subspaces of $\mathcal{H}$ (for details, see [AI]). The resolvent set, spectrum, and point spectrum of a linear operator G are denoted by $\rho(G)$, $\sigma(G)$, and $\sigma_p(G)$, respectively. Finally, for the linear operator U $(= S_1 S_2^{-1})$, we use the partition

$$U = \begin{pmatrix} U_{11} & U_{12} \\ U_{21} & U_{22} \end{pmatrix},$$

corresponding to the decomposition $\mathcal{H} = \mathcal{R}(P_+) \oplus \mathcal{R}(P_-)$, determined by $J = P_+ - P_-$. The identity operator on $\mathcal{R}(P_-)$ is denoted by I_-.

In the following theorem several necessary and sufficient conditions for the anti-tonicity of the inverse operator function have been collected. Similar conditions for the J-bicontractivity of the operator U can be found in various places in the literature, cf. Kreĭn and Shmul'yan [KS] (in particular their Theorem 3.3) and also [AI, Theorem 2.4.17]. For a complete proof, cf. [HN].

Theorem 1.2 *Let A and B be invertible selfadjoint operators, and suppose that $A \le B$ and $\mathrm{In}(A) = \mathrm{In}(B)$. Then any condition from the following two sets of conditions is necessary and sufficient for the inequality $A^{-1} \ge B^{-1}$:*

    I *(Conditions in terms of U)*

    (i) *U is J-bicontractive,*

    (iia) *$U\mathcal{L} \in \mathcal{M}^-$ for some $\mathcal{L} \in \mathcal{M}^-$,*

    (iib) *$U\mathcal{L} \in \mathcal{M}^-$ for any $\mathcal{L} \in \mathcal{M}^-$,* (iiia) $P_-U\mathcal{L} = \mathcal{R}(P_-)$ *for some* $\mathcal{L} \in \mathcal{M}^-$,

(iiib) $P_-U\mathcal{L} = \mathcal{R}(P_-)$ *for any* $\mathcal{L} \in \mathcal{M}^-$;

II *(Conditions in terms of the blocks of U)*

(iva) *for some* $K \in \mathcal{K}^-$, $(U_{21}K + U_{22})^{-1}$ *exists, is bounded, and is defined on the subspace* $\mathcal{R}(P_-)$,

(ivb) *for any* $K \in \mathcal{K}^-$, $(U_{21}K + U_{22})^{-1}$ *exists, is bounded, and is defined on the subspace* $\mathcal{R}(P_-)$,

(v) $\mathcal{R}(U_{22}) = \mathcal{R}(P_-)$,

(vi) $U_{22}U_{22}^* \geq kI_-$ *for some* $k > 0$,

(vii) $0 \notin \sigma_p(U_{22}^*)$,

(viii) $0 \in \rho(U_{22})$,

(ix) $\{\lambda \in \mathbb{C} : |\lambda| < 1\} \subset \rho(U_{22})$.

Remark 1.3 Professor Heinz Langer has pointed out to us that the reduction (1.6) can alternatively be derived from a result of Bognár and Krámli [BK] instead of using the result (1.5) by K

othe [K]. According to this result, for a continuous J-selfadjoint operator A on a J-space, there exists a continuous operator C such that $A = C^{[*]}C$ iff

$$k_A^+ \leq k^+ \quad \text{and} \quad k_A^- \leq k^-.$$

Above $C^{[*]}$ denotes the J-adjoint of C, and k_A^+ and k_A^- (resp. k^+ and k^-) denote the intrinsic dimensions of A-positivity and A-negativity (resp. J-positivity and J-negativity), cf. [BK, Theorem 1]. It follows that, for invertible selfadjoint operators A and B and a canonical symmetry J, $\mathrm{In}(A) = \mathrm{In}(B) = \mathrm{In}(J)$ gives the representations

$$A = C_1^*JC_1 \quad \text{and} \quad B = C_2^*JC_2.$$

Since A and B are invertible, C_1 and C_2 can also be taken to be invertible, and the operator U now takes the form $U = C_1C_2^{-1}$.

Using a result on J-bicontractions due to Ginzburg [G] (cf. [IKL, Lemma 11.3]) yields the following result (cf. [S, Theorems 1 and 2] or [HN, Theorem 2]).

Theorem 1.4 *Let* A *and* B *be invertible selfadjoint operators, and suppose that* $\min\{i_+(A), i_-(A)\} < \infty$. *Then*

$$A \leq B \implies A^{-1} \geq B^{-1} \quad \text{iff} \quad \mathrm{In}(A) = \mathrm{In}(B). \tag{1.2}$$

This result provides a solution to problem (a) in that it shows that the finite-dimensional result carries over to a general Hilbert space $\mathcal{H}$ as long as the spectral subspace of A corresponding to the positive or negative part of the spectrum of A is finite-dimensional, i.e., $(A\cdot, \cdot)$ [or $(B\cdot, \cdot)$] defines a Pontryagin space structure on $\mathcal{H}$. Hence, difficulties can arise only when $(A\cdot, \cdot)$ defines at least a Kreĭn space structure on $\mathcal{H}$.

2. Conditions in terms of A and B

Let C and D be operators on the Hilbert space $\mathcal{H}$. It is well known that

$$C \geq 0, \; D = D^* \implies \sigma(CD) \subset \mathbb{R}, \tag{2.1}$$

and

$$C \geq 0, \; D \geq 0 \implies \sigma(CD) \subset \mathbb{R}^+. \tag{2.2}$$

Indeed, both of these relations follow at once from $\sigma(CD) \subset \sigma(C^{1/2}DC^{1/2}) \cup \{0\}$ (compare [H, Problem 76]). Similarly,

$$C \geq 0 \text{ invertible}, \; D = D^*, \text{ and } \sigma(CD) \subset \mathbb{R}^+ \implies D \geq 0. \tag{2.3}$$

[Here $\sigma(C^{1/2}DC^{1/2}) \subset \mathbb{R}^+$ implies $D \geq 0$.]

The following theorem gives a necessary and sufficient condition for (1.1) to hold in terms of the location of the spectrum $\sigma(B^{-1}A)$ in the complex plane, cf. [S, Theorem 4] or [HN, Theorem 5]. Below we give a different proof based on the above elementary properties of the spectrum of the product of operators.

Theorem 2.1 *Let A and B be invertible selfadjoint operators, and suppose that $A \leq B$. Then*

$$A^{-1} \geq B^{-1} \quad \textit{iff} \quad \sigma(B^{-1}A) \subset \mathbb{R}^+.$$

Proof. First observe that

$$A \leq B \implies \sigma(B^{-1}A) \subset \mathbb{R} \tag{2.4}$$

(compare [S, Lemma 3] or [HN, proof of Theorem 5]). Since $0 \in \rho(B^{-1}A)$ and

$$(A^{-1} - B^{-1})(B - A) = (B^{-1}A)^{-1} + B^{-1}A - 2I,$$

the spectral mapping theorem (cf., e.g., [KR, Theorem 3.3.6]) yields

$$\sigma((A^{-1} - B^{-1})(B - A)) = \{\lambda^{-1} + \lambda - 2 : \lambda \in \sigma(B^{-1}A)\}. \tag{2.5}$$

Suppose now that $A^{-1} \geq B^{-1}$. Then, by virtue of (2.2),

$$\sigma((A^{-1} - B^{-1})(B - A)) \subset \mathbb{R}^+, \tag{2.6}$$

and (2.5) shows that $\sigma(B^{-1}A) \subset \mathbb{R}^+$.

Conversely, suppose that $\sigma(B^{-1}A) \subset \mathbb{R}^+$. Then (2.5) gives (2.6), and hence, if $0 \in \rho(B - A)$, the implication (2.3) yields $A^{-1} \geq B^{-1}$. In the case when $0 \in \sigma(B - A)$, we consider $B_\epsilon = B + \epsilon I \geq A$, $\epsilon > 0$. Since $\sigma(B^{-1}A) \subset \,]0,\infty[$, the upper semi-continuity of the spectrum (cf. [H, Problem 103]) implies that $\sigma(B_\epsilon^{-1}A) \subset \mathbb{R}^+$ for every $\epsilon > 0$ small enough. Further, $0 \in \rho(B_\epsilon - A)$ so that $A^{-1} \geq B_\epsilon^{-1}$, and letting $\epsilon \to 0$ we obtain $A^{-1} \geq B^{-1}$. (The set of positive operators is closed, cf., e.g., [KR, Theorem 4.2.2 (i)]). $\blacksquare$

Let H be an invertible selfadjoint operator on the Hilbert space $\mathcal{H}$, and let $\|\cdot\|$ denote the underlying Hilbert norm. An H-positive subspace $\mathcal{L} \subset \mathcal{H}$ is defined to be uniformly H-positive if there exists an $\epsilon > 0$ such that $(Hx,x) \geq \epsilon\|x\|^2$ for every $x \in \mathcal{L}$. Further, $\mathcal{L}$ is said to be maximal uniformly H-positive if it is not contained in a larger uniformly H-positive subspace of $\mathcal{H}$. Every maximal uniformly H-positive subspace is maximal H-nonnegative [B, Corollary V.7.2], and is closed. Similar definitions and assertions hold for H-negative subspaces of $\mathcal{H}$.

Let $\mathcal{M}$ be a maximal uniformly H-positive subspace of $\mathcal{H}$, and let $\mathcal{N} = \mathcal{M}^{\perp_H}$ be the H-orthogonal complement of $\mathcal{M}$. Then

$$\mathcal{H} = \mathcal{M}\,[\dot{+}]\,\mathcal{N} \tag{2.7}$$

and the subspace $\mathcal{N}$ is maximal uniformly H-negative [B, Corollary V.7.4]. Further, denote by $P_{\mathcal{M}}$ the (H-orthogonal) projection onto $\mathcal{M}$ corresponding to the decomposition (2.7). The operator $P_{\mathcal{M}}$ is linear and bounded (by the Closed Graph Theorem).

We shall derive a criterion for (1.1) involving maximality properties of definite subspaces, and begin with a lemma which is obtained by slightly modifying the proofs of Theorems 1.4.1 and 1.4.5 in [AI].

Lemma 2.2 *Let $\mathcal{L}$ be an H-nonnegative subspace of $\mathcal{H}$. Then the mapping $P_{\mathcal{M}}|\mathcal{L} : \mathcal{L} \to P_{\mathcal{M}}\mathcal{L}$ is a linear homeomorphism. Moreover, $\mathcal{L}$ is maximal iff $P_{\mathcal{M}}\mathcal{L} = \mathcal{M}$.*

Theorem 2.3 *Let A and B be invertible selfadjoint operators, and suppose that $A \leq B$. Then the following conditions are equivalent:*

(i) $A^{-1} \geq B^{-1}$,

(ii) there exists a maximal uniformly A-positive subspace which is maximal uniformly B-positive,

(iii) there exists a decomposition $\mathcal{H} = \mathcal{L}\,\dot{+}\,\mathcal{K}$, where $\mathcal{L}$ is an A-nonnegative subspace and $\mathcal{K}$ is a uniformly B-negative subspace.

(iv) every maximal A-nonnegative subspace is maximal B-nonnegative.

Proof. (i) $\Longrightarrow$ (ii): Suppose that $A^{-1} \geq B^{-1}$. Replacing A and B by S^*AS and S^*BS, respectively, for a suitable invertible operator S, we can assume that $B = B^{-1}$. Then we have $A \leq B \leq A^{-1}$, and it is easily verified that the positive spectral subspace $\mathcal{H}_+$ of A satisfies condition (ii).

(ii) $\implies$ (iii): Let $\mathcal{M}$ be a subspace satisfying (ii). Then (iii) follows by taking $\mathcal{L} = \mathcal{M}$ and $\mathcal{K} = \mathcal{M}^{\perp_B}$.

(iii) $\implies$ (iv): Suppose that

$$\mathcal{H} = \mathcal{L}\dotplus\mathcal{K}, \tag{2.8}$$

where the subspace $\mathcal{L}$ is A-nonnegative and $\mathcal{K}$ is uniformly B-negative. Then $\mathcal{L}$ must be maximal A-nonnegative and maximal B-nonnegative, and similarly $\mathcal{K}$ must be maximal uniformly B-negative and also maximal uniformly A-negative. (Any extension of $\mathcal{K}$ intersects $\mathcal{L}$, and is hence not a definite subspace, and similarly for $\mathcal{L}$.) Let $\mathcal{L}'$ be any maximal A-nonnegative subspace. To show that $\mathcal{L}'$ is maximal B-nonnegative, we verify that $\mathcal{H} = \mathcal{L}'\dotplus\mathcal{K}$ (compare [B, Lemma V.7.6]). Clearly $\mathcal{L}' \cap \mathcal{K} = \{0\}$, and it remains to prove that $\mathcal{L} \subset \mathcal{L}' + \mathcal{K}$. But this follows easily from $P_\mathcal{L}\mathcal{L}' = \mathcal{L}$, which holds by Lemma 1 [replace $\mathcal{L}$ in (2.8) by $\mathcal{K}^{\perp_A}$].

To complete the proof it is verified that (ii) $\implies$ (i). Let $\mathcal{M}$ be a subspace satisfying the condition (ii). We show that $]-\infty,0] \cap \sigma(B^{-1}A) = \emptyset$, or equivalently, that the operator $C_\lambda = (1-\lambda)A + \lambda B$ is invertible for every $\lambda \in [0,1]$. (See the proof of Theorem 5 in [HN].) Suppose the contrary, and let $\lambda \in]0,1[$ be a number for which the operator C_λ is not invertible. Choosing a sequence (x_n) of vectors satisfying $\|C_\lambda x_n\| \to 0$, and proceeding as in the proof of Lemma 4 in [HN], we can construct a B-nonnegative extension $\mathcal{M}' \supset \mathcal{M}$, which contradicts the fact that $\mathcal{M}$ is maximal uniformly B-positive. Hence, we must have $]-\infty,0] \cap \sigma(B^{-1}A) = \emptyset$, and then (i) follows by Theorem 2.1, cf. (2.4). ∎

Criteria similar to those in Theorem 2.3, involving maximal nonpositive (or maximal nonnegative) subspaces, were given as Corollary 1 in [HN], and were obtained there from results concerning J-bicontractions on Kreĭn spaces. As outlined in Section 1, such an approach was crucial in [S] and [HN] for solving the problem (1.1). Above Theorem 2.3 as well as Theorem 2.1 have been proved directly without relying on results on J-bicontractions.

Consider the case $i_+(A) < \infty$ [or $i_-(A) < \infty$]. Then clearly e.g. condition (ii) in Theorem 2.3 is equivalent to $\mathrm{In}(A) = \mathrm{In}(B)$, and Theorem 1.4 is hence obtained without using Ginzburg's [G] result that every J-contractive operator on a Pontryagin space is J-bicontractive.

3. Convexity and constant inertia

In Section 3 of [HN] it was seen that the antitonicity problem (1.1) is closely related both to the location of the spectrum of the product of selfadjoint operators (see also the previous section) and to the constancy of inertia on certain subsets of selfadjoint operators. In this section we explore more fully these connections, and prove inter alia a generalization of results to be found in Shmul'yan [S] and Cain [C2].

As a partial converse of (2.2), Shmul'yan [S, Theorem 6] established (2.3) under the weaker assumption that C is positive and $\text{Ker}(C) = \{0\}$. The following result can be seen as a more general inertial version of (2.3).

Theorem 3.1 *(Cain [C2, Corollary 6.15]). Let C and D be selfadjoint operators. If*

$$\sigma(CD) \cap \,] - \infty, 0] = \emptyset, \tag{3.1}$$

then $\text{In}(C) = \text{In}(D)$.

Note that the converse of Theorem 3.1 is not valid, in general. Taking

$$C = -D = \begin{pmatrix} 1 & 0 \\ 0 & -1 \end{pmatrix},$$

we have $\text{In}(C) = \text{In}(D)$, but $\sigma(CD) = \{-1\}$.

Below we shall show that the spectral condition (3.1) implies actually a stronger inertial property. We begin with a lemma. A similar result is proved by Cain [C1, Theorem 4], under slightly stronger assumptions on the operators in the set $\mathcal{U}$.

Lemma 3.2 *Let $\mathcal{U}$ be a (pathwise) connected set of invertible selfadjoint operators. Then, for any $C, D \in \mathcal{U}$, $\text{In}(C) = \text{In}(D)$.*

Proof. Let $\Gamma(t)$, $t \in [0,1]$, be a path in $\mathcal{U}$ joining C to D, and let

$$P_+(t) = -\frac{1}{2\pi i} \oint_{\gamma_+} [\Gamma(t) - zI]^{-1} \, dz$$

and

$$P_-(t) = -\frac{1}{2\pi i} \oint_{\gamma_-} [\Gamma(t) - zI]^{-1} \, dz$$

be Riesz projections associated with the positive and negative part of the spectrum of $\Gamma(t)$. Here γ_+ (γ_-) is a closed rectifiable Jordan curve lying in $\rho(\Gamma(t))$ such that the interior of γ_+ (γ_-) contains the positive (negative) part of $\sigma(\Gamma(t))$ for every $t \in [0,1]$. (Such γ_+ and γ_- can be found, since $\sigma(\Gamma(t)) \subset \mathbb{R}$ and the set $\{\|\Gamma(t)\| : t \in [0,1]\}$ is compact.) It is easily verified that $P_+(t)$ and $P_-(t)$ depend continuously on $t \in [0,1]$ (compare [DK, Theorem I.2.2]), i.e., $P_+(t)$ and $P_-(t)$ define paths of projections in $\mathcal{B}(\mathcal{H})$. By [H, Problem 57], $\dim P_+(t)\mathcal{H}$ and $\dim P_-(t)\mathcal{H}$ do not depend on t. This gives $\text{In}(C) = \text{In}(D)$ [cf. (1.3) and (1.4)]. ∎

Theorem 3.3 *Let $\mathcal{A}$ be a set of invertible selfadjoint operators. Then the following conditions are equivalent:*

(i) $\sigma(C^{-1}D) \cap \,] - \infty, 0] = \emptyset$ *for every $C, D \in \mathcal{A}$,*

(ii) *there exists a convex set of invertible selfadjoint operators containing $\mathcal{A}$,*

(iii) *there exists a convex set of selfadjoint operators with constant inertia containing $\mathcal{A}$.*

Proof. Suppose that (i) holds. Then, for any $C, D \in \mathcal{A}$ and all $\mu_1, \mu_2 \geq 0$ such that $\mu_1 + \mu_2 \neq 0$, we have

$$(\mu_2 C^{-1} D + \mu_1 I)^{-1} = (\mu_1 C + \mu_2 D)^{-1} C \in \mathcal{B}(\mathcal{H}),$$

yielding $(\mu_1 C + \mu_2 D)^{-1} \in \mathcal{B}(\mathcal{H})$. Hence, (ii) follows.

Suppose, conversely, that (ii) holds. Then, for any $C, D \in \mathcal{A}$,

$$[\lambda C + (1 - \lambda)D]^{-1} \in \mathcal{B}(\mathcal{H}) \quad \text{for every } \lambda \in [0, 1],$$

and, as in the first part of the proof, it is seen that (i) is valid.

In view of Lemma 3.2 and the definition of inertia [cf. (1.3) and (1.4)], (ii) and (iii) are equivalent. $\blacksquare$

Theorem 3.3 reveals that the spectral condition (3.1) implies not only $\text{In}(C) = \text{In}(D)$, but constant inertia on certain convex sets of selfadjoint operators determined by C and D. In this stronger form the inertial condition is hence also sufficient for (3.1). [Note that for selfadjoint C and D, invertibility of CD implies invertibility of C and D, and $\text{In}(C^{-1}) = \text{In}(C)$ for invertible selfadjoint C.]

A finite-dimensional analogue of Theorem 3.3, formulated in terms of the convex hull of Hermitian matrices, has been given by Johnson and Rodman [JR]. In particular they point out that, given m nonsingular Hermitian matrices $A_1, A_2, \ldots, A_m$, all matrices in the convex hull $\text{co}\{A_1, A_2, \ldots, A_m\}$ have the same inertia iff every matrix in $\text{co}\{A_1, A_2, \ldots, A_m\}$ is nonsingular, and conclude that the study of convex sets of Hermitian matrices with constant inertia includes the study of convex sets of nonsingular Hermitian matrices [JR, p. 353]. In view of Lemma 3.2, the equivalence of (ii) and (iii) in Theorem 3.3 continues to hold under the weaker topological assumption of connectedness of the sets. As the following example shows, convexity is, however, needed for the spectral condition (i) in Theorem 3.3.

Example 3.4 Consider the following path of operators:

$$A(t) = \begin{pmatrix} 1 - 2t & t(t - 1) \\ t(t - 1) & 2t - 1 \end{pmatrix}, \; t \in [0, 1].$$

Then $A(t) = A(t)^*$ and $\det A(t) < 0$ for every $t \in [0, 1]$, but taking $C = A(0)$ and $D = A(1)$ gives $\sigma(C^{-1}D) = \{-1\}$.

In view of (2.4), Theorem 3.3 combined with Theorem 2.1 yields the following characterizations of the antitonicity of the inverse.

Theorem 3.5 *Let A and B be invertible selfadjoint operators, and suppose that $A \leq B$. Then the following conditions are equivalent:*

(i) $A^{-1} \geq B^{-1}$,

(ii) *there exists a convex set of invertible selfadjoint operators containing A and B,*

(iii) *there exists a convex set of selfadjoint operators with constant inertia containing A and B.*

From the proof of Theorem 3.3 it is clear that, as the convex set appearing in conditions (ii) and (iii) in Theorem 3.3 (resp. Theorem 3.5), one can take either the conical hull or the convex hull of $\mathcal{A}$ (resp. $\{A, B\}$). However, in Theorem 3.5 a further example of such a convex set is the operator interval

$$[A, B] = \{C : A \leq C \leq B\},$$

cf. Lemma 4 in [HN].

Remark 3.6 Let A and B be invertible selfadjoint operators satisfying $A \leq B$ and $\text{In}(A) = \text{In}(B)$. For any $C \in [A, B]$ we then have

$$i_+(C) = i_+(C) + i_0(C) = i_+(A) \quad \text{and} \quad i_-(C) = i_-(C) + i_0(C) = i_-(A), \qquad (3.2)$$

but, in general, we cannot conclude that $i_0(C) = i_0(A) \ (= 0)$ (compare Example 1). If, however, $\min\{i_+(A), i_-(A)\} < \infty$, then (3.2) yields $i_0(C) = 0$, and constant inertia on $[A, B]$ obtains.

As a consequence of Theorem 3.5, we obtain criteria for bicontractivity in Kreĭn spaces.

Corollary 3.7 *Let U be an invertible J-contractive operator. Then the following conditions are equivalent:*

(i) U *is J-bicontractive,*

(ii) *there exists a convex set of invertible selfadjoint operators containing J and U^*JU,*

(iii) *there exists a convex set of selfadjoint operators with constant inertia containing J and U^*JU.*

In Example 1.1 the operator U does not satisfy the assertions (ii) and (iii) of Corollary 3.7, hence U^* is not J-contractive. In fact, we have $UJU^* \geq J$, i.e., the operator U^* is J-expansive.

Acknwledgments. After completion of [HN], Professor T.Ya. Azizov kindly informed us that the late Professor Shmul'yan had considered the same antitonicity problem at a lecture at the 1990 Winter School in Voronezh. We are grateful to Professor A. A. Nudel'man for informing us that Shmul'yan's results have subsequently been published, and for sending us a copy of the original article. We are also grateful to Professor H. Langer for useful discussions on this topic.

References

[AI] T. Ya. Azizov and I. S. Iokhvidov: *Linear Operators in Spaces with an Indefinite Metric.* Wiley, Chichester, 1989.

[B] J. Bognár: *Indefinite Inner Product Spaces.* Springer, Berlin, 1974.

[BK] J. Bognár and A. Krámli: Operators of the form C^*C in indefinite inner product spaces. *Acta Sci. Math. (Szeged)* **29** (1968), 19-29.

[C1] B. E. Cain: An inertia theory for operators on a Hilbert space. *J. Math. Anal. Appl.* **41** (1973), 97-114.

[C2] B. E. Cain.: Inertia theory. *Linear Algebra Appl.* **30** (1980), 211-240.

[DK] Yu. L. Daletskiĭ and M. G. Kreĭn: *Stability of Solutions of Differential Equations in Banach Space.* Transl. Math. Monogr. Vol. 43, Amer. Math. Soc., Providence, Rhode Island, 1974.

[G] Yu. P. Ginzburg: *J*-nonexpansive Analytic Operator Functions. Candidate's Dissertation, Odessa, 1958. (Russian)

[H] P. R. Halmos: *A Hilbert Space Problem Book.* Second Edition. Springer, New York, 1982.

[HN] S. Hassi and K. Nordström: Antitonicity of the inverse of selfadjoint operators. DFG–Report No. 364, Institut für Mathematik, Universität Augsburg, 1992 (submitted).

[IKL] I. S. Iokhvidov, M. G. Kreĭn and H. Langer: *Introduction to the Spectral Theory of Operators in Spaces with an Indefinite Metric.* Akademie–Verlag, Berlin, 1982.

[JR] C. R. Johnson and L. Rodman: Convex sets of Hermitian matrices with constant inertia. *SIAM J. Alg. Disc. Meth.* **6** (1985), 351-359.

[K] G. Köthe: Das Trägheitsgesetz der quadratischen Formen im Hilbertschen Raum. *Math. Z.* **41** (1936), 137-152.

[KR] R. V. Kadison and J. R. Ringrose: *Fundamentals of the Theory of Operator Algebras I*. Academic Press, New York, 1983.

[KS] M. G. Kreĭn and Yu. L. Shmul'yan: Plus-operators in a space with indefinite metric. *Amer. Math. Soc. Transl. Ser. 2* **85** (1969), 93-113.

[N] K. Nordström: Some further aspects of the Löwner-ordering antitonicity of the Moore-Penrose inverse. *Comm. Statist. Theory Methods* **18** (1989), 4471-4489.

[S] Yu. L. Shmul'yan: A question regarding inequalities between Hermitian operators. *Math. Notes* **49** (1991), 423-425 [translation of *Mat. Zametki* **49** (1991), 138-141].

Seppo Hassi
Keneth Nordstrœm

Department of Statistics
University of Helsinki
SF–10100 Helsinki
Finland

Operator Theory:
Advances and Applications, Vol. 61
© 1993 Birkhäuser Verlag Basel

Unitary Extensions of a System of Commuting Isometric Operators

María Dolores Morán

Abstract. The paper is concerned with commuting unitary extensions of a system of isometric operators. A description of the set of all commuting unitary extensions is given. The results are applied to the problem of describing all the commuting selfadjoint extensions of a pair of closed symmetric operators.

1. Introduction

Let $(U; V; \mathcal{H})$ be a pair of isometries with domains $\mathcal{D}_U, \mathcal{D}_V$ and ranges $\mathcal{R}_U, \mathcal{R}_V$, respectively, closed subspaces of the Hilbert space $\mathcal{H}$. We say that $(U'; V'; \mathcal{F})$ is a commuting unitary extension of the pair $(U; V; \mathcal{H})$ if $H \subseteq \mathcal{F}$ and U', V' are two commuting unitary operators in $\mathcal{F}$ which extend U and V, respectively. We say that $(U'; V'; \mathcal{F})$ is a minimal extension of $(U; V; \mathcal{H})$ if in addition

$$\mathcal{F} = \bigvee_{(n,m)\in \mathbf{Z}\times\mathbf{Z}} U'^n V'^m(\mathcal{H}),$$

that is, the closure of the linear hull of

$$\bigcup_{(n,m)\in \mathbf{Z}\times\mathbf{Z}} U'^n V'^m(\mathcal{H}).$$

Assume that the given pair $(U; V; \mathcal{H})$ satisfies

$$U^n \mathcal{D}_V \subseteq \mathcal{D}_U, \quad U^n \mathcal{R}_V \subseteq \mathcal{D}_U, \quad n = 0, 1, \ldots$$

Then

$$< U^n V f, V f' > \; = \; < U^n f, f' >, \quad f, f' \in \mathcal{D}_V, \quad n = 1, 2, \ldots \tag{1.1}$$

is a necessary and sufficient condition for the existence of a minimal commuting unitary extension $(U'; V'; \mathcal{F})$ of $(U; V; \mathcal{H})$ (see [1]).

In the first part of this paper we will recall a description (see [7]) of the set $\mathfrak{U}_{U,V}$ of all equivalence classes of minimal unitary extensions of $(U; V; \mathcal{H})$, where two minimal unitary extensions $(U'; V'; \mathcal{F}')$ and $(U''; V''; \mathcal{F}'')$ are said equivalent if there exists a unitary isomorphism Φ of $\mathcal{F}'$ onto $\mathcal{F}''$ which leaves invariant the elements of $\mathcal{H}$ and satisfies:

$$\Phi U'^n V'^m = U''^n V''^m \Phi, \quad n, m \in \mathbf{Z}.$$

The problem was solved by D. Z. Arov and L. Z. Grossman in case U is the identity on $\mathcal{H}$ ([3]).

In the second part, we will give some applications to the description of the self-adjoint extensions of a system of commuting symmetric operators, considered by R. S. Ismagelov ([5]) and also by A. Koranyi ([6]). Working with Cotlar-Sadosky lifting theorem ([4]), Alegría ([1]) gives some other applications.

2.

Let $\mathcal{H}$ be a Hilbert space, $\mathcal{D}_V$ and $\mathcal{R}_V$ closed subspaces of $\mathcal{H}$ and $V : \mathcal{D}_V \to \mathcal{R}_V$ an isometry. We say that $(U; \mathcal{F})$ is a unitary extension of the isometry V (defined in $\mathcal{H}$) if $\mathcal{F}$ is a Hilbert space which contains $\mathcal{H}$ as a closed subspace and U is a unitary operator on $\mathcal{F}$ which extends V. We say that $(U; \mathcal{F})$ is a minimal unitary extension of V if in addition

$$\mathcal{F} = \bigvee_{n \in \mathbf{Z}} U^n(\mathcal{H}).$$

Two minimal unitary extensions of V, $(U; \mathcal{F})$ and $(U'; \mathcal{F}')$, are equivalent if there exists a unitary operator Φ from $\mathcal{F}$ onto $\mathcal{F}'$ which leaves invariant the elements of $\mathcal{H}$ and satisfies:

$$\Phi U = U' \Phi.$$

Let $\mathcal{N}, \mathcal{M}$ be Hilbert spaces. The class of all holomorphic functions Θ in $|z| < 1$ such that $\Theta(z) \in \mathcal{L}(\mathcal{N}, \mathcal{M})$ and $\Theta(z)$ is a contraction for each z will be denoted by $\mathcal{B}(\mathcal{N}, \mathcal{M})$.

Arov and Grossman ([3]) give a bijection between the set $\mathfrak{U}_V$ of all equivalence classes of minimal unitary extensions of an isometry V defined in $\mathcal{H}$ and the unitary ball of analytic operator valued functions $\mathcal{B}(\mathcal{N}, \mathcal{M})$, where $\mathcal{N} = \mathcal{H} \ominus \mathcal{D}_V$ and $\mathcal{M} = \mathcal{H} \ominus \mathcal{D}_V$. Given $(U; \mathcal{F})$ a minimal unitary extension of V, the function

$$\Theta_U(z) = P_{\mathcal{M}}\{U(I - zP_{\mathcal{F} \ominus \mathcal{H}}U)^{-1}\}|_{\mathcal{N}} \tag{2.1}$$

is an analytic contractive operator valued function. We call $\Theta_U(z)$ the holomorphic function associated to the minimal unitary extension $(U; \mathcal{F})$. For $\Theta \in \mathcal{B}(\mathcal{N}, \mathcal{M})$, we denote by V^Θ the minimal unitary extension of V having Θ as its associated holomorphic function.

Let $(U; V; \mathcal{H})$ be a pair of isometries with domains $\mathcal{D}_U, \mathcal{D}_V$ and ranges $\mathcal{R}_U, \mathcal{R}_V$, respectively. We set

$$\mathcal{N}_U = H \ominus \mathcal{D}_U, \quad \mathcal{N}_V = \mathcal{H} \ominus \mathcal{D}_V, \quad \mathcal{M}_U = \mathcal{H} \ominus \mathcal{D}_U, \quad \mathcal{M}_V = \mathcal{H} \ominus \mathcal{D}_V.$$

Assume that for $(U; V; \mathcal{H})$ we have

$$U^n \mathcal{D}_V \subseteq \mathcal{D}_U, \quad U^n R_V \subseteq \mathcal{D}_U, \quad n = 0, 1, \ldots,$$

and condition 1.1 holds. Then the following result is true (see [7]):

Theorem 2.1 *Let $(U; V; \mathcal{H})$ be a pair of isometries as above. There exists a bijection between the set $\mathfrak{U}_{U,V}$ of all equivalence classes of minimal unitary extensions of $(U; V; \mathcal{H})$ and*

$$\bigcup_{\Theta \in B(\mathcal{N}_U, \mathcal{M}_U)} \left\{ \begin{array}{l} \beta \in B\big([\bigvee_{n\in\mathbf{Z}}(U^\Theta)^n(\mathcal{H})] \ominus [\bigvee_{n\in\mathbf{Z}}(U^\Theta)^n(\mathcal{D}_V)], \\[2mm] [\bigvee_{n\in\mathbf{Z}}(U^\Theta)^n(\mathcal{H})] \ominus [\bigvee_{n\in\mathbf{Z}}(U^\Theta)^n(\mathcal{R}_V)]\big) : \\[2mm] U^\Theta \beta = \beta U^\Theta|_{[\bigvee_{n\in\mathbf{Z}}(U^\Theta)^n(\mathcal{H})]\ominus[\bigvee_{n\in\mathbf{Z}}(U^\Theta)^n(\mathcal{D}_V)]} \end{array} \right\}$$

Since in some applications the operator U is unitary, we also state the next corollary.

Corollary 2.2 *Let $(U; V; \mathcal{H})$ be a pair of isometries which satisfies the conditions of Theorem 2.1. If in addition U is a unitary operator in $\mathcal{H}$ (i.e. $\mathcal{N}_U = \mathcal{M}_U = \{0\}$) then there exists a bijection between the set $\mathfrak{U}_{U,V}$ of all equivalence classes of minimal unitary extensions of $(U; V; \mathcal{H})$ and the set*

$$\left\{ \begin{array}{l} \beta \in B\big(\mathcal{H} \ominus [\bigvee_{n\in\mathbf{Z}} U^n(\mathcal{D}_V)], \mathcal{H} \ominus [\bigvee_{n\in\mathbf{Z}} U^n(\mathcal{R}_V)]\big) : \\[2mm] U\beta = \beta U|_{\mathcal{H}\ominus[\bigvee_{n\in\mathbf{Z}} U^n(\mathcal{D}_V)]} \end{array} \right\}$$

Finally, if only $\mathcal{N}_U = \{0\}$, we have

Corollary 2.3 *Let $(U; V; \mathcal{H})$ be a pair of isometries which satisfies the conditions of Theorem 2.1. If in addition the defect of the domain of U is zero (i.e. $\mathcal{N}_U = \{0\}$) then there exists a bijection between the set $\mathfrak{U}_{U,V}$ of all equivalence classes of minimal unitary extensions of $(U; V; \mathcal{H})$ and the set*

$$\left\{ \begin{array}{l} \beta \in B(\mathcal{H} \ominus \mathcal{D}_V, \mathcal{H} \ominus \mathcal{R}_V) : \\[2mm] P_{\mathcal{R}_U}(V P_{\mathcal{D}_V} + \beta(z) P_{\mathcal{N}_V})[I - z(V P_{\mathcal{D}_V} + \beta(z) P_{\mathcal{N}_V})]^{-1} U = \\[2mm] U(V P_{\mathcal{D}_V} + \beta(z) P_{\mathcal{N}_V})[I - z(V P_{\mathcal{D}_V} + \beta(z) P_{\mathcal{N}_V})]^{-1}|_{\mathcal{H}} \end{array} \right\}.$$

3.

Let $\{U_\lambda\}_{\lambda\in\Lambda}$ be a unitary representation of a commutative group Λ on the Hilbert space $\mathcal{H}$, and let $V : \mathcal{D} \to \mathcal{R}$ be an isometry from $\mathcal{D}$ onto $\mathcal{R}$, both of them closed subspaces of $\mathcal{H}$. We say that $(\{U'_\lambda\}_{\lambda\in\Lambda}; V'; \mathcal{H}')$ is a commuting unitary extension of $(\{U_\lambda\}_{\lambda\in\Lambda}; V; \mathcal{H})$ if $\mathcal{H}'$ is a Hilbert space which contains $\mathcal{H}$ as a closed subspace, $\{U'_\lambda\}_{\lambda\in\Lambda}$ is a unitary dilation of the unitary representation $\{U_\lambda\}_{\lambda\in\Lambda}$, while V' is a unitary extension of V in $\mathcal{H}'$, such that
$, U'_\lambda V' = V'U'_\lambda$ for all $\lambda \in \Lambda$. We say that $(\{U'_\lambda\}_{\lambda\in\Lambda}; V'; \mathcal{H}')$ is a minimal extension if in addition $\mathcal{H}' = \vee_{n\in\mathbf{Z}} V'^m(\mathcal{H})$. Koranyi ([6]) gives a necessary and sufficient condition for the existence of a commuting unitary extension of $(\{U_\lambda\}_{\lambda\in\Lambda}; V; \mathcal{H})$:

$$\|U_\lambda Vx + VU_\lambda y\| = \|x + y\| \quad \text{for all} \quad x \in \mathcal{D}, y \in U_\lambda^{-1}\mathcal{D}, \lambda \in \Lambda. \tag{3.1}$$

We are interested in giving a description of the set $\mathfrak{U}_{(\{U_\lambda\}_{\lambda\in\Lambda};V;\mathcal{H})}$ of all equivalence classes of minimal unitary extensions of $(\{U_\lambda\}_{\lambda\in\Lambda}; V; \mathcal{H})$, where two minimal unitary extensions $(\{U'_\lambda\}_{\lambda\in\Lambda}; V'; \mathcal{H}')$ and $(\{U''_\lambda\}_{\lambda\in\Lambda}; V''; \mathcal{H}'')$ are said equivalent if there exists a unitary isomorphism Φ of $\mathcal{H}'$ onto $\mathcal{H}''$ which leaves invariant the elements of $\mathcal{H}$ and satisfies

$$\Phi U'_\lambda V' = U''_\lambda V''\Phi, \quad \lambda \in \Lambda.$$

Proposition 3.1 *Let $\{U_\lambda\}_{\lambda\in\Lambda}$ be a unitary representation of a commutative group Λ on the Hilbert space $\mathcal{H}$, and let $V : \mathcal{D} \to \mathcal{R}$, $\mathcal{D}, \mathcal{R} \subset \mathcal{H}$, be an isometry from $\mathcal{D}$ onto $\mathcal{R}$. Assume that $(\{U_\lambda\}_{\lambda\in\Lambda}; V; \mathcal{H})$ satisfies condition (3). Then there is a bijection between the set $\mathfrak{U}_{(\{U_\lambda\}_{\lambda\in\Lambda};V)}$ of all equivalence classes of minimal unitary extensions of $(\{U_\lambda\}_{\lambda\in\Lambda}; V; \mathcal{H})$ and the set*

$$\mathfrak{A} = \bigcap_{\lambda\in\Lambda} \left\{ \begin{array}{l} \beta \in \mathcal{B}\left(\mathcal{H} \ominus [\vee_{\lambda\in\Lambda} U_\lambda(\mathcal{D})], \mathcal{H} \ominus [\vee_{\lambda\in\Lambda} U_\lambda(\mathcal{R})]\right) : \\[2mm] U_\lambda\beta = \beta U_\lambda|_{\mathcal{H}\ominus[\vee_{\lambda\in\Lambda} U_\lambda(\mathcal{D})]} \end{array} \right\}$$

Proof. Let us consider $(\{U'_\lambda\}_{\lambda\in\Lambda}; V'; \mathcal{H}') \in \mathfrak{U}_{(\{U_\lambda\}_{\lambda\in\Lambda};V)}$. The function

$$\beta(z) = P_{\mathcal{H}\ominus[\vee_{\lambda\in\Lambda} U_\lambda(R)]}V'(I - zP_{\mathcal{H}'\ominus\mathcal{H}}V')^{-1}|_{\mathcal{H}\ominus[\vee_{\lambda\in\Lambda} U_\lambda(\mathcal{D})]} \tag{3.2}$$

is an analytic contractive operator valued function such that

$$U_\iota\beta = \beta U_\iota|_{\mathcal{H}\ominus[\vee_{\lambda\in\Lambda}U_\lambda(\mathcal{D})]}$$

for all $\iota \in \Lambda$. Then β is the holomorphic function associated to the minimal unitary extension $(V'; \mathcal{H}')$ of the isometric operator $\bar{V} : \vee_{\lambda\in\Lambda}U_\lambda(\mathcal{D}) \to \vee_{\lambda\in\Lambda}U_\lambda(\mathcal{R})$, uniquely determined by

$$\bar{V}(\sum_{\lambda\in\Lambda} U_\lambda(f_\lambda)) = \sum_{\lambda\in\Lambda} U_\lambda(Vf_\lambda), \quad f_\lambda \in \mathcal{D}, \quad \lambda \in \Lambda.$$

Condition 3.1 insures that $\bar{V}$ is a well defined isometric extension of V which commutes with U_λ for all $\lambda \in \Lambda$. Since V' commutes with U_λ for all $\lambda \in \Lambda$, the same is true for β; thus $\beta \in A$. It is easy to check that the correspondence is injective and also surjective. ∎

In this context, it is worth noticing that the commuting unitary extension of $(\{U_\lambda\}_{\lambda\in\Lambda}; V; \mathcal{H})$ given by Koranyi in [6] corresponds to

$$\Theta(z) = z P_{\mathcal{H}\ominus[\bigvee_{\lambda\in\Lambda} U_\lambda(R)]}(I - z\bar{V}^{-1}P_{[\bigvee_{\lambda\in\Lambda} U_\lambda(R)]})^{-1}|_{\mathcal{H}\ominus[\bigvee_{\lambda\in\Lambda} U_\lambda(\mathcal{D})]}.$$

Let $\{U_i\}_{i=1}^k$ be a family of commuting unitary operators on $\mathcal{H}$; then $U_{(n_1,\ldots,n_k)} = U_1^{n_1}\ldots U_k^{n_k}$ is a unitary representation of $\mathbf{Z}^k$ in $\mathcal{H}$. We denote by $\mathfrak{U}_{(\{U_i\}_{i=1}^k;V)}$ the corresponding set of all classes of minimal unitary extensions.

As a consequence of Proposition 3.1 we have:

Proposition 3.2 *Let $\{U_i\}_{i=1}^k$ be a family of commuting unitary operators in $\mathcal{H}$ and let $V : \mathcal{D} \to \mathcal{R}$ be an isometry from $\mathcal{D}$ onto $\mathcal{R}$, both of them closed subspaces of $\mathcal{H}$. If*

$$\|U_1^{n_1}\ldots U_k^{n_k}Vx + VU_1^{n_1}\ldots U_k^{n_k}y\| = \|x + y\| \tag{3.3}$$

holds for all $x \in \mathcal{D}, y \in U_1^{-n_1}\ldots U_k^{-n_k}\mathcal{D}$ and all natural numbers $n_1, \ldots, n_k$, then there exists a bijection between the set $\mathfrak{U}_{(\{U_i\}_{i=1}^k;V)}$ and

$$\bigcap_{i=1}^k \left\{ \begin{array}{l} \beta \in \mathcal{B}\left(\mathcal{H}\ominus[\bigvee_{n_1,\ldots,n_k\in\mathbf{Z}} U_1^{n_1}\ldots U_k^{n_k}(\mathcal{D})], \mathcal{H}\ominus[\bigvee_{n_1,\ldots,n_k\in\mathbf{Z}} U_1^{n_1}\ldots U_k^{n_k}(\mathcal{R})]\right): \\[2mm] U_i\beta = \beta U_i|_{\mathcal{H}\ominus[\bigvee_{n_1,\ldots,n_k\in\mathbf{Z}} U_1^{n_1}\ldots U_k^{n_k}(\mathcal{D})]} \end{array} \right\}$$

Let $(A, B; \mathcal{H})$ be a pair of operators in $\mathcal{H}$, such that A is selfadjoint and B is closed and symmetric. We say that $(A', B'; \mathcal{H}')$ is a commuting selfadjoint extension of the pair $(A, B; \mathcal{H})$ if $\mathcal{H} \subseteq \mathcal{H}'$ and A', B' are two commuting selfadjoint operators in $\mathcal{H}'$ (i.e. their projection-valued measures commute) which extend A and B, respectively. We say that $(A', B'; \mathcal{H}')$ is a minimal extension in a similar way. Denote by $\mathfrak{A}_{A,B}$ the set of all equivalence classes of minimal selfadjoint extensions of $(A; B; \mathcal{H})$, where two minimal selfadjoint extensions $(A'; B'; \mathcal{H}')$ and $(A''; B''; \mathcal{H}'')$ are said equivalent if there exists an unitary isomorphism Φ of $\mathcal{H}'$ onto $\mathcal{H}''$ which leaves invariant the elements of $\mathcal{H}$ and satisfies:

$$\Phi A' = A''\Phi \quad \text{and} \quad \Phi B' = B''\Phi. \tag{3.4}$$

Now let us denote by $C(T)$ the Cayley transform of a closed symmetric operator T (i.e. $C(T) = (T - iI)(T + iI)^{-1}$). If we consider $U = C(A)$ and $V = C(B)$, with A,

B as above, then U is a unitary operator in $\mathcal{H}$ and V is an isometry with domain $\mathcal{D}$ equal to the domain of $(B+iI)^{-1}$ and range $\mathcal{R}$. Let us check that if $\mathfrak{U}_{U,V}$ is not empty then $\mathfrak{A}_{A,B}$ is not empty. Indeed, if $(U',V';\mathcal{H}') \in \mathfrak{U}_{U,V}$ and $< h',(I-V')g' >= 0$ for all $g \in \mathcal{H}'$, then also $< h',(I-V)g >= 0$ for all $g \in \mathcal{D}$. So $h' \in \mathcal{H}' \ominus \mathcal{H}$, and we have $V'h' = h = V'^*h'$. Let us consider the set $\mathcal{K} = \{h' \in \mathcal{H}' \ominus \mathcal{H} : V'h' = h'\}$. If $h' \in \mathcal{K}$ then

$$0 = < U'h', U'(I-V')g' > = < U'h', (I-V')U'g' >$$

Thus $0 = < U'h',(I-V)g >$ for all $g \in \mathcal{D}$, hence $U'h' \in \mathcal{H}' \ominus \mathcal{H}$ and $V'U'h' = U'V'h' = U'h'$. It follows that $U'(\mathcal{K}) \subseteq \mathcal{K}$. Since the extension is minimal, the same argument can be used for U'^{-1}; thus $\mathcal{K}$ must be $\{0\}$. We may therefore define

$$A' = i(I + U')(I - U')^{-1}, \quad B' = i(I + V')(I - V')^{-1}.$$

They are both selfadjoints operators in $\mathcal{H}'$, which extend A , respectively B. It is easy to verify that they commute.

Let us also mention that if $\mathfrak{U}_{C(A),C(B)} \neq \emptyset$ then there exists a bijection between the set $\mathfrak{U}_{C(A),C(B)}$ and the set $\mathfrak{A}_{A,B}$. For this we only have to verify that for each $(A';B';\mathcal{H}') \in \mathfrak{A}_{A,B}$ there exists $(U',V';\mathcal{H}') \in \mathfrak{U}_{C(A),C(B)}$ such that

$$A' = i(I + U')(I - U')^{-1}, \quad B' = i(I + V')(I - V')^{-1},$$

and set

$$U' = C(A') \text{ and } V' = C(B').$$

Proposition 3.3 *Let $(A,B;\mathcal{H})$ be a pair of operators in $\mathcal{H}$, such that A is selfadjoint, and B is closed symmetric. If*

$$< [(A-iI)(A+iI)^{-1}]^n Bf, f' > = < [(A-iI)(A+iI)^{-1}]^n f, Bf' > \tag{3.5}$$

for all $n \in \mathbb{N}$ and f, f' in the domain of B, then there is a bijection between $\mathfrak{A}_{A,B}$ and

$$\left\{ \begin{array}{l} \beta \in \mathcal{B}\big(\mathcal{H} \ominus V_{n\in\mathbb{Z}}[(A-iI)(A+iI)^{-1}]^n(\mathcal{D}), \mathcal{H} \ominus V_{n\in\mathbb{Z}}[(A-iI)(A+iI)^{-1}]^n(\mathcal{R})\big) : \\[2mm] (A-iI)(A+iI)^{-1}\beta = \beta(A-iI)(A+iI)^{-1}|_{\mathcal{H}\ominus V_{n\in\mathbb{Z}}[(A-iI)(A+iI)^{-1}]^n(\mathcal{D})} \end{array} \right\}$$

where $\mathcal{D}$ is the domain of $(B+iI)^{-1}$ and $\mathcal{R}$ is the range of $C(B)$.

Proof. If condition (7) holds, then condition (1) will be true for $U = C(A)$ and $V = C(B)$; thus $\mathfrak{U}_{C(A),C(B)} \neq \emptyset$. ∎

As a final remark, let $\mathcal{M}$ be a dense linear manifold in the Hilbert space $\mathcal{H}$. Let A_0, B_0 be two symmetric operators defined on $\mathcal{M}$; A and B their closures, with A selfadjoint (cf [5],[6]). Then $\mathfrak{U}_{C(A),C(B)} \neq \emptyset$ implies $\mathfrak{A}_{A,B} \neq \emptyset$. But it is not clear whether it might be possible that $\mathcal{A}_{A,B} \neq \emptyset$, while $\mathfrak{U}_{C(A),C(B)} = \emptyset$.

References

[1] R. Arocena: On the extension problem for a class of translation invariant positive forms, *J. Operator Theory*, **21**(1989), 323-347.

[2] P. Alegría: Parametrización y algoritmo de Schur para las representaciones integrales de nucleos de Toeplitz generalizados en Z y Z^2, Ph.D. Dissertation. Universidad del País Vasco, Faculdad de Ciencias, Matemáticas.

[3] D.Z. Arov; L.Z. Grossman: Scattering matrices in the theory of extensions of isometric operators, *Soviet Math. Dokl.*, **127**(1983), 508-512.

[4] M. Cotlar; C. Sadosky: Two parameters lifting theorems and double Hilbert transforms in commutative and non-commutative settings, *J. Math. Anal. Appl.*, **150**(1990), 439-480.

[5] R.S. Ismagelov: Self-adjoint extensions of a system of commuting symmetric operators, *Soviet Math.Dokl.*, **133**(1960), 867-870.

[6] A. Korányi: On some classes of analytic functions of several variables, *Trans. A.M.S.* 1961.

[7] M.D.Morán: On commuting isometries, *J. Operator Theory*, **24**(1990), 75-83.

María Dolores Morán

Universidad Central de Venezuela
Faculdad de Ciencias
Escuela de Fisica y Matematicas
Departamento de Matematicas
A.P. 20.513–Caracas 1020A
Venezuela

Operator Theory:
Advances and Applications, Vol. 61
© 1993 Birkhäuser Verlag Basel

Some Generalizations of Classical Interpolation Problems

Adolf A. Nudelman

Abstract. In this paper we consider some matrix versions of classical interpolation problems (such as the Nevanlinna-Pick, the Carathéodory or the Schur problems). We consider these generalized problems on domains defined by some polynomial inequalities and we find the necessary and sufficient conditions under which the main results of classical case may be extended over this case.

1. The (A,B,C)-problem on classical domains

Each of the classical interpolation problems includes interpolation conditions as well as a class of analytic functions. In its turn a corresponding class of analytic functions is defined by a domain and a range. The classical domains are:

(i) the unit disc $\Delta_1 = \{z \mid |z| < 1\}$,

(ii) the upper half-plane $\Delta_2 = \{z \mid \operatorname{Im} z > 0\}$,

(iii) the right half-plane $\Delta_3 = \{z \mid \operatorname{Re} z > 0\}$.

If we introduce the polynomials

$$P_1(z,\zeta) = 1 - z\zeta, \qquad P_2(z,\zeta) = (z - \zeta)/i, \qquad P_3(z,\zeta) = z + \zeta$$

then these domains can be described by the polynomial inequalities

$$\Delta_j = \{z \mid P_j(z,\bar{z}) > 0\}, \qquad j = 1, 2, 3.$$

The classical ranges for matrix-valued functions are defined by one of the following inequalities

$$I - w(z)w(z)^* \geq 0, \ (w(z) - w(z)^*)/i \geq 0, \ w(z) + w(z)^* \geq 0.$$

Here $w(z)$ is an $m \times m$ matrix-function analytic on Δ_j. Denoting

$$J_1 = \begin{pmatrix} -I_m & 0 \\ 0 & I_m \end{pmatrix}, \ J_2 = \begin{pmatrix} 0 & -iI_m \\ iI_m & 0 \end{pmatrix}, \ J_3 = \begin{pmatrix} 0 & I_m \\ I_m & 0 \end{pmatrix}$$

then these inequalities can be rewritten in the unified form

$$(w(z), I_m)J_k(w(z), I_m)^* \geq 0. \tag{1.1}$$

We say that an $m \times m$ matrix function $w(z)$ belongs to the class $\mathcal{F}(\Delta_l, J_k)$ if $w(z)$ is analytic on Δ_l and the inequality (1.1) is true for all $z \in \Delta_l$. It is known that for $w(z) \in \mathcal{F}(\Delta_l, J_k)$ the kernel

$$K_w(z, \bar{\zeta}) = \frac{(w(z), I)J_k(w(\zeta), I)^*}{P_l(z, \bar{\zeta})}$$

is positive on Δ_l, that is the block-matrix

$$(K(z_m, \bar{z}_n))_{m,n=1}^{\nu}$$

is positive semidefinite for any $\{z_n\}_1^{\nu} \subset \Delta_l$. In the sequel we shall omit the subscripts l, k in the notations of Δ_l and J_k.

Explicit formulations of the above mentioned classical problems are as follows.

The Nevanlinna-Pick problem: Let the class $\mathcal{F}(\Delta, J)$ and sequences $\{z_j\}_1^{\nu} \subset \Delta$ and $\{w_j\}_1^{\nu}$ be given, where w_j $(j = 1, 2, \ldots, \nu)$ is an $m \times m$ matrix. Find matrix-functions $w(z) \in \mathcal{F}(\Delta, J)$ such that $w(z_j) = w_j$, $j = 1, 2, \ldots, \nu$.

The Schur (respectively, *the Carathéodory) problem*: Let Δ be the unit disc and let a sequence $\{c_k\}_0^{\nu}$ of $m \times m$ matrices be given. Find matrix-functions $w(z) \in \mathcal{F}(\Delta, J)$ such that

$$w(z) = c_0 + c_1 z + \cdots + c_\nu z^\nu + O(z^\nu) \tag{1.2}$$

Here $J = J_1$ for the Schur problem and $J = J_3$ for the Carathéodory problem.

These problems can be generalized as follows. Let the following data be given: the class $\mathcal{F}(\Delta, J)$, the set $Z = \{z_j\}_1^l \subset \Delta$ and to each $z_j \in Z$ there correspond some non-negative integer numbers ν_j $(j = 1, 2, \ldots, l)$ and two sequences of $1 \times m$ matrices (rows) $\{B_k^{(j)}\}_{k=0}^{\nu_j}$ and $\{c_k^{(j)}\}_{k=0}^{\nu_j}$. Find $m \times m$ matrix functions $w(z) \in \mathcal{F}(\Delta, J)$ for which

$$(b_0^{(j)} + b_1^{(j)}(z - z_j) + \cdots + b_{\nu_j}^{(j)}(z - z_j)^{\nu_j})w(z)$$
$$= c_0^{(j)} + c_1^{(j)}(z - z_j) + \cdots + c_{\nu_j}^{(j)}(z - z_j)^{\nu_j} + O((z - z_j)^{\nu_j}), \tag{1.3}$$
$$j = 1, 2, \cdots, \nu.$$

This is the so called one side tangential problem considered in [7,8]. For further generalizations and a detailed historical survey see, for instance, [1].

In order to obtain a more compact and convenient form of the problem, we introduce the following matrices. For each of the given points z_j we construct the lower

Jordan's block

$$A_j = \begin{pmatrix} z_j & & & 0 \\ 1 & \ddots & & \\ & \ddots & \ddots & \\ 0 & & 1 & z_j \end{pmatrix}$$

of size $(\nu_j + 1) \times (\nu_j + 1)$ and we set

$$A = \begin{pmatrix} A_1 & & 0 \\ & \ddots & \\ 0 & & A_l \end{pmatrix}.$$

Thus A is a block-diagonal matrix of size $N \times N$ where $N = \sum_{j=1}^{l}(\nu_j + 1)$. Now we set

$$b_j = \begin{pmatrix} b_0^{(j)} \\ \vdots \\ b_{\nu_j}^{(j)} \end{pmatrix}, \ c_j = \begin{pmatrix} c_0^{(j)} \\ \vdots \\ c_{\nu_j}^{(j)} \end{pmatrix}, \ B = \begin{pmatrix} b_1 \\ \vdots \\ b_l \end{pmatrix}, \ C = \begin{pmatrix} c_1 \\ \vdots \\ c_l \end{pmatrix}.$$

The matrices b_j and c_j, B and C are of sizes $(\nu_j + 1) \times m$ and $N \times m$, respectively.

It is easy to check that the interpolation conditions (1.3) can be rewritten in the form

$$\frac{1}{2\pi i} \int_\gamma (\lambda I - A)^{-1} B w(\lambda) d\lambda = C \tag{1.4}$$

where γ is a contour encircling the spectrum of A, $\gamma \subset \Delta$. We remark that the matrix function $w(z)$ satisfying (1.4) also satisfies the same relation obtained by replacing A, B, C by TAT^{-1}, TB, TC respectively. In this connection we shall require of A only that its spectrum is in Δ. When this requirement is satisfied, the problem of finding all the $w(z) \in \mathcal{F}(\Delta, J)$ satisfying (1.4) is called the (A, B, C)-*problem* in the class $\mathcal{F}(\Delta, J)$.

The above mentioned classical problems are special cases of the (A, B, C)-problem. For example, we obtain the Nevanlinna-Pick problem if we put

$$A = \begin{pmatrix} z_1 I_m & & 0 \\ & \ddots & \\ 0 & & z_n I_m \end{pmatrix}, B = \begin{pmatrix} I_m \\ \vdots \\ I_m \end{pmatrix}, C = \begin{pmatrix} w_1 \\ \vdots \\ w_n \end{pmatrix}. \tag{1.5}$$

Indeed, in this case comparing block by block we can write (1.4) as

$$\frac{1}{2\pi i} \int_\gamma (\lambda I_m - z_j I_m)^{-1} I_m w(\lambda) d\lambda = w_j, \ j = 1, 2, \ldots, n,$$

that is, as

$$w(z_j) = w_j, \ j = 1, 2, \ldots, n.$$

In order to obtain the Schur or the Carathéodory problem we put

$$
A = \begin{pmatrix} 0 & & & 0 \\ I_m & 0 & & \\ & \ddots & \ddots & \\ 0 & & I_m & 0 \end{pmatrix}, \quad B = \begin{pmatrix} I_m \\ 0 \\ \vdots \\ 0 \end{pmatrix}, \quad C = \begin{pmatrix} c_0 \\ c_1 \\ \vdots \\ c_\nu \end{pmatrix}
$$

where $c_j (j = 0, 1, \ldots, \nu)$ is an $m \times m$ matrix defined by the expansion (1.2).

There are three questions in the (A, B, C)-problem as well as in classical problems: (a) find conditions under which the problem has a solution; (b) investigate whether the solution is unique or not: (c) describe all solutions. These questions were considered in [7] and [8], in the subsections 1.2 - 1.4 we indicate some facts on which the main results of [7,8] are based. Later on these results will be proved in more general situations.

1.2. The solvability of the (A, B, C)-problem depends on the positivity of some Hermitian matrix W which is a (unique) solution of the so called Fundamental Matrix Equation (FME). The right hand side of FME is equal to $(C, B)J(C, B)^*$ but the form of the left hand side is defined by the domain Δ, that is it depends on the polynomial $P(z, \zeta)$. More precisely, FME has the form

$$
W - AWA^* = (C, B)J(C, B)^* \quad \text{if} \quad \Delta = \Delta_1 \ (P(z, \zeta) = 1 - z\zeta),
$$

$$
(AW - WA^*)/i = (C, B)J(C, B)^* \quad \text{if} \quad \Delta = \Delta_2 \ (P(z, \zeta) = (z - \zeta)/i),
$$

$$
AW + WA^* = (C, B)J(C, B)^* \quad \text{if} \quad \Delta = \Delta_3 \ (P(z, \zeta) = z + \zeta).
$$

We shall suppose additionally that

$$
\operatorname{rank} B = m \tag{1.6}
$$

if either $J = J_2$ or $J = J_3$. This requirement is satisfied for classical problems, later on (Lemma 1.4) its meaning will be explained.

Theorem 1.1 *The (A, B, C)-problem in $\mathcal{F}(\Delta, J)$ is solvable if and only if the solution W of the corresponding FME is positive semidefinite.*

Note that for classical problems the solution of FME can be found very easy. For instance, in case of the Nevanlinna-Pick problem when $\Delta = \Delta_1$ and $J = J_1$, FME has the form

$$
W - AWA^* = BB^* - CC^*. \tag{1.7}
$$

where A, B, C are the same as in (1.5). If we decompose all the matrices A, B, C, W in blocks of size $m \times m$ then we obtain from (1.7) that

$$W_{ij} - z_i W \bar{z}_j = I_m - w_i w_j^*,$$

hence

$$W = \left(\frac{I_m - w_i w_j^*}{1 - z_i \bar{z}_j} \right)_{i,j=1}^n,$$

that is W is a Pick matrix.

1.3. Now we consider the so-called *extended* (A, B, C)-*problem*. Let a given (A, B, C)-problem be solvable, that is the solution W of FME is positive semidefinite, and let ζ be some point of Δ and v be some $m \times m$ constant matrix. An extended (A, B, C)-problem in $\mathcal{F}(\Delta, J)$ is $(A_\zeta, B_\zeta, C_\zeta)$-problem in $\mathcal{F}(\Delta, J)$ where

$$A_\zeta = \left(\begin{array}{cc} A & 0 \\ 0 & \zeta I_m \end{array} \right), B_\zeta = \left(\begin{array}{c} B \\ I_m \end{array} \right), C_\zeta = \left(\begin{array}{c} C \\ v \end{array} \right).$$

On the one hand, the $(A_\zeta, B_\zeta, C_\zeta)$-problem in $\mathcal{F}(\Delta, J)$ is solvable if and only if there exists the representation

$$C_\zeta = \frac{1}{2\pi i} \int_\gamma (\lambda I - A_\zeta)^{-1} B_\zeta w(\lambda) d\lambda, \quad w(z) \in \mathcal{F}(\Delta, J)$$

that is, if and only if

$$\left(\begin{array}{c} C \\ v \end{array} \right) = \frac{1}{2\pi i} \int_\gamma \left(\begin{array}{c} (\lambda I - A)^{-1} B \\ (\lambda - \zeta)^{-1} I_m \end{array} \right) w(\lambda) d\lambda, \quad w(z) \in \mathcal{F}(\Delta, J).$$

Hence $(A_\zeta, B_\zeta, C_\zeta)$-problem is solvable if and only if the initial (A, B, C)-problem is solvable and $v = w(\zeta)$ where $w(z)$ is an arbitrary solution.

On the other hand, applying Theorem 1.1 we see that the $(A_\zeta, B_\zeta, C_\zeta)$-problem is solvable if and only if its corresponding FME has a positive semidefinite solution W_ζ. We can obtain an explicit formula for W_ζ after simple calculations based on the comparison of blocks.

Both these reasons imply the next theorem.

Theorem 1.2 *A matrix function $w(z)$, which is analytic on Δ, is a solution of (A, B, C) - problem in $\mathcal{F}(\Delta, J)$ if and only if it is a solution of the matrix inequality*

$$\left(\begin{array}{cc} W & R(z)^*(w(z), I_m)^* \\ (w(z), I_m)R(z) & \frac{(w(z), I_m)J(w(z), I_m)^*}{P(z, \bar{z})} \end{array} \right) \geq 0 \tag{1.8}$$

where $R(z) = J(C, B)^* P^{-1}(z, A^*)$.

According to Potapov's terminology this inequality will be named a *Fundamental Matrix Inequality* (FMI). FMI plays an essential role in Potapov's approach to classical interpolation problems.

1.4. Here and in the sequel we suppose that the solution W of FME is positive definite. In this case inequality (1.8) is equivalent to the following one

$$\frac{(w(z), I_m) J (w(z), I_m)^*}{P(z, \bar{z})} - (w(z), I_m) R(z) W^{-1} R(z)^* (w(z), I_m)^* \geq 0.$$

Since $P(z, \bar{z}) > 0$ for $z \in \Delta$, the last inequality can be rewritten as

$$(w(z), I_m)(J - P(z, \bar{z}) R(z) W^{-1} R(z)^*)(w(z), I_m)^* \geq 0.$$

Remember that $R(z) = J(C, B)^* P^{-1}(z, A^*)$. The key to the description of all solutions of (A, B, C) - problem is the next theorem.

Theorem 1.3 *There exists a factorization*

$$J - P(z, \bar{z}) R(z) W^{-1} R(z)^* = \mathfrak{A}(z) J \mathfrak{A}(z)^*. \tag{1.9}$$

where $\mathfrak{A}(z)$ is a $2m \times 2m$ rational matrix function.

It is clear that $\mathfrak{A}(z)$ is J-inner, that is

$$\begin{aligned} J - \mathfrak{A}(z) J \mathfrak{A}(z)^* &\geq 0 \quad \text{in} \quad \Delta, \\ J - \mathfrak{A}(z) J \mathfrak{A}(z)^* &= 0 \quad \text{on} \quad \partial\Delta. \end{aligned} \tag{1.10}$$

Using an analyticitiy argument we can deduce from (1.9) that for any z and ζ

$$J - P(z, \bar{\zeta}) R(z) W^{-1} R(\zeta)^* = \mathfrak{A}(z) J \mathfrak{A}(\zeta)^*. \tag{1.11}$$

Since $W > 0$, this equality implies that the kernel

$$\frac{J - \mathfrak{A}(z) J \mathfrak{A}(\zeta)^*}{P(z, \bar{\zeta})}$$

is positive on Δ. Note that the factor $\mathfrak{A}(z)$ in (1.11) is not uniquely defined. In order to normalize $\mathfrak{A}(z)$ we choose some $z_0 \in \partial\Delta$ and introduce the additional condition $\mathfrak{A}(z_0) = I_{2m}$. Then we obtain from (1.11) the explicit expression for $\mathfrak{A}(z)$:

$$\mathfrak{A}(z) = I_{2m} - P(z, \bar{z}_0) J(C, B)^* P^{-1}(z, A^*) W^{-1} P^{-1}(A, \bar{z}_0)(C, B).$$

1.5. It is useful to introduce a set $\mathcal{F}(\Delta, J)$ of pairs of $m \times m$ matrix functions $(p(z), q(z))$ for which the next conditions are satisfied

(i) $p(z), q(z)$ are analytic on Δ;

(ii) $\operatorname{rank}(p(z), q(z)) = m$ for all $z \in \Delta$;

(iii) $(p(z), q(z))J(p(z), q(z))^* \geq 0$, $z \in \Delta$.

It turns out that for any pair $(p(z), q(z)) \in \mathcal{F}(\Delta, J)$ the kernel

$$\frac{(p(z), q(z))J(p(\zeta), q(\zeta))^*}{P(z, \bar{\zeta})}$$

is positive on Δ if P and J are the above introduced P_j and J_k. If $(p(z), q(z)) \in \mathcal{F}(\Delta, J)$ and the matrix function $q(z)$ is invertible, then $q^{-1}(z)p(z) \in \mathcal{F}(\Delta, J)$.

1.6. As it has already been mentioned, FMI is equivalent to

$$(w(z), I_m)\mathfrak{A}(z)J\mathfrak{A}(z)^*(w(z), I_m)^* \geq 0, \quad z \in \Delta$$

From the last inequality it is easy to deduce that the pair $(p(z), q(z))$ defined by

$$(w(z), I_m)\mathfrak{A}(z) = (p(z), q(z))$$

belongs to the class $\tilde{\mathcal{F}}(\Delta, J)$. Conversely, for any pair $(p(z), q(z)) \in \tilde{\mathcal{F}}(\Delta, J)$ let us denote

$$(P(z), Q(z)) = (p(z), q(z))\mathfrak{A}^{-1}(z). \tag{1.12}$$

It is obvious that $P(z), Q(z)$ are analytic on Δ, and rank $(P(z), Q(z) = m$. Moreover,

$$(P(z), Q(z))J(P(z), Q(z))^* \geq 0, z \in \Delta, \tag{1.13}$$

because (1.10) implies that on Δ

$$(P(z), Q(z))J(P(z), Q(z))^* \geq (P(z), Q(z))\mathfrak{A}(z)J\mathfrak{A}(z)^*(P(z), Q(z))^*,$$

or taking into account (1.12), that

$$(P(z), Q(z))J(P(z), Q(z))^* \geq (p(z), q(z))J(p(z), q(z))^* \geq 0, z \in \Delta$$

Lemma 1.4 *Let the matrix function $Q(z)$ be defined by (1.12) where $p(z), q(z)) \in \mathcal{F}(\Delta, J)$. If rank $B = m$ then $Q(z)$ is invertible.*

Proof. If $(p(z), q(z)) \in \mathcal{F}(\Delta, J)$ then

$$(P(z), Q(z))^*\mathfrak{A}(z)J\mathfrak{A}(z)^*(P(z), Q(z))^* \geq 0, z \in \Delta.$$

This inequality is equivalent to

$$\begin{pmatrix} W & R(z)^*(P(z),Q(z))^* \\ (P(z),Q(z))R(z) & \dfrac{(P(z),Q(z))J(P(z),Q(z))^*}{P(z,\bar{z})} \end{pmatrix} \geq 0 \qquad (1.14)$$

where $R(z) = J(C,B)^* P^{-1}(z,A^*)$. Assume that $f^*Q(z_0) = 0$ for some $z_0 \in \Delta$ and $f \in \mathbb{C}^m$. Let us show that this is possible only if $f = 0$. From (1.13) we deduce first that $(f^*P(z_0), f^*Q(z_0))J(f^*P(z_0), f^*Q(z_0) \geq 0$, i.e.

$$(f^*P(z_0), 0)J(f^*P(z_0), 0)^* \geq 0. \qquad (1.15)$$

If $J = J_1$ then (1.15) can be rewritten as $-f^*P(z_o)P(z_0)^*f \geq 0$ hence $f^*P(z_0) = 0$. This fact together with $f^*Q(z_0) = 0$ and $\operatorname{rank}(P(z_0), Q(z_0)) = m$ mean that $f = 0$. If $J = J_2$ or $J = J_3$ then the left hand side of (1.15) is null. From (1.14) we deduce that for every $g \in \mathbb{C}^m$

$$\begin{pmatrix} g^*Wg & g^*R(z_0)^*(P(z_0),Q(z_0))^*f \\ f^*(P(z_0),Q(z_0))R(z_0)g & 0 \end{pmatrix} \geq 0. \qquad (1.16)$$

This implies

$$f^*(P(z_0),Q(z_0))R(z_0)g = 0,$$

or since g is arbitrary, $f^*Q(z_0) = 0$,

$$(f^*P(z_0), 0)R(z_0) = 0.$$

From this we obtain, using the expression $R(z) = J(C,B)^*P^{-1}(z,A^*)$, that

$$(f^*P(z_0), 0)J(C,B)^* = 0.$$

Since $J = J_2$ or $J = J_3$ this can be rewritten as

$$f^*P(z_0)B^* = 0.$$

But we suppose that $\operatorname{rank} B^* = m$, so $f^*P(z_0) = 0$. As it was shown above, this implies $f = 0$. $\blacksquare$

The proof of the next theorem is the combination of Lemma 1.4 and the preceding reasonings.

Theorem 1.5 *If $W > 0$, then all solutions $w(z)$ of the (A,B,C)-problem in the class $\mathcal{F}(\Delta, J)$ are given by the formulas*

$$w(z) = Q^{-1}(z)P(z)$$

where

$$(P(z),Q(z) = (p(z),q(z))\mathfrak{A}^{-1}(z)$$

and the pairs $(p(z),q(z))$ run through the class $\mathcal{F}(\Delta, J)$.

2. The (A,B,C)-problem on algebraic domains. The necessity

2.1. Let now $P(z, \zeta)$ be an arbitrary polynomial

$$P(z, \zeta) = \sum_{i,j=0}^{M} \gamma_{ij} z^i \zeta^j$$

with a Hermitian matrix of coefficients

$$\Gamma_p = (\gamma_{ij}) = \Gamma_p^*.$$

Put

$$\Delta = \{z \mid P(z, \bar{z}) > 0\}$$

and denote by $\mathcal{F}(\Delta, J)$ the class of those matrix functions $w(z)$ analytic on Δ for which the kernel

$$K_w(z, \bar{\zeta}) = \frac{(w(z), I_m) J (w(\zeta), I_m)^*}{P(z, \bar{\zeta})}$$

is positive on Δ.

Assume additionally that $\mathcal{F}(\Delta, J)$ contains constant matrices. If $v \in \mathcal{F}(\Delta, J), v = $ const, then $(v, I_m) J (v, I_m)^*$ is positive since for any $z \in \Delta$ we have $K_v(z, \bar{z}) \geq 0$ and $P(z, \bar{z}) > 0$. Thus the positivity of the kernel $K_v(z, \bar{\zeta})$ implies the positivity of $1/P(z, \bar{\zeta})$. Conversely, if for a constant matrix v we have $(v, I_m) J (v, I_m)^* \geq 0$ and the kernel $1/P(z, \bar{\zeta})$ is positive on Δ, then $v \in \mathcal{F}(\Delta, J)$.

A.Mazko and V.Kharitonov [4]–[7] proved that the kernel $1/P(z, \bar{\zeta})$ is positive on Δ if and only if the matrix Γ_P has only one positive eigenvalue. If this condition holds, we have $P(z, \bar{\zeta}) \neq 0$ for any $z, \zeta \in \Delta$.

We say that the domain Δ defined by the polynomial P is a *domain of classical type* if Theorems 1.1,1.2,1.3, and 1.5 are true for any (A, B, C)-problem in the class $\mathcal{F}(\Delta, J)$. Later on we determine the conditions on polynomial $P(z, \zeta)$ under which Δ is a domain of classical type.

2.2. If $P(z, \zeta) = \sum_{i,j=0}^{M} \gamma_{ij} z^i \zeta_j$, then for the given (A, B, C)-problem in $\mathcal{F}(\Delta, J)$ FME has a form

$$\sum_{0}^{M} \gamma_{ij} A^i W (A^*)^j = (C, B) J (C, B)^*.$$

This equation has the unique solution W since the spectrum of A is situated in Δ and $P(z, \bar{\zeta}) \neq 0$ for $z, \zeta \in \Delta$, [8]. Let the given (A, B, C)-problem be solvable and $w(z)$ be its solution. Let us verify that in this case

$$W = \frac{1}{4\pi^2} \int_\gamma \int_\gamma (\lambda I - A)^{-1} B K_w(\lambda, \bar{\mu}) B^* (\bar{\mu} I - A^*)^{-1} d\lambda d\bar{\mu}. \tag{2.1}$$

180   A.A. Nudelman

Indeed, note that

$$A^i(\lambda I - A)^{-1} = \lambda^i(\lambda I - A)^{-1} + \varphi(\lambda)$$

where $\varphi(\lambda)$ is a matrix polynomial. From this fact we can easily deduce that

$$A^i W(A^*)^j = \frac{1}{4\pi^2} \int_\gamma \int_\gamma \lambda^i \bar{\mu}^j (\lambda I - A)^{-1} B K_w(\lambda, \bar{\mu}) B^*(\bar{\mu} I - A^*)^{-1} d\lambda d\bar{\mu}$$

and

$$\sum_{i,j=0}^{M} \gamma_{ij} A^i W(A^*)^j$$

$$= \frac{1}{4\pi^2} \int_\gamma \int_\gamma P(\lambda, \bar{\mu})(\lambda I - A)^{-1} B K_w(\lambda, \bar{\mu}) B^*(\bar{\mu} I - A^*)^{-1} d\lambda d\bar{\mu}$$

$$= \frac{1}{4\pi^2} \int_\gamma \int_\gamma (\lambda I - A)^{-1} B(w(\lambda), I_m) J(w(\mu), I_m)^* B^*(\bar{\mu} I - A^*)^{-1} d\lambda d\bar{\mu}.$$

The last expression equals $(C, B)J(C, B)^*$ since

$$\frac{1}{2\pi i} \int_\gamma (\lambda I - A)^{-1} B w(\lambda) d\lambda = C, \quad \frac{1}{2\pi i} \int_\gamma (\lambda I - A)^{-1} B d\lambda = B.$$

Now the positivity of the kernel $K_w(\lambda, \bar{\mu})$ implies the positive semidefiniteness of W. The part "only if" of Theorem 1.1 is proved.

2.3. Let $w(z)$ be some solution of (A, B, C)-problem in $\mathcal{F}(\Delta, J)$. As we know, the extended $(A_\zeta, B_\zeta, C_\zeta)$-problem, where

$$A_\zeta = \begin{pmatrix} A & 0 \\ 0 & \zeta I_m \end{pmatrix}, B_\zeta = \begin{pmatrix} B \\ I_m \end{pmatrix}, C_\zeta = \begin{pmatrix} C \\ w(\zeta) \end{pmatrix},$$

is solvable, hence $W_\zeta \geq 0$. Since

$$W_\zeta = \frac{1}{4\pi^2} \int_\gamma \int_\gamma (\lambda I - A_\zeta)^{-1} B_\zeta K_w(\lambda, \bar{\mu}) B_\zeta^*(\bar{\mu} I - A^*)^{-1} d\lambda d\bar{\mu},$$

it follows that W_ζ can be expressed by a matrix

$$W_\zeta = \begin{pmatrix} W & X \\ X^* & Z \end{pmatrix}$$

where

$$X = \frac{1}{4\pi^2} \int_\gamma \int_\gamma (\lambda I - A)^{-1} B K_w(\lambda, \bar{\mu})(\bar{\mu} - \bar{\zeta})^{-1} d\lambda d\bar{\mu},$$

and

$$Z = \frac{1}{4\pi^2} \int_\gamma \int_\gamma (\lambda - \zeta)^{-1} K_w(\lambda, \bar{\mu})(\bar{\mu} - \bar{\zeta})^{-1} d\lambda d\bar{\mu}$$

Let us calculate, for instance, one of the entries X^*

$$\frac{1}{4\pi^2} \int_\gamma \int_\gamma (\lambda - \zeta)^{-1} P^{-1}(\lambda, \bar\mu)(w(\lambda), I_m) J(w(\mu), I_m)^* B^*(\bar\mu I - A^*)^{-1} d\lambda d\bar\mu$$

$$= -\frac{1}{2\pi i} \int_\gamma P^{-1}(\zeta, \bar\mu)(w(\zeta), I_m) J(w(\mu), I_m)^* B^*(\bar\mu I - A^*)^{-1} d\bar\mu \tag{2.2}$$

$$= (w(\zeta), I_m) J(C, B)^* P^{-1}(\zeta, A^*) = (w(\zeta), I_m) R(\zeta).$$

Similarly we obtain

$$\frac{1}{4\pi^2} \int_\gamma \int_\gamma (\lambda - \zeta)^{-1} K_w(\lambda, \bar\mu)(\bar\mu - \bar\zeta)^{-1} d\lambda d\bar\mu$$

$$= K_w(\zeta, \bar\zeta) = \frac{(w(\zeta), I_m) J(w(\zeta), I_m)^*}{P(\zeta, \bar\zeta)} \tag{2.3}$$

We have obtained thus

$$\begin{pmatrix} W & R(\zeta)^*(w(\zeta), I_m)^* \\[2mm] (w(\zeta), I_m) R(\zeta) & \frac{(w(\zeta), I_m) J(w(\zeta), I_m)^*}{P(\zeta, \bar\zeta)} \end{pmatrix} \geq 0 \tag{2.4}$$

and the part "only if" of Theorem 1.2 is proved.

2.4. Now we find the necessary conditions for the factorization (1.9) to hold. We know already that in this case one can take

$$\mathfrak{A}(z) = I_{2m} - P(z, \bar z_0) J(C, B)^* P^{-1}(z, A^*) W^{-1} P^{-1}(A, \bar z_0)(C, B).$$

Put $A = \alpha I$ where $\alpha \in \Delta$ and choose B and C such that $U = (C, B) J(C, B)^* \geq 0$. The FME for this (A, B, C)-problem has the form

$$P(\alpha, \bar\alpha) W = U$$

so that

$$W = U/P(\alpha, \bar\alpha).$$

Put for brevity $K = (C, B)$. Then $U = KJK^*$ and

$$\mathfrak{A}(z) = I_{2m} - \frac{P(z, \bar z_0) P(\alpha, \bar\alpha)}{P(z, \bar\alpha) P(\alpha, \bar z_0)} JV$$

where $V = K^* U^{-1} K = K^*(KJK^*)^{-1} K > 0$.

The identity (1.9) can be written in this case as

$$J - \frac{P(z, \bar z) P(\alpha, \bar\alpha)}{P(z, \bar\alpha) P(\alpha, \bar z)} JVJ = \left(I_{2m} - \frac{P(z, \bar z_0) P(\alpha, \bar\alpha)}{P(z, \bar\alpha) P(\alpha, \bar z_0)} JV\right) J \left(I_{2m} - \frac{P(z_0, \bar z) P(\alpha, \bar\alpha)}{P(\alpha, \bar z) P(z_0, \bar\alpha)} VJ\right).$$

182   A.A. Nudelman

Since $VJV = (K^*(KJK^*)^{-1}K)J(K^*(KJK^*)^{-1}K) = K^*(KJK^*)^{-1}K = V$, the following identity must hold for the polynomial $P(z,\zeta)$

$$\frac{P(z,\bar{z})}{P(z,\bar{\alpha})P(\alpha,\bar{z})} = \frac{P(z,\bar{z}_0)}{P(z,\bar{\alpha})P(\alpha,\bar{z}_0)} + \frac{P(z_0,\bar{z})}{P(\alpha,\bar{z})P(z_0,\bar{\alpha})}$$
$$- \frac{P(z,\bar{z}_0)P(z_0,\bar{z})P(\alpha,\bar{\alpha})}{P(z,\bar{\alpha})P(\alpha,\bar{z})P(z_0,\bar{\alpha})P(\alpha,\bar{z}_0)}$$

or

$$P(z,\bar{z}) = P(z,\bar{z}_0)\frac{P(\alpha,\bar{z})}{P(\alpha,\bar{z}_0)} + \frac{P(z,\bar{\alpha})}{P(z_0,\bar{\alpha})}P(z_0,\bar{z})$$
$$- \frac{P(\alpha,\bar{\alpha})}{P(\alpha,\bar{z}_0)P(z_0,\alpha)}P(z,\bar{z}_0)P(z_0,\bar{z})$$

Lemma 2.1 *The identity (1.9) is true if and only if the polynomial $P(z,\zeta)$ admits a representation of type*

$$P(z,\bar{\zeta}) = r(z)\overline{s(\zeta)} + s(z)\overline{r(\zeta)}, \tag{2.5}$$

where $r(z)$ and $s(z)$ are polynomials.

Proof. Let $P(z,\bar{\zeta})$ admit (2.4). Since $P(z_0,\bar{z}_0) = 0$, there are three possibilities: either $s(z_0) = 0$, or $r(z_0) = 0$, or $r(z_0) = i\tau s(z_0)$ where $\bar{\tau} = \tau$. We can consider only the first case because the second one can be considered in the similar way, and the third case will be reduced to the first one by substituting $s(z)$ by $r(z) - i\tau s(z)$. For the first case the right hand side of (1.9) can be written as

$$s(z)\overline{r(z_0)}\frac{r(\alpha)\overline{s(z)} + s(\alpha)\overline{r(z)}}{s(\alpha)\overline{r(z_0)}} + \frac{r(z)\overline{s(\alpha)} + s(z)\overline{r(\alpha)}}{r(z_0)\overline{s(\alpha)}}r(z_0)\overline{s(\alpha)}$$
$$- \frac{r(\alpha)\overline{s(\alpha)} + s(\alpha)\overline{r(\alpha)}}{s(\alpha)\overline{r(z_0)}r(z_0)\overline{s(\alpha)}}s(z)\overline{r(z_0)}r(z_0)\overline{s(z)}$$
$$= s(z)\left(\frac{r(\alpha)}{s(\alpha)}\overline{s(z)} + \overline{r(z)}\right) + \left(r(z) + s(z)\frac{\overline{r(\alpha)}}{\overline{s(\alpha)}}\right)\overline{s(z)}$$
$$- \left(\frac{r(\alpha)}{s(\alpha)} + \frac{\overline{r(\alpha)}}{\overline{s(\alpha)}}\right)s(z)\overline{s(z)}$$
$$= s(z)\overline{r(z)} + r(z)\overline{s(z)} = P(z,\bar{z})$$

Conversely, let (2.3) be true for some $z_0 \in \partial\Delta$ and $\alpha \in \Delta$. Put

$$u(z) = P(z,\bar{z}_0), \quad v(z) = \frac{P(z,\bar{\alpha})}{P(z_0,\bar{\alpha})}, \quad 2x = \frac{P(\alpha,\bar{\alpha})}{P(\alpha,\bar{z}_0)P(z_0,\bar{\alpha})}$$

It is easy to see that $\bar{x} = x$ and according to (2.3)

$$P(z,\bar{z}) = u(z)\overline{v(z)} + v(z)\overline{u(z)} - 2xu(z)\overline{u(z)}.$$

Hence

$$P(z,\bar{z}) = r(z)\overline{s(z)} + s(z)\overline{r(z)} \tag{2.6}$$

where $r(z) = u(z), s(z) = v(z) - xu(z)$. Now (2.5) implies (2.4). ∎

Lemma 2.2 *The polynomial $P(z,\zeta)$ admits the representation (2.4) if and only if the matrix Γ_P has only one positive and only one negative eigenvalue (so rank $\Gamma_P = 2$ and signat $\Gamma_P = 0$).*

Proof. First we remark that (2.4) is equivalent to

$$P(z,\bar{\zeta}) = p(z)\overline{p(\zeta)} - q(z)\overline{q(\zeta)} \tag{2.7}$$

since one can take

$$p(z) = (r(z) + s(z))/\sqrt{2}, \quad q(z) = (r(z) - s(z))/\sqrt{2}.$$

Put $\Lambda(z) = (1, z, \ldots, z^M)$ so that $P(z,\bar{\zeta}) = \Lambda(z)\Gamma_P\Lambda(\zeta)^*$. Let $p(z) = \sum_0^M p_j z^j, q(z) = \sum_0^M q_j z^j$, put $p = col(p_0, p_1, \ldots, p_M)$, $q = col(q_0, q_1, \ldots, q_M)$, then

$$p(z) = \Lambda(z)p, \quad q(z) = \Lambda(z)q \tag{2.8}$$

and from (2.6) we see that

$$\Gamma_p = pp^* - qq^*. \tag{2.9}$$

Thus Γ_P has only one positive and only one negative eigenvalue.

Conversely, if Γ_P has only one positive and only one negative eigenvalue, then Γ_P admits the representation (2.8) whence it follows (6.2). ∎

3. The (A,B,C)-problem on classical domains. The necessity

The next theorem gives a condition under which a domain Δ is the domain of classical type.

Theorem 3.1 *A polynomial*

$$P(z,\zeta) = \sum_0^M \gamma_{ij} z^i \zeta^j$$

defines the domain

$$\Delta = \{z \mid P(z,\bar{z}) > 0\}$$

of classical type if and only if the matrix $\Gamma_p = (\gamma_{ij})_0^M$ has only one positive and only one negative eigenvalue.

The part "only if" is proved in sections 2.2–2.4. Now we come to the proof of the parts "if" of Theorems 1.1,1.2,1.3, and 1.5. So in the sequel we suppose

$$P(z,\bar\zeta) = r(z)\overline{s(\zeta)} + s(z)\overline{r(\zeta)} \tag{3.1}$$

and for definiteness we take

$$J = J_3 = \begin{pmatrix} 0 & I_m \\ I_m & 0 \end{pmatrix}.$$

In this case FME has a form

$$r(A)Ws(A)^* + s(A)Wr(A)^* = BC^* + CB^* \tag{3.2}$$

3.1. To prove the part "if" of the Theorem 1.1 we suppose $W > 0$ (the case $W \geq 0$ is reduced to this case) and put

$$w(z) = C^*\psi^{-1}(z)C$$

where

$$\psi(z) = \left(\frac{r(z)}{s(z)}s(A) - r(A) \right) Ws(A)^* + BC^*.$$

a) $w(\lambda) \in \mathcal{F}(\Delta, J_3)$. Indeed, $s(z) \neq 0$ for $z \in \Delta$, so $s(A)$ is invertible and

$$\psi(z) + \psi(\zeta)^* =$$
$$= \left(\frac{r(z)}{s(z)} + \frac{\overline{r(\zeta)}}{\overline{s(\zeta)}} \right) s(A)Ws(A)^* - r(A)Ws(A)^* - s(A)Wr(A)^* + BC^* + CB^*.$$

Taking into account (3.1) and (3.2) we obtain

$$(\psi(z) + \psi(\zeta)^*)/P(z,\bar\zeta) = s(A)Ws(A)^*/s(z)\overline{s(\zeta)}$$

so that this kernel is positive on Δ. In particular this implies that the real part of $\psi(z)$ is positive definite for any $z \in \Delta$ and hence $\psi(z)$ is invertible for $z \in \Delta$. Now it is easy to deduce the positivity of the kernel

$$\begin{aligned}
K_w(z,\bar\zeta) &= (w(z) + w(\zeta)^*)/P(z,\bar\zeta) \\
&= C^*\psi^{-1}(z)\frac{\psi(z) + \psi(\zeta)^*}{P(z,\bar\zeta)}\psi^{-1}(\zeta)^*C \\
&= C^*\psi^{-1}(z)\frac{s(A)Ws(A)^*}{s(z)\overline{s(\zeta)}}\psi^{-1}(\zeta)^*C.
\end{aligned}$$

b) $w(z)$ is a solution of the given (A, B, C)-problem. Indeed,

$$\frac{1}{2\pi i}\int_\gamma (\lambda I - A)^{-1}Bw(\lambda)d\lambda = \frac{1}{2\pi i}\int_\gamma (\lambda I - A)^{-1}BC^*\psi^{-1}(\lambda)Cd\lambda$$

$$= \frac{1}{2\pi i}\int_\gamma (\lambda I - A)^{-1}(\psi(\lambda) - (r(\lambda)s(A) - s(\lambda)r(A))W\frac{s(A)^*}{s(\lambda)})\psi^{-1}(\lambda)Cd\lambda$$

$$= \frac{1}{2\pi i}\int_\gamma (\lambda I - A)^{-1}Cd\lambda$$

$$-\frac{1}{2\pi i}\int_\gamma (\lambda I - A)^{-1}(r(\lambda)s(A) - s(\lambda)r(A))W\frac{s(A)^*}{s(\lambda)})\psi^{-1}(\lambda)Cd\lambda.$$

The first integral is equal to C and the second one is equal to zero since the product

$$(\lambda I - A)^{-1}(r(\lambda)s(A) - s(\lambda)r(A))$$

$$= (\lambda\Gamma - A)^{-1}((r(\lambda)I - r(A))s(A) - (s(\lambda)I - s(A))r(A))$$

is analytic on Δ. $\blacksquare$

3.2. The proof of Theorem 1.2 is based on Theorem 1.1; the argument of section 1.3 can be repeated again.

3.3. Now we shall prove the factorization (1.11) under conditions (3.1) where

$$\mathfrak{A}(z) = I_{2m} - P(z, z_0)J_3(C, B)^*P^{-1}(z, A^*)W^{-1}(A, z_0)(C, B)$$

$$= I_{2m} - J_3K^*T(z)s^{-1}(A)^*W^{-1}s^{-1}(A)K,$$

and

$$K = (C, B),\ T(z) = (v(A)^* + v(z)I)^{-1},\ v(z) = r(z)/s(z)$$

so that

$$T(z)v(A)^* = I - v(z)T(z). \tag{3.3}$$

Let us compute

$$\mathfrak{A}(z)J_3\mathfrak{A}(\zeta)^* = J_3 - J_3K^*T(z)s^{-1}(A)^*W^{-1}s^{-1}(A)J_3$$

$$-J_3K^*s^{-1}(A)^*W^{-1}s^{-1}(A)T(\zeta)^*KJ_3 \tag{3.4}$$

$$+J_3K^*T(z)s^{-1}(A)^*W^{-1}s^{-1}(A)KJ_3K^*s^{-1}(A)^*W^{-1}s^{-1}(A)T(\zeta)^*KJ_3.$$

Taking into account FME

$$KJ_3K^* = r(A)Ws(A)^* + s(A)Wr(A)^*$$

we get

$$W^{-1}s^{-1}(A)KJ_3K^*s^{-1}(A)^*W^{-1} = W^{-1}v(A) + v(A)^*W^{-1}$$

and the last term of (3.4) can be transformed into

$$J_3 K^* T(z) v(A) s^{-1}(A)^* W^{-1} s^{-1}(A) T(\zeta)^* K J_3$$
$$+ \quad J_3 K^* T(z) s^{-1}(A)^* W^{-1} s^{-1}(A) v(A) T(\zeta)^* K J_3,$$

and using (3.3) it can be transformed into

$$J_3 K^* s^{-1}(A)^* W^{-1} s^{-1}(A) T(\zeta)^* K J_3$$
$$+ \quad J_3 K^* T(z) s^{-1}(A)^* W^{-1} s^{-1}(A) K J_3$$
$$- \quad (v(z) + \overline{v(\zeta)}) J_3 K^* T(z) s^{-1}(A)^* W^{-1} s^{-1}(A) K J_3$$

Now (3.4) transforms to

$$\begin{aligned}
\mathfrak{A}(z) J_3 \mathfrak{A}(\zeta)^* &= J_3 - (v(z) + v(\bar{\zeta})) J_3 K^* T(z) s^{-1}(A)^* W^{-1} s^{-1}(A) K J_3 \\
&= J_3 - P(z, \bar{\zeta}) J_3 K^* P^{-1}(z, A^*) W^{-1} P^{-1}(A, \bar{\zeta}) K J_3.
\end{aligned}$$

Thus (1.11) is proved.

3.4. To complete the proof it remains to verify the validity of Theorem 1.5 under the condition (3.1).

Lemma 3.2 *If the polynomial $P(z, \zeta)$ admits the representation (3.1), then a pair $(p(z), q(z))$ of matrix functions which are analytic on Δ belongs to $\tilde{\mathcal{F}}(\Delta, J)$ if and only if*

$$(p(z), q(z)) J (p(z), q(z))^* \geq 0, \quad z \in \Delta \tag{3.5}$$

Proof. According to the definition, $(p(z), q(z)) \in \tilde{\mathcal{F}}(\Delta, J)$ if and only if the kernel

$$\frac{(p(z), q(z)) J (p(\zeta), q(\zeta))^*}{P(z, \bar{\zeta})}$$

is positive in Δ. Evidently, if $(p(z), q(z)) \in \tilde{\mathcal{F}}(\Delta, J)$, then (3.5) is true. Conversely, let $(p(z), q(z))$ satisfy (3.5). The inequality

$$P(z, \bar{z}) = r(z) \overline{s(z)} + s(z) \overline{r(z)} > 0, \quad z \in \Delta$$

implies that the function $t(z) = r(z)/s(z)$ has positive real part on Δ. Hence any branch of the inverse function $z = z(t)$ maps the right half-plane Δ_3 onto Δ so that

$$(p(z(t)), q(z(t))) \in \tilde{\mathcal{F}}(\Delta_3, J)$$

Thus the kernel

$$\frac{(p(z(t)), q(z(t))) J (p(z(\tau)), q(z(\tau)))}{t + \bar{\tau}}$$

is positive in Δ_3. Put $z = z(t)$, $\zeta = z(\tau)$. The kernel

$$\frac{(p(z),q(z))J(p(\zeta),q(\zeta))^*}{t(z)+\overline{t(\zeta)}}$$

is positive in Δ, together with the kernel

$$s(z)\frac{(p(z),q(z))J(p(\zeta),q(\zeta))^*}{r(z)\overline{s(\zeta)}+s(z)\overline{r(\zeta)}}\overline{s(\zeta)}$$

and hence with the kernel

$$\frac{(p(z),q(z))J(p(\zeta),q(\zeta))^*}{P(z,\bar\zeta)}.$$

The proof is complete. $\blacksquare$

Now in order to prove Theorem 1.5 for the domain considered here we can repeat reasonings of section 1.6.

References

[1] J. Ball; I.Gohberg; L. Rodman: *Interpolation of rational matrix functions*, Birkhäuser Verlag, Basel 1990.

[2] Ju. L. Daletskii; M. G. Krein: *Stability of solutions of differential equations on Banach space*, AMS Transl. Math. Monographs, Vol.43, 1974.

[3] V. L. Kharitonov: The problem of the distribution of roots for the characteristic polynomial of a control system, *Avtomatika i telemekhanika*, 5 (1981), 42-47 (in Russian).

[4] A. G. Mazko; V. L. Kharitonov: On the Routh-Hurvitz problem for a class of algebraic domains, *Dinamika i ustoychivost mekhanicheskikh system*, Kiev, Inst. mathem., 1980, 124-128 (in Russian).

[5] A. G. Mazko: A matrix algorithm for design of optimal linear system with specified spectral properties, *Avtomatika i telemekhanika*, 5(1981), 33-37 (in Russian).

[6] A. G. Mazko: A generalization of Lyapunov theorem for a class of domains bounded by algebraic curves, *Avtomatika*, Kiev, 1 (1982), 85-87 (in Russian).

[7] A. A. Nudelman: On a new problem of moment type, *Dokl. Akad. Nauk* SSSR **233**:5(1977), 792-795; English transl. in *Soviet Math. Dokl.*, **18**(1977), 507-510.

[8] A. A. Nudelman: On a generalization of classical interpolation problems, *Dokl. Akad. Nauk* SSSR, **256**(1981), **4**, 790-793; English transl. in *Soviet Math. Dokl.* **23**(1981), 125-128.

A. A. Nudelman

Odessa Civil Engineering Institute
Ul. Didrikhsona 4
270029 Odessa
Ukraine

Operator Theory:
Advances and Applications, Vol. 61
© 1993 Birkhäuser Verlag Basel

The Kobayashi Distance between two Contractions

Ion Suciu

Abstract. The Kobayashi pseudodistance can be defined on each complex Banach manifold. Using contractive analytic functions from the unit disk into $\mathcal{B}(\mathcal{H})$ we can define the Kobayashi pseudodistance on the closed unit ball $\mathcal{B}_1$ of $\mathcal{B}(\mathcal{H})$, where $\mathcal{B}(\mathcal{H})$ is the algebra of all bounded linear operators on the Hilbert space $\mathcal{H}$. The main result of the paper asserts the fact the Kobayashi pseudodistance is a true distance on the *Shmulyan parts* of $\mathcal{B}_1$. Some connections between Pick condition for the two-nodes Nevanlinna-Pick interpolation problem and the Kobayashi distance are established.

1. The classical Nevanlinna–Pick interpolation problem and the Schwarz–Pick Lemma

Throughout this paper **D** will stand for the open unit disc in the complex plane **C**. The classical *Nevanlinna–Pick interpolation problem* is the following: given $\{z_1, z_2, \ldots, z_n\}$ a finite set of n distinct points in **D** and $\{w_1, w_2, \ldots, w_n\}$ a set of n complex numbers find necessary and sufficient conditions for the existence of an analytic function f from **D** into **C** such that $\|f\| = \sup_{z \in \mathbf{D}} |f(z)| \le 1$ and $f(z_j) = w_j$ for any $j = 1, 2, \ldots n$. In case such an f exists, give recurrent formulas to produce, by free parameters, the set of all solutions.

We shall call $[\{z_1, z_2, \ldots, z_n\}, \{w_1, w_2, \ldots, w_n\}]$ a *n-nodes system of data* for the Nevanlinna–Pick problem. Clearly we have necessarily $|w_j| \le 1$ for $1 \le j \le n$.

A necessary and sufficient condition for the existence of a solution in the Nevanlinna–Pick problem is the *Pick condition* on the system of data which asserts that the matrix

$$
\begin{bmatrix}
\frac{1-\bar{w}_1 w_2}{1-\bar{z}_1 z_2} & \cdots & \frac{1-\bar{w}_1 w_n}{1-\bar{z}_1 z_n} \\
\vdots & \ddots & \vdots \\
\frac{1-\bar{w}_n w_1}{1-\bar{z}_n z_1} & \cdots & \frac{1-\bar{w}_n w_n}{1-\bar{z}_n z_n}
\end{bmatrix}
\tag{1.1}
$$

on $\mathbf{C}^n$ is positive.

In the particular case of a system of data of the form $[\{0, z\}, \{0, w\}]$ the Pick condition reads

$$
\begin{bmatrix}
1 & 1 \\
1 & \frac{1-|w|^2}{1-|z|^2}
\end{bmatrix} \ge 0
\tag{1.2}
$$

and by computing the eigenvalues we can easily prove that (1.2) is equivalent to

$$|w| \leq |z|. \tag{1.3}$$

So if $|z| < 1$ and $|w| \leq 1$ there exists an analytic function in $\mathbf{D}$ with $\|f\| \leq 1$, $f(0) = 0$ and $f(z) = w$ if and only if (1.3) holds. Hence the Pick condition in this special case is the inequality which apears in the Schwarz Lemma.

The general two nodes case $[\{z_1, z_2\}, \{w_1, w_2\}]$ can be reduced to the above special case by composing with a Moebius transform on $\mathbf{D}$. From (1.3) it follows that in this case the Pick condition is equivalent to

$$\left| \frac{w_2 - w_1}{1 - \bar{w}_1 w_2} \right| \leq \left| \frac{z_2 - z_1}{1 - \bar{z}_1 z_2} \right| \tag{1.4}$$

Recall that the *hyperbolic (Poincaré) distance* on $\mathbf{D}$ is given by

$$\delta(z, w) = \tanh^{-1} \left(\left| \frac{z - w}{1 - \bar{w}z} \right| \right), \quad z, w \in \mathbf{D} \tag{1.5}$$

and that the *Schwarz–Pick Lemma* asserts that for any analitic function f from $\mathbf{D}$ into $\mathbf{D}$ and any $z, w \in \mathbf{D}$ we have

$$\delta(f(z), f(w)) \leq \delta(z, w) \tag{1.6}$$

So, in the two nodes case the Pick condition for the existence of a solution in the Nevanlinna–Pick problem is equivalent to the inequality contained in the Schwarz–Pick Lemma.

The general n-nodes case can be reduced inductively, under certain conditions, to the two nodes case, by composing succesively with adequate Moebius transforms. This was the way in which Pick proved its condition and gave recursive solutions to the Nevanlinna-Pick problem (cf. [8], [11]). The uniqueness parts from Schwarz and Schwarz–Pick Lemma produce corresponding uniqueness conditions in the Nevanlinna–Pick problem.

2. The operator Nevanlinna–Pick interpolation problem

The matrix or operator variants for the classical interpolation problems were intensively studied: remarkable results give to operator extrapolation theory a consistency close to the classical one. Most of these results are now encompassed in the elegant operator extrapolation theory generated by the famous Sz.-Nagy-Foiaş theorem concerning the dilations of intertwining operators. The monograph *The commutant lifting aproach to interpolation problems*, by C. Foiaş and A. E. Frazho ([5]), presents a unified approach, based on the geometric framework of the commutant lifting theorem, of the classical and modern interpolation problems and their recent applications in engineering and geophysics.

The operator Nevanlinna–Pick interpolation problem (as well as the Carathéodory problem) appears as the standard test case of the whole theory. The formulation of the problem is the same as the classical one.

Let $\mathcal{H}, \mathcal{K}$ be two Hilbert spaces, $\mathcal{B}(\mathcal{H}, \mathcal{K})$ be the Banach space of all linear bounded operators from $\mathcal{H}$ to $\mathcal{K}$ and $\mathcal{B}_1(\mathcal{H}, \mathcal{K})$ its closed until ball. By $H^\infty(\mathcal{H}, \mathcal{K})$ we denote the set of all norm bounded analytic functions from $\mathbf{D}$ into $\mathcal{B}(\mathcal{H}, \mathcal{K})$. With pointwise operations and supnorm $H^\infty(\mathcal{H}, \mathcal{K})$ is a Banach space. $H_1^\infty(H, \mathcal{K})$ stands for the closed unit ball in $H^\infty(\mathcal{H}, \mathcal{K})$ and $H^\infty(\mathcal{H}) = H^\infty(\mathcal{H}, \mathcal{H})$.

We formulate the operator Nevanlinna–Pick problem in the following special form:

Given a finite set of n distinct points $z_1, z_2, \ldots, z_n$ in $\mathbf{D}$ and a set of n contractions $A_1, A_2, \ldots, A_n$ on $\mathcal{H}$, find necessary and sufficient conditions for the existence of a function $F \in H_1^\infty(\mathcal{H})$ such that

$$F(z_j) = A_j, \quad j = 1, 2, \ldots, n.$$

The Pick condition can be written exactly as in the scalar (classical) case: the *operator Nevanlinna–Pick system of data* $[\{z_1, z_2, \ldots, z_n\}, \{A_1, A_2, \ldots A_n\}]$ satisfies the *Pick conditions* provided the $n \times n$ operator matrix

$$\begin{bmatrix} \frac{I - A_1^* A_1}{1 - \bar{z}_1 z_1} & \cdots & \frac{I - A_1^* A_n}{1 - \bar{z}_1 z_n} \\ \vdots & \ddots & \vdots \\ \frac{I - A_n^* A_1}{1 - \bar{z}_n z_1} & \cdots & \frac{I - A_n^* A_n}{1 - \bar{z}_n z_n} \end{bmatrix} \tag{2.1}$$

defines a positive operator on the direct sum $\mathcal{H}^n$ of n copies of $\mathcal{H}$.

Using intertwining dilation theory we can prove that the Pick condition is necessary and sufficient for the existence of a solution in the operator Nevanlinna–Pick problem. This theory also produces a complete Schur analysis of the set of all solutions, including algorithms for obtaining them recursively from several different types of free parameters (cf.[5]).

We write here the Schur type formula in a very special case; namely, the system of data $[\{0\}, \{A\}]$. Clearly $[\{0\}, \{A\}]$ verifies the Pick condition. Setting, as usually, for a contraction T on $\mathcal{H}$, $D_T^2 = I - T^*T$ and $\mathcal{D}_T = \overline{D_T \mathcal{H}}$, the formula

$$F(z) = A + z D_{A^*} F_1(z)[I + z A^* F_1(z)]^{-1} D_A \tag{2.2}$$

with $F_1 \in H_1^\infty(\mathcal{D}_A, \mathcal{D}_{A^*})$, describes all solutions of the Nevanlinna–Pick problem with the data $[\{0\}, \{A\}]$.

If $A = 0_{\mathcal{H}}$, then $F(z) = zF_1(z)$ with $F_1 \in H_1^{\infty}(\mathcal{H})$; this is the first part of Schwarz Lemma.

If we take the two nodes system of data of the form $[\{0, z\}, \{0_{\mathcal{H}}, A\}]$ then the Pick condition is

$$\begin{bmatrix} I_H & I_H \\ I_H & \frac{1}{1-|z|^2}D_A^2 \end{bmatrix} \geq 0. \tag{2.3}$$

Since the 2×2 operator matrix of the form

$$\begin{bmatrix} R & X \\ X^* & Q \end{bmatrix}$$

on $\mathcal{H} \oplus \mathcal{H}$ is positive if and only if R, Q are positive and

$$X = R^{1/2}\Gamma Q^{1/2}$$

with Γ a contraction from $\overline{Q^{1/2}\mathcal{H}}$ into $\overline{R^{1/2}\mathcal{H}}$ (cf. [5]) it results that the system of data $[\{0, z\}, \{0_{\mathcal{H}}, A\}]$ satisfies the Pick condition if and only if $\|A\| \leq |z|$. Thus again the Pick condition is equivalent to the inequality contained in the Schwarz Lemma.

The following question is now natural. *Does there exists a distance on $\mathcal{B}_1(\mathcal{H})$, or on parts of it, such that the Schwarz–Pick Lemma remains valid and the Pick condition for the systems of two nodes data in the Nevanlinna–Pick problem is equivalent with the inequality contained in the Schwarz–Pick Lemma?*

This paper contains some results connected to this question.

3.   Schwarz–Pick systems of pseudodistances

The universal nature of Schwarz Lemma is one of the wonderful things in mathematics. The excellent survey contained in the book *The Schwarz Lemma* by Sean Dineen [3] presents, in the contents and in the historical notes and remarks, the evolution of the ideas contained in the Schwarz Lemma with special orientation to infinite dimensional (Banach spaces) holomorphy.

I extract from this book the notion of the Schwarz–Pick systems of pseudodistances.

A *pseudodistance* on a nonempty set M is a function $\rho\colon M \times M \to [0, \infty)$ verifying $\rho(p, p) = 0$, $\rho(p, q) = \rho(q, p)$ and the triangle inequality $\rho(p, r) \leq \rho(p, q) + \rho(q, r)$, $p, q, r \in M$. Sometimes we allow ρ to take the value ∞ and think on ρ as a pseudodistance on each part of M (equivalence class) modulo the equivalence relation $p \sim q$ provided $\rho(p, q) < \infty$.

A system which assigns a pseudodistance ρ_M to each complex Banach manifold M (in particular to any domains in a Banach space) is called a *Schwarz–Pick system of pseudodistances* if the following conditions hold:

(i) The pseudodistance $\rho_{\mathbf{D}}$ asigned to $\mathbf{D}$ is the hyperbolic (Poincaré) distance on $\mathbf{D}$.

(ii) If M_1 and M_2 are two complex manifolds and f is a holomorphic function from M_1 to M_2 then

$$\rho_{M_2}(f(p), f(q)) \le \rho_{M_1}(p, q) \qquad p, q \in M_1.$$

Among the Schwarz–Pick systems there are two of maximal interest:
The *Carathéodory pseudodistance* on M is defined by

$$C_M(p, q) = \sup\{\delta(f(p), f(q)) \mid f \colon M \to \mathbf{D} \text{ is holomorphic}\}. \tag{3.1}$$

The *Kobayashi pseudodistance* on M is defined by

$$K_M(p, q) = \inf \sum_{k=1}^{n} \delta(0, z_k) \tag{3.2}$$

where the infimum in (3.3) is taken over all finite sets $z_1, z_2, \ldots, z_n$ of distinct points in $\mathbf{D}$ for which there exist holomorphic functions $f_1, f_2, \ldots, f_n$ from $\mathbf{D}$ into M such that

$$f_1(0) = p, \quad f_j(z_{j-1}) = f_j(0), \quad i \le j \le n-1, \quad f_n(z_n) = q.$$

The Carathéodory and the Kobayashi pseudodistances form two Schwarz–Pick systems; the following important extremal property is valid: for any other Schwarz–Pick system ρ_M we have

$$C_M(p, q) \le \rho_M(p, q) \le K_M(p, q), \quad p, q \in M \tag{3.3}$$

In general they are not true distances on M (for $M = \mathbf{C}$, for example, $C_M \equiv K_M \equiv 0$). However if M is a convex bounded domain in a Banach space they are true distances and generate the original topology on M (cf.[3]).

The importance of these two systems of Schwarz–Pick pseudodistance comes from (3.3). While the definition of Carathéodory pseudodistance is quite natural, the case of Kobayashi pseudodistance is more intricate, since the natural candidate

$$\delta_M(p, q) = \inf\{\delta(z, w)\mid \quad z, w \in \mathbf{D}, f \colon \mathbf{D} \to M, \quad \text{holomorphic with} \atop f(z) = p, f(w) = q\} \tag{3.4}$$

does not satisfy the triangle inequality. If δ_M satisfies the triangle inequality then $\delta_M = K_M$; this is the same as assuming that the Pick condition for two nodes data is equivalent to the inequality contained in the Schwarz–Pick Lemma.

All these considerations can be applied to the open unit ball of the Banach algebra $\mathcal{B}(\mathcal{H})$. But the most important extrapolation problems involve data with norm one contractions. Since on the closed unit ball $\mathcal{B}_1(\mathcal{H})$ of $\mathcal{B}(\mathcal{H})$ there is no canonical structure of complex Banach manifold, we can not apply the general theory in the most relevant operator extrapolation problems.

4. The Kobayashi distance between two contractions

We shall define the *Kobayashi pseudodistance* on $\mathcal{B}_1(\mathcal{H})$ setting for two contractions A, B on $\mathcal{H}$.

$$\delta_K(A, B) = \inf \sum_{j=1}^{n} \delta(0, z_j), \tag{4.1}$$

where the infimum is taken over all finite systems $z_1, z_2, \ldots, z_n$ of points in $\mathbf{D}$ for which there exist $F_1, F_2, \ldots, F_n \in H_1^{\infty}(\mathcal{H})$ such that

$$F_1(0) = A, \qquad F_j(z_j) = F_{j+1}(0) \text{ for } 1 \le j \le n-1, \qquad F_n(0) = B, \tag{4.2}$$

if such a system exists and is taken to be ∞ otherwise.

For $A, B, C \in \mathcal{B}_1(\mathcal{H})$ we have $\delta_K(A, A) = 0$, $\delta_K(A, B) = \delta_K(B, A)$ and $\delta_K(A, C) \le \delta_K(A, B) + \delta_K(B, C)$.

The equivalence relation ' $A \overset{K}{\sim} B$ provided $\delta_K(A, B) < \infty$ ' splits $\mathcal{B}_1(\mathcal{H})$ into disjoint parts called Δ_K-parts; δ_K restricted to each Δ_K-part is a finite valued pseudodistance. For a contraction A we shall denote by $\Delta_K(A)$ the Δ_K-part containing A.

From the discussion in section 2 it follows that $\Delta_K(O_{\mathcal{H}})$ is the open unit ball in $\mathcal{B}(\mathcal{H})$. Also it is clear that for any $F \in H_1^{\infty}(\mathcal{H})$, $F(\mathbf{D})$ is contained into a certain Δ_K-part and for the pseudodistance δ_K the Schwarz–Pick Lemma holds: for any $F \in H_1^{\infty}(\mathcal{H})$ we have

$$\delta_K(F(z), F(w)) \le \delta(z, w), \quad z, w \in \mathbf{D}. \tag{4.3}$$

The following theorem gives the first information on the structure of Δ_K-parts.

Theorem 4.1 *Let A, B be two contractions on $\mathcal{H}$. The following are equivalent:*

(i) *There exist z_0 in $\mathbf{D}$ and $F \in H_1^{\infty}(\mathcal{H})$ such that $F(0) = A$, $F(z_0) = B$.*

(ii) *There exist bounded operators R from $\mathcal{D}_A$ to $\mathcal{D}_{A^*}$ and Q from $\mathcal{D}_B$ to $\mathcal{D}_{B_*}$ such that*

$$B = A + D_{A^*}RD_A, \qquad A = B + D_{B^*}QD_B. \tag{4.4}$$

(iii) *A and B belong to the same Δ_K-part i.e.*

$$\delta_K(A, B) < \infty. \tag{4.5}$$

Proof. Let $F \in H_1^{\infty}(\mathcal{H})$ and $z_0 \in \mathbf{D}$ such that $F(0) = A$ and $F(z_0) = B$. Setting

$$G(z) = F\left(\frac{z + z_0}{1 + \bar{z}_0 z}\right)$$

then clearly $G \in H_1^\infty(\mathcal{H})$, $\quad G(0) = F(z_0) = B$ and $G(-z_0) = F(0) = A$.

Writing F and G under the form (2.1) we obtain (4.4) in (ii). Hence $(i) \to (ii)$ is proved.

In order to prove $(ii) \to (i)$ we have to show that there exists $z_0 \in \mathbf{D}$ such that the Pick condition

$$\begin{bmatrix} D_A^2 & I - A^*B \\ I - B^*A & \dfrac{D_B^2}{1-|z_0|^2} \end{bmatrix} \geq 0, \tag{4.6}$$

for the data $[\{0, z_0\}, \{A, B\}]$, holds.

Using the second equality in (4.4) we obtain

$$\|Ah\|^2 = \|Bh + D_B \cdot QD_B h\|^2 = \|h\|^2 + \|QD_B h\|^2 - \|D_B h - B^* QD_B h\|^2.$$

We used the fact that for a contraction T the operator matrix $\begin{bmatrix} T & D_{T^*} \\ D_T & -T^* \end{bmatrix}$ defines an isometric operator.

Hence

$$\|D_A h\|^2 = \|(I - B^*Q)D_B h\|^2 - \|QD_B h\|^2 \leq \|(I - B^*Q)\|^2 \|D_B h\|^2.$$

It follows that there exists a bounded operator X from $\mathcal{D}_B$ in $\mathcal{D}_A$ such that

$$D_A = XD_B \tag{4.7}$$

From the first equality in (4.4) we obtain

$$\begin{aligned} I - A^*B &= I - A^*A + A^*(A - B) = D_A^2 - A^* D_A \cdot RD_A \\ &= D_A^2 - D_A A^* RD_A = D_A(I - A^*R)D_A \\ &= D_A(I - A^*R)XD_B. \end{aligned}$$

Let $z_0 \in \mathbf{D}$ such that $\Gamma = (1 - |z_0|^2)^{1/2}(I - A^*R)X$ is a contraction from D_B into D_A. Then

$$I - A^*B = D_A \Gamma \frac{D_B}{(1 - |z_0|^2)^{1/2}}$$

and this implies the positivity condition (4.6). So the implication $(ii) \to (i)$ is proved. Since the implication $(i) \to (iii)$ is obvious, it remains to prove $(iii) \to (ii)$.

Let us remark first that the equalities (4.4) define an equivalence relation on $\mathcal{B}_1(\mathcal{H})$.

Let for

$$C = B + D_B \cdot R_1 D_B,$$

$$B = C + D_C \cdot Q_1 D_C.$$

Then

$$C = A + D_A \cdot RD_A + D_B \cdot R_1 D_B$$

196   *I. Suciu*

$$A = C + D_C \cdot Q_1 D_C + D_B \cdot R D_B$$

and using factorizations of type (4.7) we clearly obtain

$$C = A + D_A \cdot R_2 D_A$$

$$A = C + D_C \cdot Q_2 D_C$$

Suppose now that $\delta_K(A, B) < \infty$. Then there exist $z_1, z_2, \ldots z_n \in \mathbf{D}$ and functions $F_1, F_2, \ldots F_n \in H_1^\infty(\mathcal{H})$ such that $F_1(0) = A$, $F_j(z_j) = F_{j+1}(0), 1 \leq j \leq n - 1$, $F_j(z_n) = B$. Then for any $j = 1, 2, \ldots n$ the pair $A_j = F_j(0)$, $B_j = F_j(z_j)$, verifies (i) and consequently (ii). Since $A = A_1, B_j = A_{j+1}$ for $1 \leq j \leq n - 1$ and $B_n = B$ clearly it follows that for the pair A, B (4.4) holds. ∎

The equivalence relation defined by (4.4) on $\mathcal{B}_1(\mathcal{H})$ was introduced by Yu. L. Shmulyan in [10]. In [6] this equivalence relation was denoted by $\overset{o}{\sim}$ (o from Odessa) and it was proved that the corresponding o-parts of $\mathcal{B}_1(\mathcal{H})$ are the open C- faces of the C-convex subset $\mathcal{B}_1(\mathcal{H})$ of $B(\mathcal{H})$. So each Δ_K-part (o-part) is contained in a R-face of $\mathcal{B}_1(\mathcal{H})$ and it would be interesting to establish the connection between the Kobayashi pseudodistance and the part-distance (cf. [2]) on Δ_K.

Our next goal is to prove that the Kobayashi pseudodistance is a true distance on Δ_K-parts.

5. Harnack distance on Harnack parts of contractions

We say that the contractions A, B are *Harnack equivalent* if there exists a positive constant a such that.

$$1/a^2 \,\Re p(B) \leq \Re p(A) \leq a^2 \,\Re p(B), \tag{5.1}$$

for any analytic polynomial p satisfying $\Re p(z) \geq 0$ for $z \in \mathbf{D}$.

Clearly (5.1) defines an equivalence relation on $\mathcal{B}_1(\mathcal{H})$. It splits $\mathcal{B}_1(\mathcal{H})$ into the equivalence classes, called *Harnack parts* or Δ_H-parts. $\Delta_H(A)$ stands for the Harnack part containing A.

If $\mathcal{H} = \mathbf{C}$ then $\mathcal{B}_1(\mathcal{H}) = \bar{\mathbf{D}}$ and the classical Harnack inequalities show that in this case $\Delta_H(0) = \mathbf{D} = \Delta_H(z)$ for $|z| < 1$ and $\Delta_H(z) = \{z\}$ for $|z| = 1$.

In [6] it was proved that each Δ_O-part is contained in a Δ_H-part and if dim $\mathcal{H} = \infty$ there exists a Δ_H part which is not a Δ_O-part.

We shall define on each Δ_H-part the *Harnack distance* δ_H (cf. [15]) setting for $A, B \in \Delta_H$

$$\delta_H(A, B) = \inf\{\log a|\ (5.1) \text{ holds for } a\}. \tag{5.2}$$

It is easy to see that δ_H is a true distance on each Δ_H-part of $\mathcal{B}_1(\mathcal{H})$ and that on $\mathbf{D}$, as the only nontrivial Harnack part of $\mathcal{B}_1(\mathbf{C})$, it coincides with the hyperbolic (Poincaré) distance δ.

It was proved in [13] that $A \overset{H}{\sim} B$ if and only if the formula

$$S_{AB} \sum U_B^j h_j = \sum U_A^j h_j$$

defines a boundedly invertible operator from $\mathcal{K}_B$ to $\mathcal{K}_A$; here, for a contraction T on $\mathcal{H}$, U_T, acting on $\mathcal{K}_T$, is the minimal unitary dilation of T (cf. [16]). Moreover

$$\max\{\|S_{AB}\|, \|S_{BA}\|\} = \inf\{a|\ (5.1)\ \text{holds for}\ a\}.$$

Hence

$$\delta_H(A,B) = \log\max\{\|S_{AB}\|, \|S_{BA}\|\}. \tag{5.3}$$

In [14], using (5.3) and the Nagy–Foiaş model for contractions, some analytic formulas for computing $\delta_H(A,B)$ were given ; as a consequence, the following Schwarz–Pick Lemma for δ_H-distance was proved:

If F is a contractive $\mathcal{B}(\mathcal{H})$-valued analytic function on $\mathbf{D}$ then $F(\mathbf{D})$ is contained in a Harnack part and

$$\delta_H(F(z), F(w)) \le \delta(z,w), \qquad z, w \in \mathbf{D}. \tag{5.4}$$

Summing up we have

Theorem 5.1 *Any Δ_K-part of $\mathcal{B}_1(\mathcal{H})$ is contained in a Δ_H-part and for $A, B \in \Delta_K$ we have*

$$\delta_H(A,B) \le \delta_K(A,B).$$

Since δ_H is a true metric on a Harnack part we obtain

Corollary 5.2 *On each Δ_K-part of $B_1(\mathcal{H})$ the Kobayashi pseudodistance is a true distance.*

It was proved in [15] that δ_H is complete on Δ_H-parts. We do not know whether the corresponding result for δ_K holds.

In case $\dim \mathcal{H} \ge 2$ the condition $\delta_H(A,B) \le \delta(0, z_0)$ does not imply the Pick condition for the data $[\{0, z_0\}, \{A, B\}]$ (cf.[14]). For the Kobayashi distance δ_K we do not know if this is true. As in the case of the complex Banach manifolds the problem is if

$$\delta_K(A,B) = \inf\{\delta(0,z), z \in \mathbf{D}|(\exists)F \in H_1^\infty, F(0) = A, F(z) = B\}$$

holds.

6. Final remarks

For $\Delta_0 = \Delta_H(O_\mathcal{H}) = \Delta_K(O_\mathcal{H})$ – the open unit ball in $B(\mathcal{H})$ – the Kobayashi distance δ_K coincides with K_{Δ_0} defined in Section 3. In this case the Carathéodory distance C_{Δ_0} is also defined and we have (cf. [3])

$$\delta_K(A, B) = K_{\Delta_0}(A, B) = \delta_{\Delta_0}(A, B) = C_{\Delta_0}(A, B). \qquad (6.1)$$

So we have

Theorem 6.1 *For $A_1, A_2 \in \Delta_0$ the Pick condition for the two nodes Nevanlinna-Pick data $[\{z_1, z_2\}, \{A_1, A_2\}]$ is equivalent to the inequality contained in the Schwarz-Pick Lemma for the Kobayashi distance, i.e. with*

$$\delta_K(A_1, A_2) \leq \delta(z_1, z_2) \qquad (6.2)$$

We also have (cf. [17])

$$\delta_K(0_\mathcal{H}, A) = K_{\Delta_0}(0_\mathcal{H}, A) = C_{\Delta_0}(0_\mathcal{H}, A) = \delta(0, \|A\|) = \frac{1}{2} \log \frac{1 + \|A\|}{1 - \|A\|}. \qquad (6.3)$$

If dim $\mathcal{H} \geq 2$ it was shown in [14] that for A of the form

$$A = \begin{bmatrix} 0 & \rho \\ 0 & 0 \end{bmatrix}$$

with $0 < \rho < 1$ we have

$$\delta_H(0_\mathcal{H}, A) = 1/2 \log \frac{1}{1 - \|A\|}$$

and hence

$$\delta_H(0_\mathcal{H}, A) < \delta_K(0_\mathcal{H}, A).$$

If A is normal it was proved in [14] that

$$\delta_H(0, A) = \delta_K(0, A) = \frac{1}{2} \log \frac{1 + \|A\|}{1 - \|A\|}.$$

If Δ_K is an arbitrary nontrivial Δ_K-part we can look for a complex Banach manifold structure on Δ_K such that $K_{\Delta_K} = \delta_K$.

A natural candidate for the topology on Δ_K is the so-called operator quasi-ball topology generated by the family of neighbourhoods of the form:

$$\mathcal{V}_{A,\epsilon} = \{B \in B_1(\mathcal{H}) | B = A + D_{A^*} Q D_A, Q \in B(D_A, D_{A^*}), \|Q\| < \epsilon\}.$$

It was shown in [14] that Δ_H and Δ_K-parts are open and closed set in this topology, Δ_K-parts being the maximal connected components of the Δ_H-part containing them.

In [10] it was shown that for $A \in B_1(\mathcal{H})$ the Moebius transform

$$Z \to A + D_A \cdot Z(I + A^*Z)^{-1}D_A$$

maps $\Delta_0 = \Delta_K(O_H)$ on $\Delta_K(A)$.

We hope that following this line we can endow the Δ_K-parts with a complex Banach manifold structure such that the Harnack and Kobayashi distances are complete and compatible. This would enable to continue the study of Δ_H and Δ_K parts of $B_1(H)$ using results from hyperbolic geometry.

References

[1] T. Ando; I. Suciu and D. Timotin: Characterization of some Harnack parts of contractions, *J. Operator Theory* **2**(1979), 233-245.

[2] H. S. Bear: *Lecture on Gleason parts, Lecture Notes in Math.* **12**(1970)

[3] S. Dineen: *The Schwarz Lemma*, Clarendom Press, Oxford, 1989.

[4] C.Foiaş: On Harnack parts of contractions, *Rev. Roumaine Math. Pures Appl.***19**(1974, 315-318.

[5] C.Foiaş and A. E. Frazho: *The Commutant Lifting Aproach to Interpolation Problems*, Birkhäuser Verlag, Basel-Boston-Berlin, (1990).

[6] V. A. Khatskievich; Yu. L. Shmulyan and V. S. Shulman: Pre-orders and equivalences in the operator balls, *Sibirsk. Mat. J.* **32**(1991), No.32, 172-183.

[7] S. Kobayashi: *Hyperbolic manifolds and holomorphic mappings*, Marcel Dekker, New York, Pure and Appl. Math. **2** 1970.

[8] G. Pick: Über die Beschränkungen analytischen Functionen, welche durch vorgegebene Funktionswarte bewirkt sind., *Math. Ann.* **77**(1916), 7-23.

[9] H. A. Schwarz: Zur theorie der Abbildung (1869), *Gesammelte Abhandlungen* Vol. II, Springer, Berlin, 1890.

[10] Yu. L. Shmulyan: Generalized Möbius Transforms of Operator Balls, *Sibirsk. Math. Zh.* **21**(1980), No.5.

[11] S. Stoilow: *Theory Functions of Complex Variable I, II* [Romanian], Ed. Academiei Bucuresti (1954-1962).

[12] I. Suciu: Harnack Inequalities for a Functional Calculus, *Coll. Math. Soc. Janos Bolyai, Hilbert space operators* Tihany(1970), 499-511.

[13] I. Suciu: Analytic Relations between Functional Models for Contractions, *Acta Sci. Math.* (Szeged) **33**(1973), 359-365.

[14] I. Suciu: Schwarz Lemma for Operator Contractive Analytic Functions , *Preprint IMAR no.5,*1992.

[15] I. Suciu and I. Valuşescu: On the Hyperbolic Metric on Harnack Parts, *Studia Mathematica* T.**LV**(1976), no.1, 97-109.

[16] B. Sz-Nagy and C. Foiaş: *Harmonic Analysis of Operators in Hilbert space*, North Holland Publishing Co., Amsterdam-Budapest(1970)

[17] H. Upmeier: *Symetric Banach Manifolds and Jordan C*- algebras*, North Holland, Amsterdam, Math. Studies, 104, 1985.

Ion Suciu

Institute of Mathematics of
the Romanian Academy
P.O.Box 1-764, 70700 Bucharest
Romania

Operator Theory:
Advances and Applications, Vol. 61
© 1993 Birkhäuser Verlag Basel

The Category of Quotient Bornological Spaces

Lucien Waelbroeck

Abstract. Quotient spaces are useful. They will be part of Functional analysis as soon as Functional Analysts understand that they are useful. I have explained how I arrived in spaces with a boundedness, then in quotient spaces.

Exactness is important in algebra. In Functional Analysis, we use exactness. I describe an abelian category q which contains the category b of b-spaces and linear bounded mappings. Its objects are couples $E^1|E^0$. Morphisms should be defined. Let **Cat** be an abelian category. Exact functors $\Phi_1 : \mathbf{b} \to \mathbf{Cat}$ are defined; they can be extended in a unique way to exact functors $\mathbf{q} \to \mathbf{Cat}$. The computation is easy.

Exact functors: For U a topological paracompact space, the functor $\mathcal{C}(U, \cdot)$; for U a smooth manifold, $\mathcal{E}(U, \cdot)$; if $r \in \mathbf{R}_+ \setminus \mathbf{N}$ and U a smooth manifold, $C^r(U, \cdot)$; if $1 \leq p \leq \infty$ and Ω is a measure space, the functor $L^p(\Omega, \cdot)$.

My Doctorate (1953) [10] was done under Professor J. Leray. In 1950-51 (my first year after I graduated), Leray had spoken of Gårding's symbolic functional n-dimension Heaviside calculus.

O. Heaviside was an electric engineer (from the 19th century); he used his symbolic calculus and solved differential equations with constant coefficients. He did not construct his symbolic calculus in a rigorous way. At present, one constructs it with the Laplace transformation. In 1950, L. Gårding [4] used the n-dimension Laplace transformation and solved hyperbolic equations with constant coefficients.

1.   About my Doctorate

In my Doctorate, I have constructed the holomorphic functional calculus in an associative, commutative, unital complete locally convex algebra $\mathcal{A}$ with a joint continuous multiplication. The unit element 1 is identified with the number 1, the product of 1 by a complex number s is identified with s, $\mathbf{C} \subset \mathcal{A}$.

Definition 1.1 *Let* $\mathcal{A} = (\underline{A}, \mathcal{T}_{\mathcal{A}})$ *be an associative, topological, unital algebra and* $a \in \underline{A}$. *The element* $a \in \underline{A}$ *is regular if* $\exists\, M \in \mathbf{R}_+ \forall\, s \in \mathbf{C}, |s| \geq M : \exists\, (s-a)^{-1},$ *and* *the set $\{(s-a)^{-1} \mid |s| \geq M\}$ is bounded.* [1] [2]

My Doctorate did not contain Gårding's symbolic calculus, the operators $\frac{\partial}{\partial x_i}$ are not regular. In 1953, I felt that the construction of holomorphic functional calculus in function of non regular elements was a topic of research. Such of this can be found in my Thèse d'Agrégation.

In the Spring 1951, I. Jacobson gave a sequence of talks at Paris. Among others, he spoke of I.M. Gelfand [5], [6], [7], I.M. Gelfand-G. Shilov [8]. My subject was not the theory of Banach algebras, but was not too far from it. The papers due to Gelfand and Gelfand-Shilov were published in the Math. Sbornik, July 1941. The U.S.S.R. had been invaded in June, 1941. I did not find the 1940-1944 Math. Sbornik between 1950 and 1953 in the Western Continental European libraries to which I went. I read the Mathematical Reviews and reconstructed the proofs I needed.

2.   How I have arrived at the bornological spaces

H. Cartan (1956) spoke of my Doctorate at the Séminaire Bourbaki. I had considered a complete, locally convex algebra with a joint continuous multiplication.

Cartan reminded me that the product of two bounded sets is bounded if A is quasi-complete and its multiplication is separately continuous. It follows that A_b, (the set $\underline{A}$, with its von Neumann boundedness) is a b-algebra if A is a quasi-complete locally convex algebra with a separately continuous multiplication. It was then that I found the b-algebras. Bourbaki suggested that I consider a quasi-complete algebra with a hypo-continuous multiplication. I prefer b-algebras to algebras with a quasi-complete topology and a hypo-continuous multiplication.

In 1956, in the bornological situation, I could only prove that the holomorphic functional calculus is a homomorphism $\mathcal{Q}([\mathrm{sp}\ a]) \rightarrow \underline{A}$. I do not consider this a complete result. A homomorphism of the b-algebra $\mathcal{O}(\mathrm{sp}\ a)$ into a b-algebra A should map a bounded subset of $\mathcal{O}([\mathrm{sp}\ a])$ onto a bounded subset of A. In my Doctorate, I proved that the holomorphic functional calculus is continuous. In 1971 [12], using an easy application of the Buchwalter theorem [2], I have proved that the mapping maps a bounded subset into a bounded subset. (If E is a b-space whose boundedness has a countable basis, F is another b-space, $u : E \rightarrow F$ is linear bounded, $u(\underline{E}) = \underline{F}$, then u is bornologically surjective). We have a homomorphism of a b-algebra into another one.

---

[1] I define regular elements of an associative unital topological algebras. Further, the reader will find b-algebra. With the same definition, one defines the regular operators of the algebra.

[2] Since a few years, my topological vector spaces E are $(\underline{E}, \mathcal{T}_E)$, my bornological vector spaces are $E = (\underline{E}, \mathcal{B}_E)$, etc.

3. Bornological spaces

Definition 3.1 *Let $\underline{X}$ be a set, $\mathcal{P}(\underline{X})$ the set of subsets of $\underline{X}$. A boundedness $\mathcal{B}_X$ is a subset of $\mathcal{P}(\underline{X})$ such that*

a. $\mathcal{B}_X \subset \mathcal{P}(\underline{X})$.

b. $B \in \mathcal{B}_X$ *if* $B_1 \in \mathcal{B}_X$, $B_2 \in \mathcal{B}_X$, *and* $B \subset B_1 \cup B_2$.

c. $\forall\, x \in \underline{X} : \{x\} \in \mathcal{B}_X$.

$\underline{E}$ is a real or a complex vector space, **D** is the unit disc if the scalars are from **C**, it is the interval [-1,1] if the scalars are from **R**.

Definition 3.2 *A boundedness $\mathcal{B}_E$ on the vector space $\underline{E}$ is a vector boundedness if*

d. $B \in \mathcal{B}_X$ *if* $B \subset \mathbf{D}B_1 + \mathbf{D}B_2$ *if* $B_i \in \mathcal{B}_E$.

Definition 3.3 *Let $\underline{E}$ be a real or a complex space on the field **C** or **R**. Then a boundedness $\mathcal{B}_E$ on $\underline{E}$ is separated if*

e. *A vector subspace $\underline{E}_0$ of the vector space $\underline{E}$ is zero if $\underline{E}_0 \in \mathcal{B}_E$.*

4. b-spaces and b-subspaces

Definition 4.1 *A subset $B \subset \underline{E}$ is completant if it is absolutely convex, does not contain a bounded non zero vector subspace, and E_B, the vector subspace absorbed by B, with the Minkowski functional of B is a Banach space. A b-space E is a bornological vector space such that every bounded subset is contained in a bounded completant subset.*

In my Thèse d'Agrégation [11], I have considered b-ideals. I assumed that the reader should understand what b-subspaces are. The majority of functional analysts do not consider the same b-subspaces as mine, and do not consider the same b-ideals as mine.

Definition 4.2 *Let $E = (\underline{E}, \mathcal{B}_E)$ be a b-space. Then $F = (\underline{F}, \mathcal{B}_F)$ is a b-subspace of E if $\underline{F}$ is a vector subspace of $\underline{E}$ and $\mathcal{B}_F \subset \mathcal{B}_E$.*

For most present functional analysts, a b-subspace F of E is a closed vector space with the boundedness, topology, or some other structure induced by that of E. My definition is not theirs. (My closed b-subspaces are their b-subspaces).

204   L. Waelbroeck

I consider usually the Cauchy integral, and use also the Bochner integral. Let E be a Banach space, K a compact space, m a Baire measure on K, and $f \in \mathcal{C}(K, E)$. One can construct

$$\int_K f(x) dm(x) \in \underline{E}$$

If F is a Banach subspace of E (the norm of F is finer than the norm induced by that of E) and $f \in \mathcal{C}(U, F)$, then the integral belongs to $\underline{F}$. Of course, if E is a Banach space and F is a genuine [3] Banach subspace of E, a function $f \in \mathcal{C}([0,1], E)$ exists such that $\forall\, x \in [0,1] : f(x) \in \underline{F}$ but the integral

$$\int_0^1 f(x) dm(x)$$

does not belong to $\underline{F}$. With my definition, one must check that the function $f \in \mathcal{C}(K, F)$ is really a continuous function taking values in E. The integral belongs to $\mathrm{cl}\,(\underline{F})$ [4] (See Example 9.3).

Consider next the Bochner integral. Consider a finite measure space Ω, and $f \in \underline{L}^1(\Omega, E)$; again $\int_{\underline{\Omega}} f(x) dm(x) \in \underline{E}$. If $f \in \underline{L}^1(\Omega, F)$, $\int_{\underline{\Omega}} f(x) dm(x) \in \underline{F}$. There is not any reason that the vector subspace be closed.

5. b-algebras, b-ideals and applications

Definition 5.1 *A b-algebra $\mathcal{A}$ is a b-space with a bilinear multiplication such that $B_1 \cdot B_2$ is bounded if B_1 and B_2 are bounded.*

Definition 5.2 *A b-ideal α of the b-algebra $\mathcal{A}$ is an ideal $\underline{\alpha}$ of the algebra with a completant boundedness $\mathcal{B}_\alpha$ such that all bounded subsets of α are bounded in $\mathcal{A}$ and the product of a bounded subset of α by a bounded subset of $\mathcal{A}$ is bounded in α.*

Definition 5.3 *A topological vector space E is complete enough if its closed bounded absolutely convex subsets are completant.*

Example 5.4 Let $\mathcal{A}$ be a complete enough locally convex algebra with a separately continuous multiplication. Then $\mathcal{A}_b$ is a b-algebra[5].

Example 5.5 Let E be a complete enough locally convex space. Then $\mathcal{L}(E)$, the set of continuous linear mappings $E \to E$ with its equicontinous boundedness is a b-algebra.

---

[3] A genuine Banach subspace is a Banach subspace which is not closed.

[4] If $X = (\underline{X}, \mathcal{T}_X)$ is a topological space and $\underline{A} \subset \underline{X}$, then $\mathrm{cl}\,(\underline{A})$ or $\mathrm{cl}\,_X(\underline{A})$ is the closure of $\underline{A}$ in the topological space X.

[5] If $E = (\underline{E}, \mathcal{T}_E)$ is a topological vector space, then E_b is the bornological vector space, whose elements are $\underline{E}$, and B is bounded in E_b if it is absorbed by all neighbourhoods of the origin. This is the von Neumann boundedness of E.

Example 5.6 Let $\mathcal{A}$ be an associative, commutative b-algebra, and $(a_1, \ldots, a_n)$ elements of the algebra. The b-ideal

$$\mathrm{idl}\,(a_1, \ldots, a_n)$$

is the set $\{\sum_{i=1}^n a_i \cdot b_i \mid \forall\, i : a_i \in \mathcal{A}\}$, and $B \subset \underline{\mathrm{idl}}(a_1, \ldots, a_n)$ is bounded in $\mathrm{idl}\,(a_1, \ldots, a_n)$ if $\exists\, B_1, \ldots, B_n$ bounded in $\mathcal{A}$ such that $B \subset \sum_{i=1}^n a_i \cdot B_i$.

Example 5.7 Let H be a Hilbert space and $\mathcal{L}(H)$ the Banach algebra of continuous linear operators on H; τ_1 is the vector space of operators with a finite trace on H with its Banach norm. It is a Banach ideal of $\mathcal{L}(H)$.

Example 5.8 Let $\mathcal{A}$ be a complete enough locally convex algebra with a separately continuous multiplication and $\alpha = (\underline{\alpha}, \mathcal{T}_\alpha)$ be an ideal α of $\underline{A}$ with a complete enough topology, such that the inclusion $\underline{\alpha} \subset \underline{A}$ is continuous, while the multiplication $\cdot|_{A \times \alpha} : \underline{A} \times \alpha \to \alpha$ is separately continuous. Then $\mathcal{A}_b \times \alpha_b \to \alpha_b$ is bounded.

In other words, α_b is a right b-ideal of $\mathcal{A}_b$.

Example 5.9 Let E and F be two complete enough topological spaces, such that $\underline{F}$ is a vector subspace of $\underline{E}$ and the inclusion $\underline{F} \subset \underline{E}$ is continuous. Then we can take $\mathcal{A}$ to be the set of linear continuous mappings $E \to E$ which map $\underline{F}$ into itself and whose restriction to $\underline{F}$ is continuous $F \to F$. A subset B of $\mathcal{A}$ is bounded (for the boundedness of $\mathcal{A}$) if it is equicontinuous $E \to E$, and its restriction to $\underline{F}$ is equicontinuous $F \to F$. We see that $\mathcal{A}$ is a b-algebra. The b-space $\mathcal{L}(E, F)$, the space of linear bounded mappings $E \to F$ with the equibounded boundedness is a b-ideal of $\mathcal{A}$.

6. The quotient bornological spaces

I have promised that I shall speak of the category **q**.

The objects of **q** are couples (E^1, E^0), where $E^1 = (\underline{E}^1, \mathcal{B}_{E^1})$ is a b-space, and $E^0 = (\underline{E}^0, \mathcal{B}_{E^0})$ is a b-subspace of E^1, ($\underline{E}^0$ is a vector subspace of $\underline{E}^1$, $\mathcal{B}_{E^0}$ is a completant boundedness on $\underline{E}^0$ and $\mathcal{B}_{E^0} \subset \mathcal{B}_{E_1}$). I write $E = E^1|E^0 = (E^1, E^0)$.

Next to objects, we must find morphisms. As morphisms we can consider Vasilescu's [9] lifted graphs: A lifted graph (here) is a b-subspace $G(u)$ of $E^1 \times F^1$ such that $G(u) \cap (E^0 \times F^1) = E^0 \times F^0$ and the first projection $G(u) \to E^1$ is bornologically surjective[6].

---

[6]If X and Y are b-subspaces of the b-space E, the intersection $X \cap Y$ has as elements $\underline{X} \cap \underline{Y}$, its bounded sets are $\mathcal{B}_x \cap \mathcal{B}_x$. A linear bounded mapping $f : U \to V$ is bornologically surjective if $\forall\, B \in \mathcal{B}_v, \exists\, C \in \mathcal{B}_v$ such that $f(C) = B$.

I use strict morphisms and pseudo-isomorphisms. If $E = E^1|E^0$ and $F = F^1|F^0$ are q-spaces, then $u_1 : E^1 \to F^1$ induces a strict morphism if $u^1 : E^1 \to F^1$ is linear, bounded, and $u_{1|E^0} : E^0 \to F^0$ is bounded. Two linear mappings, u_1 and u'_1, each inducing a strict morphism, induce the same strict morphism if $u_1 - u'_1$ is linear bounded, $E^1 \to F^0$. The composition of strict morphisms is defined in an obvious natural way. A category $\tilde{q}$ is defined. It is the strict category.

A strict morphism s is a pseudo-isomorphism if it is induced by a bornologically surjective linear bounded mapping $s_1 : E^1 \to F^1$, and $(s_1)^{-1}(F^0) = E^0$.

Proposition 6.1 *A category* q *exists, such that all pseudo–isomorphisms of* $\tilde{q}$ *are isomorphisms of* q. *If* **Cat** *is a category, a strict functor* $\tilde{\Phi} : \tilde{q} \to$ **Cat** *can be extended to a functor* $\Phi : q \to$ **Cat** *iff* $\forall\, s$, *pseudo–isomorphism of* $\tilde{q}$, $\tilde{\Phi}(s)$ *is an isomorphism of* **Cat**. *The extension of the strict functor* $\tilde{\Phi}$ *is unique (modulo isomorphisms of functors) if it exists.*

Vasilescu's definition looks better than mine. I prefer mine. Let **Cat** be a category. To compute in a category, we need functors, here functors $\Phi : q \to$ **Cat**. With my definition, one first shows that the restriction $\tilde{\Phi}$ is a functor $\tilde{q} \to$ **Cat**. This is usually easy. Next, we show that for all pseudo-isomorphism, $\tilde{\Phi}(s)$ is an isomorphism. This is a Functional Analysis problem. The two categories defined are isomorphic.

7. Exactness in the category b and exact functors b → Cat

Definition 7.1 *A complex* $(u,v) : E \to F \to G$ *of* b *is exact if* $\forall\, B$, *bounded in* F, $v(B) = 0, \exists\, C$ *bounded in* E *such that* $u(B) = C$.

The exact complexes of b ressemble the exact complexes of the category **Fré**.

Definition 7.2 *Let* **Cat** *be an abelian category. A functor* $\Phi_1 : b \to$ **Cat** *is exact if* $(\Phi_1(u), \Phi_1(v))$ *is exact in* **Cat** *as soon as* (u,v) *is exact in* b.

It is equivalent to say that the functor Φ maps a short exact complex of b into a short exact complex of b, and maps an injective linear mapping into a monic morphism. I cannot work with the definition: "A functor is exact if it changes a short exact complex into a short complex". This does not prove that this is not equivalent to my definition of an exact functor. I do not believe that it is.

Proposition 7.3 *An exact functor* $\Phi_1 : b \to$ **Cat** *has a unique exact extension* $\Phi : q \to$ **Cat**.

We identify completant bornological spaces E^1, E^0 with the q-spaces $E^1|\{0\}$, $E^0|\{0\}$. The inclusion $E^0 \subset E^1$ is a monic morphism of q. We have several exact functors b → b ; we let $\Phi(E^1|E^0) \simeq \Phi_1(E^1)|\Phi_1(E^0)$.

8. Some exact functors

I give some exact functors $\mathbf{b} \to \mathbf{b}$ and $\mathbf{q} \to \mathbf{q}$. They are useful when we study problems in the category $\mathbf{b}$ and can be extended to exact functors $\mathbf{b} \to \mathbf{q}$.

Example 8.1 Let $E = (\underline{E}, \mathcal{B}_E)$ be a b-space. We let $\sigma(E) \simeq \underline{E}$. If $E = E^1|E^0$, we let $\sigma E \simeq \underline{E}^1/\underline{E}^0$.

Example 8.2 If E is a b-space and X is a set, then $\underline{\beta}(X, E)$ is the vector space of mappings $f : X \to \underline{E}$ such that $f(X)$ is bounded in E. A subset $B \subset \underline{\beta}(X, E)$ is bounded if it is "equibounded", that is, the set $B(B_1) = \{b(b_1) | b \in B, b_1 \in B_1\}$ is bounded in E for all B_1, bounded subset of E. If U a manifold that is countable at infinity, $\underline{\beta}'(U, E)$ is the vector space of the mappings $U \to E$ which are locally bounded. A subset B of $\underline{\beta}'(U, E)$ is bounded in $\beta'(U, E)$ if $\forall\, V \subset\subset U$, the set $B(V) = \{b(v) | b \in B, v \in V\}$ is bounded in E. If $E = (E^1|E^0)$, we let $\beta'(U, E) \simeq \beta'(U, E^1)|\beta'(U, E^0)$.

$\sigma(\beta(X, E))$ is a vector space if E is a q-space and X is a set. A large amount of mathematics done in the category $\mathbf{q}$ can be done using the functor $\sigma\beta(X, \cdot)$. I call it the "miracle functor". It is exact. A q-space E vanishes (i.e. is isomorphic to the q-space 0) iff $\sigma(\beta(X, E)) = \{0\}$ for all set X. A morphism $u : E \to F$ is monic in the category $\mathbf{q}$ ($E = E^1|E^0, F = F^1|F^0$ are q-spaces) iff $\forall\, X : \sigma(\beta(X, u))$ is injective. A complex $(u, v) : E \to F \to G$ of $\mathbf{q}$ is exact iff for all sets X, the complex $(\sigma(\beta(X, u)), \sigma(\beta(X, v)))$ is exact in the category $\mathbf{EV}$. I use the Miracle Functor, believing that it is useful in category $\mathbf{q}$. If $E = E^1|E^0$ is a q-space and X is a set, I write $E_X = \sigma(\beta(X, E))$. Instead of writing $a \in E_X$, I write $a \in_X E$. The set E_X is a "realisation" of the q-space E.

The following example is an application of the Bartle and Graves theorem [1]:

Example 8.3 Let K be a compact or a locally compact space which is "countable at infinity". The functor $C(K, \cdot)$ is exact, $\mathbf{b} \to \mathbf{b}$. Its continuation is such that

$$C(K, E^1|E^0) \simeq C(K, E^1)|C(K, E^0)$$

If E is a b-space, a subset $C \subset \underline{E}$ is relatively compact if $\exists\, B$, bounded completant, such that C is relatively compact in the Banach space E_B. If E is a Fréchet space, one knows what is relatively compact. In both cases, we let C_E be the sets that are relatively compact.

Example 8.4 E_c is the b-space as elements $\underline{E}$ and as boundedness C_E if E is either a b-space or a Fréchet space. The functor is exact $\mathbf{b} \to \mathbf{b}$. If $E = E^1|E^0$ is a q-space, $E_c \simeq E_c^1|E_c^0$.

These functors are defined $\mathbf{b} \to \mathbf{b}, \mathbf{Fr\acute{e}} \to \mathbf{b}$ and are exact. They have extensions $\mathbf{q} \to \mathbf{q}, \mathbf{qFr\acute{e}} \to \mathbf{q}$. All these functors are called $(\cdot)_c$. They allow us to translate Fréchet and quotient Fréchet problems in problems in the categories $\mathbf{b}, \mathbf{q}$.

If E is a Banach space, and U is a manifold countable at infinity, a mapping $f : U \to E$ belongs to $\mathcal{E}(U, E)$ if it is of the class C^∞. The space of these mappings is a priori a Fréchet space. On it we consider its von Neumann boundedness. This space is now a b-space if E is a Banach space.

Example 8.5 Let E be a b-space and U a manifold that is countable at infinity. Then $\mathcal{E}(U, E)$ is the space of mappings $U \to \underline{E}$ such that $\forall\, x \in U, \exists\, V_x$ neighbourhood of $x, \exists\, B_x$, bounded completant in E such that $f_{|V_x} \in \mathcal{E}(V_x, E^1_{B_x})$. A subset $B \subset \mathcal{E}(U, E)$ is bounded if $\forall\, x \in U, \exists\, V_x$ neighbourhood of $x, \exists\, B_x$, bounded completant in E such that $B_{|V_x}$ is bounded in $\mathcal{E}(V_x, E^1_{B_x})$.

If $r \in \mathbf{R} \setminus \mathbf{N}$ and E is a Banach space, a mapping $U \to E$ is of the class C^r taking its values in a E if it is of the class $C^{[r]}$ on U and its derivatives of order $[r]$ are of Hölder exponent $r - [r]$. On the space $\mathcal{C}^r(U, E)$, one can place its Fréchet locally convex topology. We replace this topology by its von Nemann boundedness. This a bornological space if E is a Banach space.

Example 8.6 Let U be a manifold that is countable at infinity. Let E be a b-space, and let $r \in \mathbf{R}_+ \setminus \mathbf{N}$. f is of the class C^r on U taking values in E if each $x \in U$ has a neighbourhood V_x and a bounded completant B_x of the b-space E exists such that $f_{|V_x} \in \mathcal{C}^r(V_x, E_{B_x})$ is bounded in $C^r(V_x, E_{B_x})$. A subset B_1 of $\mathcal{C}^r(U, E)$ is bounded if each $x \in U$ has a neighbourhood V_x and a bounded completant B_x of the b-space E exists such that $B_{|V_x}$ is bounded in the b-space $C^r(V_x, E_{B_x})$.

The functors are exact, $\mathbf{b} \to \mathbf{b}$, they can be extended to exact functors $\mathbf{q} \to \mathbf{q}$. If $E = E^1|E^0$, we see that $\mathcal{E}(U, E) \simeq \mathcal{E}(U, E^1)|\mathcal{E}(U, E^0); C^r(U, E) \simeq C^r(U, E^1)|C^r(U, E^0)$.

In both examples 8.4 and 8.5, we begin with a Banach space E. In the example 8.4, $\mathcal{E}(U, E)$, defined in the example, it is isomorphic to the b-space defined a few lines earlier. In the example 8.5, $C^r(U, E)(r \in \mathbf{R} \setminus \mathbf{N})$ in the example is isomorphic to the b-space defined a few lines earlier. (I mean, the examples generalize the definitions given in the Banach situation).

It seems good to define what I call the functor $C^\infty(U, \cdot)$ when E is a q-space. The functor is left exact, in the category $\mathbf{q}$, it is not exact:

Example 8.7 Let U be a manifold and E be a quotient bornological space. Then $C^\infty(U, E) \simeq \varprojlim_{r \to \infty} C^r(U, E)$.

Whenever E is a quotient Fréchet space, the q-spaces $C^\infty(U, E_c)$, $\mathcal{E}(U, E_c)$ and $(\mathcal{E}(U, E))_c$ are naturally isomorphic. When we consider b-spaces, the functor $C^\infty(U, \cdot)$ is left exact and is not exact. The functor $\mathcal{E}(U, \cdot)$ is exact. These restrictions of the functors to **Fré** are isomorphic, but the two functors are not isomorphic. Quotient Fréchet mathematics is easier than q-space mathematics.

Let $\Omega = (\underline{\Omega}, A_\Omega, m_\Omega)$ be a measure space. Our definition of measurable function taking values in a Banach space will be the one given by Diestel and Uhl [3]. When E is a b-space, we say that $f : \underline{\Omega} \to \underline{E}$ is measurable if $\exists\, B$, bounded completant such that $f(B) \subset \underline{E}_B$ and $f : \Omega \to E_B$ is measurable.

Example 8.8 Let $p \in \mathbf{R}$ be such that $1 \leq p \leq \infty$. A mapping $f : \Omega \to \underline{E}$ belongs to $\underline{L}^p(\Omega, E)$ (or, if φ is an Orlicz function, $f : \Omega \to \underline{E}$ belongs to $L_\varphi(\Omega, E)$) if it is measurable and, when E is a Banach space, the mapping $\omega \mapsto \|f(\omega)\|_E$ belongs to $L^p(\Omega)$ (respectively to $L_\varphi(\Omega)$), while, if E a b-space, then a bounded completant B exists such that the mapping $\omega \mapsto \|f(\omega)\|_{E_B}$ belongs to $L^p(\Omega)$ (or to $L_\varphi(\Omega)$). This is a Banach space if we place on it the norm $\|\|f\|\|_p = \|(\|f(\cdot)\|_E)\|_p$, $\|\|f\|\|_p = \|(\|f(\cdot)\|)_{E_B}\|_p$, or $\|\|f\|\|_\varphi = \|(\|f(\cdot)\|_E)\|_\varphi$, $\|\|f\|\|_\varphi = \|(\|f(\cdot)\|_{E_B})\|_\varphi$ according to the circumstances.

9. The continuous functions and the integral

I shall speak of the "Cauchy integral". At the end of this paragraph, I shall consider the integral of a continuous function taking values in E^1 such that $\forall\, t : f(t) \in E^0$. I do not construct the Bochner integral, but shall use it when I describe what happens when $E = E^1|E^0$ is a genuine Banach space.

My space is not locally convex. I must define continuous functions taking values in b-spaces.

Definition 9.1 *Let $X = (\underline{X}, T_X)$ be a completely regular topological space and E a b-space; $f \in \underline{C}(X, E)$ if $\forall\, x \in \underline{X}, \exists\, V_x$ neighbourhood of $x, \exists\, B_x$, bounded completant in E such that $f_{|V_x} \in \underline{C}(V_x, E_{B_x})$; B is bounded in $C(X, E)$ if $\forall\, x \in \underline{X}, \exists\, V_x$ neighbourhood of $x, \exists\, B_x$, bounded completant in E, such that $B_{|V_x}$ is bounded in $C(V_x, E_{B_x})$.*

This gives a functor $\mathbf{b} \to \mathbf{b}$. I can prove that the functor is exact $\mathbf{b} \to \mathbf{b}$ if the topological space X is paracompact. We consider a compact space, X; I have said that the functor $C(X, \cdot)$ is exact $\mathbf{b} \to \mathbf{b}$. (Example 8.3).

We use the following fact.

Proposition 9.2 *Let K be a compact space and $E = \bigcup_B E_B$ be a b-space, B ranging over the bounded completant subsets. Then*

$$\mathcal{C}(K, E) \simeq \bigcup_B \mathcal{C}(K, E_B)$$

Let $f \in \mathcal{C}(K, E)$. For all $x \in \underline{K}$, a neighbourhood V_x of x and B_x a bounded completant subset of E exist such that $f_{|V_x} \in \mathcal{C}(V_x, E_{B_x})$. The compact space K is contained in a finite union of sets $V_{x_n}, n = 1, \ldots, N$. We consider the bounded set $B = \sum_{i=1}^N B_i$. Every $f \in \mathcal{C}(X, E)$ belongs to some $\mathcal{C}(K, E_B)$. In a similar way, if C is bounded in $\mathcal{C}(K, E), \exists\, B$, bounded completant such that C is bounded in $\mathcal{C}(K, E_B)$.

The aim of the paragraph is the Cauchy integral. The Heine theorem shows that a continuous function on a compact space K is uniformly continuous. We assume that K is a metric compact and E is a Banach space. (The construction can be done in all compact spaces and locally convex topological spaces).

$$\forall\, \varepsilon > 0, \exists\, \eta > 0, \forall\, (x, y) \in \underline{E}, d(x, y) < \eta \Rightarrow \ \|f(x) - f(y)\|_E < \varepsilon.$$

On K we consider a finite Baire (or Borel, the space is metrisable) measure m. We can consider a finite "semi-open" partition of $\underline{K}$, each subset $\underline{K}_i$ is the intersection of an open subset and a closed subset and $\exists\, i : x \in K_{i,y} \in K_i \Rightarrow \ d(x, y) < \eta$, so $\|f(x) - f(y)\| < \varepsilon$. For all i we choose $k_i \in \underline{K}_i$. We can write

$$\sum_{i \in I} f(k_i) \cdot m(K_i).$$

This is an approximation of the Cauchy integral of $f \in C(K, E)$. If $(L_j)_j$ is a partition finer than $(K_i)_i$ for all j, we choose $l_j \in L_j$. We obtain a new approximation of the integral, the norm of difference between of the approximations,

$$\|\sum_{i \in I} f(k_i) \cdot m(K_i) - \sum_{i \in J} f(l_j) \cdot m(L_j)\|$$

is less than $m(K)\varepsilon$. We obtain a Cauchy sequence of approximations of the Cauchy integral. The limit exists, it is the integral.

We study quotient bornological spaces. Let $E = E^1 | E^0$ be a quotient Banach space, K is compact, m is a Baire measure and $f \in \mathcal{C}(K, E^1)$

$$\int_K f(x) dm(x) \in E^1.$$

If $f \in \mathcal{C}(K, E^0)$ then

$$\int_K f(x) dm(x) \in \underline{E}^0.$$

For the integral to belong to $\underline{E}^0$, it is not sufficient to check that the function is continuous and $\forall\, x : f(x) \in \underline{E}^0$. We must check that $f \in \mathcal{C}(K, E_0)$.

Example 9.3 Let $E = E^1|E^0$ be a genuine bornological Banach space. For all x, element of $\mathrm{cl}\,(E^0)$, a function $f \in \mathcal{C}([0,1], E^1)$ exists such that $\int_0^1 f(t)dm(t) = x, \forall\, t \in [0,1], f(t) \in \underline{E}^0$, though $x \in \mathrm{cl}\,(\underline{E}^0) \setminus E^0$.

Let $x \in \mathrm{cl}\,(\underline{E}^0)$. A sequence y_n of elements of $\underline{E}^0$ exists which converges to x and $\|y_n - x\|_{E^0} \le n^{-1}2^{-n}$. We let next $z_n = y_n - y_{n+1}, \|z_n\| \le n^{-1}2^{-n+1}$. We choose a continuous function $\bar{\omega}$ on the interval $[0,1]$ whose support is in the interval $[1/2, 1]$, it vanishes at the point 1 and its integral is equal to 1. We let

$$f(t) = \sum_{n=1}^{\infty} \varpi(2^n t) \cdot 2^n z_n$$

This function belongs to $\underline{\mathcal{C}}([0,1], E^1)$. For each $t \in [0,1], f(t) \in \underline{E}^0$. The integral is equal to x. The restriction of f to $[0,1]$ belongs to $\mathcal{C}([0,1], E^0)$. If we accept the Bochner definition, the function is measurable $[0,1] \to E^0$, it is not summable, does not belong to $\underline{L}^1([0,1], E^0)$, i.e.

$$\int_0^1 \|f(t)\|_{E^0}\,dt = \infty$$

(If the function were summable, the integral would belong to $\underline{E}^0$).

References

[1] R.G. Bartle and L.M. Graves: *Trans. of the Amer. Math. Soc.*, **72**(1952), 400-413.

[2] H. Buchwalter: Topologies, bornologies et compactologies. *Doctorat à l'Université de Lyon* (1968).

[3] J. Diestel and J. J. Uhl: Vector measures. *Mathematical Surveys American Mathematical Society.* **15**(1977).

[4] L. Gårding: Linear Hyperbolic Partial Differential Equations with Constant Co-efficients. *Acta Math.* **85**(1951), 1-62.

[5] I. M. Gelfand: Normierte Ringe. *Math. Sbornik,***9(51)** (1941), 3-24.

[6] I. M. Gelfand: Ideale und primäre Ideale in normierten Ringen. .*Math. Sbornik,* **9(51)** (1941), 41-48.

[7] I. M. Gelfand: Über absolut konvergente trigonometrische Reihen und Integrale. *Math. Sbornik,* **9(51)** (1941), 51-66.

[8] I. M. Gelfand and G. Shilov: Über verschiedene Methoden der Einführung der Topologie in die Menge der maximalen Ideale einer normierten Ringes. *Math. Sbornik*,**9(51)** (1941), 25-39.

[9] F.-H. Vasilescu: Spectral theory in quotient Fréchet spaces I. *Rev. Roumaine Math. et App.* **32**(1987), 561-571.

[10] L. Waelbroeck: Le calcul symbolique dans les algèbres commutatives. *J. Math. P. et App.* **9:33**(1954), 147-186.

[11] L. Waelbroeck: Etude spectrale des algèbres complètes. *Acad. R. de Belgique. Cl. des Sc. Mém. Coll. in 8° **7**(1960),2-31.

[12] L. Waelbroeck: Topological Vector Spaces and Algebras. *Lecture Notes in Mathematics* Vol. **230**, Springer-Verlag, Berlin–Heidelberg–New-York 1971.

L. Waelbroeck

Université Libre de Bruxelles
Faculté des Sciences
Département des Mathématiques
Campus Plaine, C.P. 214
1050 Bruxelles, Belgique

19. **H. Bart, I. Gohberg, M.A. Kaashoek** (Eds.): Operator Theory and Systems, 1986, (3-7643-1783-3)

20. **D. Amir:** Isometric characterization of Inner Product Spaces, 1986, (3-7643-1774-4)

21. **I. Gohberg, M.A. Kaashoek** (Eds.): Constructive Methods of Wiener-Hopf Factorization, 1986, (3-7643-1826-0)

22. **V.A. Marchenko:** Sturm-Liouville Operators and Applications, 1986, (3-7643-1794-9)

23. **W. Greenberg, C. van der Mee, V. Protopopescu:** Boundary Value Problems in Abstract Kinetic Theory, 1987, (3-7643-1765-5)

24. **H. Helson, B. Sz.-Nagy, F.-H. Vasilescu, D. Voiculescu, Gr. Arsene** (Eds.): Operators in Indefinite Metric Spaces, Scattering Theory and Other Topics, 1987, (3-7643-1843-0)

25. **G.S. Litvinchuk, I.M. Spitkovskii:** Factorization of Measurable Matrix Functions, 1987, (3-7643-1843-X)

26. **N.Y. Krupnik:** Banach Algebras with Symbol and Singular Integral Operators, 1987, (3-7643-1836-8)

27. **A. Bultheel:** Laurent Series and their Pade Approximation, 1987, (3-7643-1940-2)

28. **H. Helson, C.M. Pearcy, F.-H. Vasilescu, D. Voiculescu, Gr. Arsene** (Eds.): Special Classes of Linear Operators and Other Topics, 1988, (3-7643-1970-4)

29. **I. Gohberg** (Ed.): Topics in Operator Theory and Interpolation, 1988, (3-7634-1960-7)

30. **Yu.I. Lyubich:** Introduction to the Theory of Banach Representations of Groups, 1988, (3-7643-2207-1)

31. **E.M. Polishchuk:** Continual Means and Boundary Value Problems in Function Spaces, 1988, (3-7643-2217-9)

32. **I. Gohberg** (Ed.): Topics in Operator Theory. Constantin Apostol Memorial Issue, 1988, (3-7643-2232-2)

33. **I. Gohberg** (Ed.): Topics in Interplation Theory of Rational Matrix-Valued Functions, 1988, (3-7643-2233-0)

34. **I. Gohberg** (Ed.): Orthogonal Matrix-Valued Polynomials and Applications, 1988, (3-7643-2242-X)

35. **I. Gohberg, J.W. Helton, L. Rodman** (Eds.): Contributions to Operator Theory and its Applications, 1988, (3-7643-2221-7)

36. **G.R. Belitskii, Yu.I. Lyubich:** Matrix Norms and their Applications, 1988, (3-7643-2220-9)

37. **K. Schmüdgen:** Unbounded Operator Algebras and Representation Theory, 1990, (3-7643-2321-3)

38. **L. Rodman:** An Introduction to Operator Polynomials, 1989, (3-7643-2324-8)

39. **M. Martin, M. Putinar:** Lectures on Hyponormal Operators, 1989, (3-7643-2329-9)

40. **H. Dym, S. Goldberg, P. Lancaster, M.A. Kaashoek** (Eds.): The Gohberg Anniversary Collection, Volume I, 1989, (3-7643-2307-8)

41. **H. Dym, S. Goldberg, P. Lancaster, M.A. Kaashoek** (Eds.): The Gohberg Anniversary Collection, Volume II, 1989, (3-7643-2308-6)

42. **N.K. Nikolskii** (Ed.): Toeplitz Operators and Spectral Function Theory, 1989, (3-7643-2344-2)

43. **H. Helson, B. Sz.-Nagy, F.-H. Vasilescu, Gr. Arsene** (Eds.): Linear Operators in Function Spaces, 1990, (3-7643-2343-4)

44. **C. Foias, A. Frazho:** The Commutant Lifting Approach to Interpolation Problems, 1990, (3-7643-2461-9)

45. **J.A. Ball, I. Gohberg, L. Rodman:** Interpolation of Rational Matrix Functions, 1990, (3-7643-2476-7)

46. **P. Exner, H. Neidhardt** (Eds.): Order, Disorder and Chaos in Quantum Systems, 1990, (3-7643-2492-9)

47. **I. Gohberg** (Ed.): Extension and Interpolation of Linear Operators and Matrix Functions, 1990, (3-7643-2530-5)

48. **L. de Branges, I. Gohberg, J. Rovnyak** (Eds.): Topics in Operator Theory. Ernst D. Hellinger Memorial Volume, 1990, (3-7643-2532-1)

49. **I. Gohberg, S. Goldberg, M.A. Kaashoek:** Classes of Linear Operators, Volume I, 1990, (3-7643-2531-3)

50. **H. Bart, I. Gohberg, M.A. Kaashoek** (Eds.): Topics in Matrix and Operator Theory, 1991, (3-7643-2570-4)

51. **W. Greenberg, J. Polewczak** (Eds.): Modern Mathematical Methods in Transport Theory, 1991, (3-7643-2571-2)

52. **S. Prössdorf, B. Silbermann:** Numerical Analysis for Integral and Related Operator Equations, 1991, (3-7643-2620-4)

53. **I. Gohberg, N. Krupnik:** One-Dimensional Linear Singular Integral Equations, Volume I, Introduction, 1992, (3-7643-2584-4)

54. **I. Gohberg, N. Krupnik:** One-Dimensional Linear Singular Integral Equations, Volume II, General Theory and Applications, 1992, (3-7643-2796-0)

55. **R.R. Akhmerov, M.I. Kamenskii, A.S. Potapov, A.E. Rodkina, B.N. Sadovskii:** Measures of Noncompactness and Condensing Operators, 1992, (3-7643-2716-2)

56. **I. Gohberg** (Ed.): Time-Variant Systems and Interpolation, 1992, (3-7643-2738-3)

57. **M. Demuth, B. Gramsch, B.W. Schulze** (Eds.): Operator Calculus and Spectral Theory, 1992, (3-7643-2792-8)

58. **I. Gohberg** (Ed.): Continuous and Discrete Fourier Transforms, Extension Problems and Wiener-Hopf Equations, 1992, (ISBN 3-7643-2809-6)

59. **T. Ando, I. Gohberg** (Eds.): Operator Theory and Complex Analysis, 1992, (3-7643-2824-X)

60. **P.A. Kuchment:** Floquet Theory for Partial Differential Equations, 1993, (3-7643-2901-7)

61. **A. Gheondea, D. Timotin, F.-H. Vasilescu** (Eds.): Operator Extensions, Interpolation of Functions and Related Topics, 1993, (3-7643-2902-5)

Integral Equations and Operator Theory

ISSN 0378-620X

Editor:
I. Gohberg
School of Mathematical Sciences
Tel-Aviv University
Ramat-Aviv
Israel

Editorial Office:
School of Mathematical
Sciences, Tel-Aviv University,
Ramat-Aviv, Israel

Editorial Board:
A. Atzmon, Tel-Aviv;
J. A. Ball, Blacksburg;
L. de Branges, West Lafayette;
K. Clancey, Athens, USA;
L. A. Coburn, Buffalo;
R. G. Douglas, Stony Brook;
H. Dym, Rehovot;
A. Dynin, Columbus;
P. A. Fillmore, Halifax;
C. Foias, Bloomington;
P. A. Fuhrmann, Beer Sheva,
S. Goldberg, College Park;
B. Gramsch, Mainz;
J. W. Helton, La Jolla;
M. A. Kaashoek, Amsterdam;
T. Kailath, Stanford;
H. G. Kaper, Argonne;
S. T. Kuroda, Tokyo;
P. Lancaster, Calgary;
L. E. Lerer, Haifa;
E. Meister, Darmstadt;
B. Mityagin, Columbus;
J. D. Pincus, Stony Brook;
M. Rosenblum, Charlottesville;
J. Rovnyak, Charlottesville;
D. E. Sarason, Berkeley;
H. Widom, Santa Cruz;
D. Xia, Nashville

Honorary and Advisory Editorial Board:
P.R. Halmos, Santa Clara;
T. Kato, Berkeley;
P.D. Lax, New York;
M.S. Livsic, Beer Sheva;
R. Phillips, Stanford;
B. Sz.-Nagy, Szeged

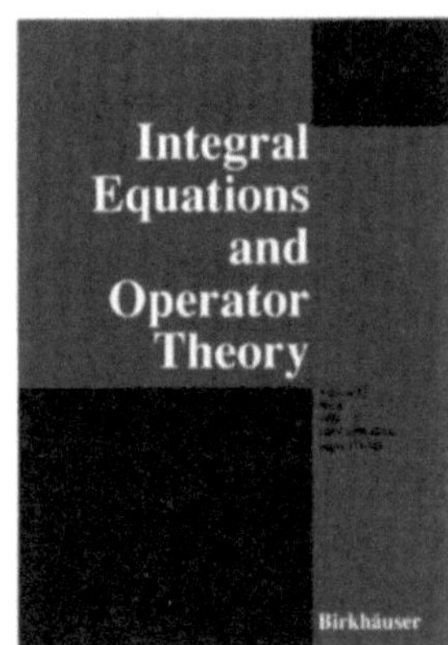

The journal is devoted to the publication of current research in integral equations, operator theory and related topics, with emphasis on the linear aspects of the theory. The very active and critical editorial board takes a broad view of the subject and puts a particularly strong emphasis on applications. The journal contains two sections, the main body consisting of refereed papers, and the second part containing short announcements of important results, open problems, information, etc.. Manuscripts are reproduced directly by a photographic process, permitting rapid publication.

Subscription information:

1993 subscription, volume 16+17 (8 issues)
sFr. 560.– / DM 658.– / US$ 378.00
(plus postage and handling)
Single copy:
sFr. 84.–/DM102.–/US$ 60.–
(plus postage and handling)

You may order through your bookseller or subscription agency, or directly from the publisher:

Birkhäuser Verlag AG
Subscription Department
P.O. Box 133
CH-4010 Basel / Switzerland
Fax ++41 / (0)61 / 271 76 66

Birkhäuser 

Birkhäuser Verlag AG
Basel · Boston · Berlin

5/93 all prices are object to changes without notice